T0214114

Lecture Notes
in Business Information Processing

392

Series Editors

Wil van der Aalst ⓘ
  RWTH Aachen University, Aachen, Germany
John Mylopoulos ⓘ
  University of Trento, Trento, Italy
Michael Rosemann ⓘ
  Queensland University of Technology, Brisbane, QLD, Australia
Michael J. Shaw
  University of Illinois, Urbana-Champaign, IL, USA
Clemens Szyperski
  Microsoft Research, Redmond, WA, USA

More information about this series at http://www.springer.com/series/7911

Dirk Fahland · Chiara Ghidini ·
Jörg Becker · Marlon Dumas (Eds.)

Business Process Management Forum

BPM Forum 2020
Seville, Spain, September 13–18, 2020
Proceedings

 Springer

Editors
Dirk Fahland 🆔
Eindhoven University of Technology
Eindhoven, The Netherlands

Chiara Ghidini 🆔
FBK-irst
Trento, Italy

Jörg Becker 🆔
University of Muenster
Münster, Germany

Marlon Dumas 🆔
University of Tartu
Tartu, Estonia

ISSN 1865-1348   ISSN 1865-1356 (electronic)
Lecture Notes in Business Information Processing
ISBN 978-3-030-58637-9   ISBN 978-3-030-58638-6 (eBook)
https://doi.org/10.1007/978-3-030-58638-6

This Springer imprint is published by the registered company Springer Nature Switzerland AG
The registered company address is: Gewerbestrasse 11, 6330 Cham, Switzerland

Preface

This volume contains the papers presented at the BPM Forum of the 18th International Conference on Business Process Management (BPM 2020). Since its introduction, the aim of the BPM Forum has been to host innovative research contributions that have high potential of stimulating discussion and scientific debate, but that do not (yet) meet the rigorous technical quality criteria to be presented at the main conference. The papers selected for the forum showcase fresh ideas from exciting and emerging topics in business process management. We received a total of 138 submissions, from which we took 125 into review. Each submission was reviewed by at least three Program Committee members and one Senior Program Committee member. At the end, 27 papers were accepted for the main conference, while 19 were invited to the BPM Forum and are compiled in this volume.

This year, the BPM conference was marked by the impact of the COVID-19 pandemic. While the conference's audience was looking forward to meeting in Seville, Spain, fate had another plan and the conference had to be held online. Under the leadership of the general chairs, Manuel Resinas and Antonio Ruiz Cortés, the conference was held online during September 13–18, 2020. On behalf of the BPM conference community, we thank them for accepting the challenge of organizing the conference fully online.

We also thank the members of the Program Committees and Senior Program Committees of the three conference tracks, as well as the external reviewers, who worked diligently to ensure a rigorous selection process both for the main conference and for the BPM Forum. We thank our sponsors: Signavio (Platinum), Celonis (Platinum), AuraPortal (Gold), DCR Solutions (Gold), Papyrus (Silver), Springer, and University of Seville. And last but not least, we thank the Organizing Committee, including Adela del Río Ortega, Amador Durán, Alfonso Márquez, Bedilia Estrada, and Beatriz Bernárdez, who, together with the general chairs, sacrificed a tremendous amount of time to overcome the challenges of switching from a physical to an online conference.

September 2020

Dirk Fahland
Chiara Ghidini
Jörg Becker
Marlon Dumas

Organization

The 18th International Conference on Business Process Management (BPM 2020) was organized by the Research Group of Applied Software Engineering (ISA Group) at the University of Seville, Spain, with the collaboration of SCORE Lab and the Instituto de Investigación en Ingeniería Informática (I3US). It took place online due to the restrictions imposed because of the COVID-19 pandemic. Originally, it was going to take place in Seville, Spain.

Steering Committee

Mathias Weske (Chair)	HPI, University of Potsdam, Germany
Boualem Benatallah	University of New South Wales, Australia
Jörg Desel	Fernuniversität in Hagen, Germany
Shazia Sadiq	The University of Queensland, Australia
Marlon Dumas	University of Tartu, Estonia
Wil van der Aalst	RWTH Aachen University, Germany
Jan Mendling	Vienna University of Economics and Business, Austria
Barbara Weber	University of St. Gallen, Switzerland
Stefanie Rinderle-Ma	University of Vienna, Austria
Manfred Reichert	Ulm University, Germany
Michael Rosemann	Queensland University of Technology, Australia

Executive Committee

General Chairs

Manuel Resinas	University of Seville, Spain
Antonio Ruiz-Cortés	University of Seville, Spain

Main Conference Program Chairs

Dirk Fahland (Track Chair, Track I)	Eindhoven University of Technology, The Netherlands
Chiara Ghidini (Track Chair, Track II)	Fondazione Bruno Kessler-IRST, Italy
Jörg Becker (Track Chair, Track III)	University of Münster, ERCIS, Germany
Marlon Dumas (Consolidation)	University of Tartu, Estonia

Workshop Chairs

Henrik Leopold	Kühne Logistics University, Germany
Adela del Río Ortega	University of Seville, Spain
Flavia Maria Santoro	Rio de Janeiro State University, Brazil

Demonstration and Resources Chairs

Marco Comuzzi	Ulsan National Institute of Science and Technology, South Korea
Claudio Di Ciccio	Sapienza University of Rome, Italy
Luise Pufahl	Technische Universität Berlin, Germany

Industry Forum Chairs

Gero Decker	Signavio, Germany
Manuel Lama	Universidade de Santiago de Compostela, Spain
Pedro Robledo	ABPMP Spain, Spain

Blockchain Forum Chairs

José María García	University of Seville, Spain
Agnes Koschmider	Kiel University, Germany
Jan Mendling	Vienna University of Economics and Business, Austria
Giovanni Meroni	Politecnico di Milano, Italy

RPA Forum Chairs

Aleksandre Asatiani	University of Gothenburg, Sweden
Nina Helander	Tampere University, Finland
Andrés Jiménez-Ramírez	University of Seville, Spain
Hajo A. Reijers	Utrecht University, The Netherlands

Doctoral Consortium Chairs

Félix García	University of Castilla-La Mancha, Spain
Manfred Reichert	Ulm University, Germany
Jan vom Brocke	University of Liechtenstein, Liechtenstein

Tutorial Chairs

Josep Carmona	Universitat Politècnica de Catalunya, Spain
Hajo A. Reijers	Utrecht University, The Netherlands
Minseok Song	Pohang University of Science and Technology, South Korea

Panel Chairs

Ernest Teniente	Universitat Politècnica de Catalunya, Spain
Mathias Weidlich	Humboldt-Universität zu Berlin, Germany

BPM Dissertation Award Chair

Jan Mendling	Vienna University of Economics and Business, Austria

Organizing Committees Chairs

Adela del Río Ortega	University of Seville, Spain
Amador Durán	University of Seville, Spain

Publicity Chairs

Cristina Cabanillas	University of Seville, Spain
Artem Polyvyanyy	The University of Melbourne, Australia
Kate Revoredo	Vienna University of Economics and Business, Austria
Armin Stein	University of Münster, Germany

Web Chair

Alfonso Márquez	University of Seville, Spain

Proceedings Chair

Bedilia Estrada-Torres	University of Seville, Spain

Supporting Staff Coordination Chair

Beatriz Bernárdez	University of Seville, Spain

Track I – Foundations

Senior Program Committee

Florian Daniel	Politecnico di Milano, Italy
Jörg Desel	Fernuniversität in Hagen, Germany
Chiara Di Francescomarino	Fondazione Bruno Kessler-IRST, Italy
Thomas Hildebrandt	University of Copenhagen, Denmark
Fabrizio Maria Maggi	Free University of Bozen-Bolzano, Italy
Marco Montali	Free University of Bozen-Bolzano, Italy
John Mylopoulos	University of Toronto, Canada
Oscar Pastor Lopez	Universitat Politècnica de València, Spain
Artem Polyvyanyy	The University of Melbourne, Australia
Manfred Reichert	Ulm University, Germany
Arthur Ter Hofstede	Queensland University of Technology, Australia
Hagen Voelzer	IBM Research, Switzerland
Mathias Weske	HPI, University of Potsdam, Germany

Program Committee

Lars Ackermann	University of Bayreuth, Germany
Daniel Amyot	University of Ottawa, Canada
Ahmed Awad	University of Tartu, Estonia
Irene Barba	University of Seville, Spain
Søren Debois	IT University of Copenhagen, Denmark
Claudio Di Ciccio	Sapienza University of Rome, Italy
Rik Eshuis	Eindhoven University of Technology, The Netherlands
Peter Fettke	German Research Center for Artificial Inteilligence (DFKI), Saarland University, Germany
Hans-Georg Fill	University of Fribourg, Switzerland
Luciano García-Bañuelos	Tecnológico de Monterrey, Mexico
María Teresa Gómez López	University of Seville, Spain
Guido Governatori	Data61, CSIRO, Australia
Gianluigi Greco	University of Calabria, Italy
Richard Hull	New York University, USA
Jetty Kleijn	LIACS, Leiden University, The Netherlands
Sander J. J. Leemans	Queensland University of Technology, Australia
Irina Lomazova	National Research University Higher School of Economics, Russia
Xixi Lu	Utrecht University, The Netherlands
Andrea Marrella	Sapienza University of Rome, Italy
Werner Nutt	Free University of Bozen-Bolzano, Italy
Wolfgang Reisig	Humboldt-Universität zu Berlin, Germany
Daniel Ritter	SAP, Germany
Andrey Rivkin	Free University of Bozen-Bolzano, Italy
Stefan Schönig	University of Regensburg, Germany
Arik Senderovich	University of Toronto, Canada
Tijs Slaats	University of Copenhagen, Denmark
Ernest Teniente	Universitat Politècnica de Catalunya, Spain
Sebastian Uchitel	University of Buenos Aires, Argentina, and Imperial College London, UK
Roman Vaculín	IBM Research, USA
Jan Martijn van der Werf	Utrecht University, The Netherlands
Francesca Zerbato	University of Verona, Italy

Track II – Engineering

Senior Program Committee

Andrea Burattin	Technical University of Denmark, Denmark
Josep Carmona	Universitat Politècnica de Catalunya, Spain
Remco Dijkman	Eindhoven University of Technology, The Netherlands
Avigdor Gal	Technion - Israel Institute of Technology, Israel
Marcello La Rosa	The University of Melbourne, Australia
Jorge Munoz-Gama	Pontificia Universidad Católica de Chile, Chile

Luise Pufahl	Technische Universität Berlin, Germany
Hajo A. Reijers	Utrecht University, The Netherlands
Stefanie Rinderle-Ma	University of Vienna, Austria
Pnina Soffer	University of Haifa, Israel
Wil van der Aalst	RWTH Aachen University, Germany
Boudewijn van Dongen	Eindhoven University of Technology, The Netherlands
Ingo Weber	Technische Universität Berlin, Germany
Matthias Weidlich	Humboldt-Universität zu Berlin, Germany

Program Committee

Marco Aiello	University of Stuttgart, Germany
Abel Armas Cervantes	The University of Melbourne, Australia
Boualem Benatallah	University of New South Wales, Australia
Cristina Cabanillas	University of Seville, Spain
Fabio Casati	ServiceNow, USA
Massimiliano de Leoni	University of Padua, Italy
Jochen De Weerdt	Katholieke Universiteit Leuven, Belgium
Joerg Evermann	Memorial University of Newfoundland, Canada
Walid Gaaloul	Télécom SudParis, France
Daniela Grigori	Laboratoire LAMSADE, Paris-Dauphine University, France
Georg Grossmann	University of South Australia, Australia
Anna Kalenkova	The University of Melbourne, Australia
Dimka Karastoyanova	University of Groningen, The Netherlands
Agnes Koschmider	Karlsruhe Institute of Technology, Germany
Henrik Leopold	Kühne Logistics University, Germany
Felix Mannhardt	SINTEF Digital, Norway
Elisa Marengo	Free University of Bozen-Bolzano, Italy
Jan Mendling	Vienna University of Economics and Business, Austria
Rabeb Mizouni	Khalifa University, UAE
Hye-Young Paik	University of New South Wales, Australia
Cesare Pautasso	University of Lugano, Switzerland
Pierluigi Plebani	Politecnico di Milano, Italy
Pascal Poizat	Université Paris Nanterre, LIP6, France
Barbara Re	University of Camerino, Italy
Manuel Resinas	University of Seville, Spain
Shazia Sadiq	The University of Queensland, Australia
Marcos Sepúlveda	Pontificia Universidad Católica de Chile, Chile
Natalia Sidorova	Eindhoven University of Technology, The Netherlands
Nick van Beest	Data61, CSIRO, Australia
Han van der Aa	Humboldt-Universität zu Berlin, Germany
Sebastiaan J. van Zelst	Fraunhofer FIT, Germany
Barbara Weber	University of St. Gallen, Switzerland
Moe Thandar Wynn	Queensland University of Technology, Australia
Nicola Zannone	Eindhoven University of Technology, The Netherlands

Track III – Management

Senior Program Committee

Daniel Beverungen	Universität Paderborn, Germany
Adela del Río Ortega	University of Seville, Spain
Patrick Delfmann	Universität Koblenz-Landau, Germany
Paul Grefen	Eindhoven University of Technology, The Netherlands
Susanne Leist	University of Regensburg, Germany
Peter Loos	IWi at DFKI, Saarland University, Germany
Martin Matzner	Friedrich-Alexander-Universität Erlangen-Nürnberg, Germany
Jan Recker	University of Cologne, Germany
Maximilian Roeglinger	FIM Research Center Finance & Information Management, Germany
Michael Rosemann	Queensland University of Technology, Australia
Flavia Maria Santoro	Rio de Janeiro State University, Brazil
Peter Trkman	University of Ljubljana, Slovenia
Amy van Looy	Ghent University, Belgium
Jan Vom Brocke	University of Liechtenstein, Liechtenstein

Program Committee

Wasana Bandara	Queensland University of Technology, Australia
Alessio Maria Braccini	University of Tuscia, Italy
Friedrich Chasin	Universität Münster, Germany
Ann-Kristin Cordes	Universität Münster, Germany
Rod Dilnutt	The University of Melbourne, Australia
Michael Fellmann	University of Rostock, Germany
Elgar Fleisch	ETH Zurich, Switzerland
Frederik Gailly	Ghent University, Belgium
Bernd Heinrich	Universität Regensburg, Germany
Mojca Indihar Štemberger	University of Ljubljana, Slovenia
Christian Janiesch	Julius-Maximilians-Universität Würzburg, Germany
Peter Kawalek	Loughborough University, UK
Ralf Knackstedt	University of Hildesheim, Germany
John Krogstie	Norwegian University of Science and Technology, Norway
Michael Leyer	University of Rostock, Germany
Alexander Mädche	Karlsruhe Institute of Technology, Germany
Monika Malinova	Wirtschaftsuniversität Wien, Austria
Fredrik Milani	University of Tartu, Estonia
Juergen Moormann	Frankfurt School of Finance & Management, Germany
Markus Nüttgens	University of Hamburg, Germany
Sven Overhage	University of Bamberg, Germany
Ralf Plattfaut	Fachhochschule Südwestfalen, Germany
Geert Poels	Ghent University, Belgium

Jens Pöppelbuß	University of Bochum, Germany
Hans-Peter Rauer	Universität Münster, Germany
Dennis M. Riehle	Universität Münster, Germany
Stefan Sackmann	University of Halle-Wittenberg, Germany
Werner Schmidt	Technische Hochschule Ingolstadt Business School, Germany
Theresa Schmiedel	University of Applied Sciences and Arts Northwestern Switzerland, Switzerland
Minseok Song	Pohang University of Science and Technology, South Korea
Oktay Türetken	Eindhoven University of Technology, The Netherlands
Jan Vanthienen	Katholieke Universiteit Leuven, Belgium
Axel Winkelmann	Julius-Maximilians-Universität Würzburg, Germany

Additional Reviewers

Mehdi Acheli
Ivo Benke
Yevgen Bogodistov
Silvano Colombo Tosatto
Jasper Feine
Rick Gilsing
Sebastian Halsbenning
Markus Heuchert
Benedikt Hoffmeister
Felix Härer
Florian Johannsen
Julian Koch
Safak Korkut
Ingo Kregel
Martin Käppel
Fabienne Lambusch
Susanne Leist

Jonas Lutz
Hugo-Andrés López
Steven Mertens
Miguel Meza
Roman Nesterov
Xavier Oriol
Nadine Ostern
Guido Perscheid
Michael Poppe
Tim Rietz
Lorenzo Rossi
Alexander Schiller
Fatemeh Shafiee
Peyman Toreini
Michael Trampler
Jonas Wanner
Anna Wilbik

Contents

BPM Adoption and Maturity

Standardization, Change and Handoffs

Process Modeling

Cross-Case Data Objects in Business Processes: Semantics and Analysis

Stephan Haarmann$^{(\boxtimes)}$ and Mathias Weske

Hasso Plattner Institute, University of Potsdam, Potsdam, Germany
{stephan.haarmann,mathias.weske}@hpi.de

Abstract. Business Process Management (BPM) provides methods and techniques to design, analyze, and enact business processes. An assumption in BPM has been that data objects are not shared among cases. Surprisingly, this often unquestioned assumption is violated in many real-world business processes. For instance, a budget data object can be read and modified by all ordering processes. These cross-case data objects have significant consequences on process modeling and verification. This paper provides a framework to describe and reason about cross-case data objects by presenting a dedicated execution semantics. Based on this framework, k-soundness is extended to cover multiple cases that share data. The paper reports on an implementation that translates BPMN process models extended with cross-case data objects to Coloured Petri nets, to properly capture their semantics.

Keywords: Inter-instance relationships · Coloured petri nets · Execution semantics.

1 Introduction

The success of organizations depends on how well their business processes are designed and implemented. Business Process Management (BPM) provides methods and tools to support process modeling, analysis, enactment, and post-execution analysis, i.e., process mining. It is a common assumption that processes operate on a local set of data objects and that data objects are never shared among different process instances, i.e., cases. This assumption enables modeling and verifying business processes as independent, decoupled entities.

However, many real-world processes defy this assumption. Rather than being isolated, cases have mutual dependencies, for example, shared data objects. We refer to such objects as *cross-case data objects*. Multiple cases may access the same cross-case data object and can, therefore, not be treated as independent, decoupled processes.

To illustrate cross-case data objects, consider a research group with an annual budget. We assume that the budget is used for travel expenses, equipment, office supplies, and also for salaries. From a process modeling perspective, the budget represents a data object that is accessed and modified by a variety of business

© Springer Nature Switzerland AG 2020
D. Fahland et al. (Eds.): BPM Forum 2020, LNBIP 392, pp. 3–17, 2020.
https://doi.org/10.1007/978-3-030-58638-6_1

processes. A hiring process allocates an annual salary on the budget; the planning and booking of conference fees and travel expenses are also allocated in the budget. Obviously, multiple cases represented by the same or by different process models can run concurrently, accessing the same cross-case data object.

In this paper, we investigate cross-case data objects in business processes. It provides a formal execution semantics based on Coloured Petri nets (CPN) [14] that is well integrated with existing semantics of BPMN process models. Based on this mapping, different correlation mechanisms are defined that relate data objects to cases. Taking into account one or multiple cross-case data objects, this formalization lays the foundation for verifying the k-soundness (no deadlocks, no dead transitions, and no remaining tokens) property, where k is the number of cases. The verification shows the validity of our approach.

The remainder of this paper is organized as follows. After an overview on related work, Sect. 3 introduces a motivating example. A semantics for processes with cross-case data is developed in Sect. 4. We employ the formalization to verify k-soundness of multiple interconnected cases of different processes in Sect. 5, before concluding the paper.

2   Related Work

There is agreement in the research community that data plays an important role in business process management. The BPMN standard [20] supports data objects and data stores, where data objects are generally volatile and are deleted after a process instance terminates, while data stores are persistent. Input and output data objects are mechanisms to define persistent data objects that are input and output, respectively, to a given process. However, BPMN fails to provide the expressive means to define shared data, which may be accessed by different cases, nor does it define how instances of persistent data are correlated to cases.

In [17], the authors extend BPMN data objects with primary keys and foreign keys. This extension enables the generation of SQL queries that return the relevant data from a relational database. Similarly, the relationships among data objects have been modeled with UML class diagrams [6]. However, neither of these approaches addresses the execution semantics in situations where multiple cases access the same object.

Cross-case data objects are one type of dependencies among different cases. Recently, the gap between processes that share nothing and real-world processes with various dependencies has been addressed. Fdhila et al. refer to such dependencies as instance spanning constraints [9]. These may involve resources, data, time, and more and are common in real world processes, as pointed out by [21]. Instance spanning constraints can be specified formally as rules using event calculus [13]. While such rules can be monitored at run-time [13] and mined from real world event logs [26], they are decoupled from process models. This makes it difficult to assess the rules' impact prior to execution. In our work, we define and integrate the semantics of cross-case data objects in process models to enable, among other things, model-based verification.

To overcome the lack of expressive means for data-related dependencies, different modeling paradigms have been developed. Data-centric process modeling languages (see Steinau et al. for a recent survey [24]), such as Proclets [2] and PHILharmonicFlows [15], define processes from a "data-first point of view". Rather than defining activities and their control flow dependencies, data-centric approaches define the life cycles of different objects as well as synchronization among them. Such approaches are very flexible and can express different cardinalities [8,23]. The Object-Centric Behavioral Constraints (OCBC) approach reported in [1] defines activities, data classes, and two types of constraints: temporal and cardinalities. However, they do not define the boundaries of a case.

What these approaches have in common: They address cross-case data mostly by abandoning a traditional notion of a case. Furthermore, data-centric approaches are not established in organizations, as activity-centric ones are. This paper aims at a common, precise understanding of cross-case data objects that is integrated with BPMN, the de-facto industry standard process modeling. Similarly, Montali and Rivkin introduce DB nets that represent persistent data that can be accessed from multiple cases [18]. While Montali and Rivkin define transactions with explicit updates of the persistent data, we treat cross-case objects similar to local objects and detect flaws without transactions.

The formal semantics proposed in this paper is based on works that map process models to Petri nets [4,7,10,16]. One of the first Petri-net based semantics for BPMN processes is [7] by Dijkman et al. The specified mapping has been extended to fit different purposes, i.e., to handle stateful data objects [4] or decisions including data object's attributes for verification [10,16]. We extend [7] by introducing cross-case data objects and unique identifiers.

3 Motivating Example

A business process comprises activities and their logical ordering necessary to achieve a business goal. A process model describes these activities, their interdependencies as well as additional requirements, for example data and resources.

The example in Fig. 1 depicts two process models that serve as motivating and illustrative examples throughout this paper. The upper process describes the procedure of buying office supplies. The process model in the lower part depicts the booking of a business trip. Notice that both processes access the budget data object, which is represented by a data store. More concretely, the data store is read and written by activities of both process models.

The process on ordering office supplies behaves as follows: A request for supplies is received, the order is placed. Receiving the goods and paying for the goods can occur concurrently. Once the goods are received and paid for, they are distributed. The travel booking process exposes an analogous behavior; it reads and writes the very same budget data object.

To differentiate cross-case data objects from traditional data objects in BPMN, we use data stores. This deviation from the BPMN pragmatics is appropriate to represent the novel concept of cross-case data objects.

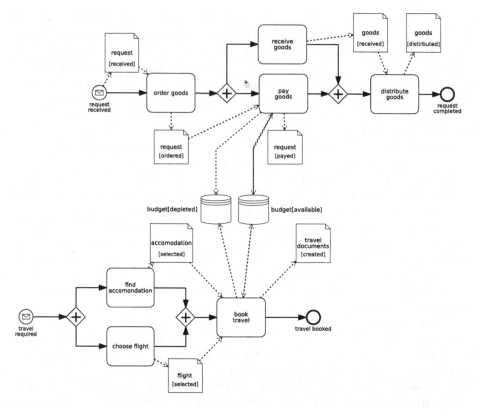

Fig. 1. Example depicting two process models: The upper one shows the process of ordering office supplies while the lower one describes a travel booking process.

In Fig. 1 we assume that goods can be paid for and travel can be booked only, if the budget data object is in state available. To illustrate cross-case data dependencies, let $c1$ be a case of the ordering process and $c2$ of the travel booking process, both of which access budget data object b. Assuming $c1$ changes the state of b from $b[available]$ to $b[depleted]$, $c2$ cannot execute the book travel activity, since its precondition is not met.

In general, if one case depletes the budget, future cases will deadlock, because neither process can handle this situation. Unfortunately, this property cannot be checked by existing verification techniques. To provide the basis for solving this problem, the next section takes an in-depth look on cross-case data objects.

4   Cross-Case Data

Traditionally, process modeling and verification either abstract from data objects [3] or rely on the assumption that data is local to a single case [4].

To incorporate cross-case data objects in process models, we introduce a formal execution semantics for BPMN processes using CPNs, which covers cross-case data objects and their impact on process execution. As in BPMN, we assume that data objects have states and that activities can only be enabled if the required data objects are in the respective states.

4.1 Formalizing Processes with Multiple Cases

The formalization is based on Coloured Petri nets, where places and tokens are typed. A type is called colorset, and a value is called color, which can be scalar or structured. When a transition consumes a token it may constrain the color, and it produces tokens of certain colors.

The formalization distinguishes between control flow tokens and data object tokens, and it introduces the respective colorsets, as shown in Fig. 2. A control flow token is passed along control flow edges, while data object tokens, obviously, represent data objects. For example, every token on a place labeled budget represents a separate budget data object. The control flow tokens carry case IDs, while each data object has a unique ID and a state. Two types of data objects are distinguished:

- *Case-specific data objects* belong to exactly one case; they store the respective caseID, which is used to correlate the data object to the case.
- *Cross-case data objects* are, in general, accessed by multiple cases.

Since states in BPMN are optional, we include the state BLANK to represent the absence of state information.

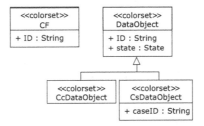

Fig. 2. Colorsets for the formalization of BPMN models with multiple cases. CF stands for control flow, CcDataObjects for cross-case data objects, and CsDataObjects for case-specific data objects. State enumerates all possible states including BLANK.

To formalize a set of process models with cross-case data objects, the process models are translated to a hierarchical CPN. We start by creating an abstract net that resembles the process structure, as shown by the mapping in Fig. 3. Then, details are added by defining a sub-net for each transition.

- Each data object is translated to a place with colorset CsDataObject, where Cs stands for case-specific. Even if a data object is represented multiple times in a given process model, the mapping contains only one place. Each instance of an object is represented by a single token in that place.

- Data stores are mapped analogously to places of type `CcDataObject` (row 2), where `Cc` stands for cross-case.
- Places with colorset `CF` (CF-places) are created (row 3–4). We add one place for each control flow in the process models (row 3) and one place for each end event (row 4), the latter of which collects CF-tokens of terminated cases.
- Hierarchical transitions are added for each activity, event (start, end, intermediate), and gateway (row 5).
- Transitions are linked to CF-places according to the control flow (rows 6–7).
- Arcs between data object places, both case-specific and cross-case, are added according to the data flow (rows 8–9). If an activity or event reads or updates an existing data object, a bidirectional arc is added linking the transition and the data object place (row 8). Exclusive gateways whose outgoing arcs have conditions of type *dataObjectName[State]* are connected with bidirectional arcs to corresponding places (row 8). If an activity or event creates a new object, the transition only produces a token on the respective place (row 9).

Mapping the example processes of travel booking and ordering office supplies results in the CPN in Fig. 4. The place `budget` holds tokens representing cross-case objects; thus, it links cases. If there is only one budget data object, the CPN has a single token on the `budget` place, which is consumed whenever `pay goods` or `book travel` fires. Any case of either process requires this single token, which accordingly presents a cross-case dependency.

The CPN in Fig. 4 represents the structure of the process models but lacks important details. To provide these details, each transition in the top-level net represents a sub-net that implements the detailed execution semantics as follows:

- New identifiers are created when a transition starts a new case or creates a new data object.
- Cases remain separated by asserting that all non-`CcDataObject` tokens consumed by a transition have the same caseID.
- Activities, events, and gateways fire only if the available data objects satisfy the respective state requirements.
- Variants of control flow nodes are represented by separate transitions. Variants of an activity or event differ in their input or output set, and variants of an exclusive gateway handle different incoming or outgoing control flow.

To create unique identifiers, an auxiliary place of type `int` is created for each process model and each data object. We name the place `count data object/process`. Such a place always holds exactly one token with the initial value 0. It counts the number of corresponding instances. Whenever a new case or data object is created, a transition consumes the token and increments it. The token's value is combined with a specific prefix to create a unique ID. We use the name of the process or data object. Note that this mechanism implements ν-transitions introduced in [22].

Figure 5 shows the detailed sub-net for the activity *book travel*. The activity reads a *flight* data object and an *accommodation* data object, which need to be in state *selected*. Tokens are consumed respectively by the transitions in Fig. 5. Each

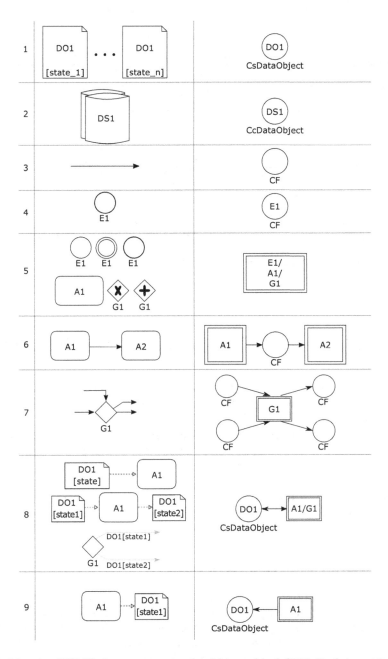

Fig. 3. Mapping BPMN elements to a top level hierarchical CPN. Each transition with double-stroke border is represented by a complete sub-net.

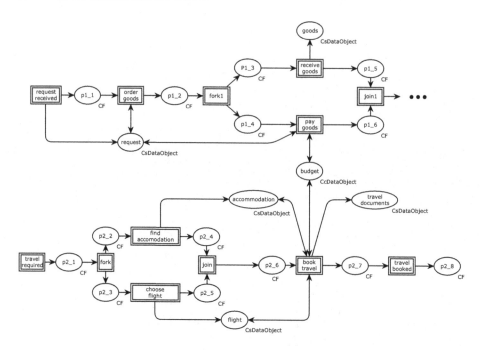

Fig. 4. CPN for the purchasing process (last activities and end events missing) and travel booking process. The place **budget** connects the sub-nets of the two processes.

transition represents a different outcome. Arc inscription require that the state of both objects is *selected* and that the caseID stored in the objects corresponds to the consumed control flow token.

Since the objects are not updated, the same data-object tokens that are consumed are produced. Furthermore, the activity reads a cross-case budget data object, which has no modeled relationship to the case. A token is consumed and produced, respectively. The activity creates a new *travel document* object. Therefore, it creates a unique ID using the counter stored in the place **count travel documents**. Additionally, control flow proceeds by producing a token on the CF-place in the transition's post set.

4.2   Correlating Data Objects to Cases

In typical BPM scenarios, correlation has to be addressed between incoming messages and cases. For instance, it is essential to correlate an incoming response from a customer to the process instance of that very customer. This is often realized through specific attributes, for instance customerID. In our setting, correlation between cases and cross-case data objects has to be covered, and the required correlation mechanisms are more complex.

If multiple cross-case data objects of the same type exist, activities can access any of these objects. However, for some settings, this arbitrary assignment is

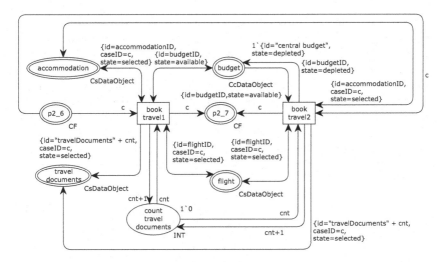

Fig. 5. Low-level CPN formalization for the activity *book travel* in the travel booking process. The two transitions describe the two possible outcomes (budget[available] or budget[depleted]). Arc inscriptions assert that all control flow tokens and case-specific data object tokens consumed and produced belong to the same case.

undesirable. In our example, we may have different budgets for different departments. More sophisticated correlation mechanisms are required.

Business process modeling standards such as BPMN [20] and WS-BPEL [19] support correlation of events to cases. BPMN supports key-based and context-based correlation, as sketched in the customer example above. In key-based correlation, each case defines a predicate on static properties of the event, the predicate determines whether or not the event belongs to a case.

In context-based correlation, events are correlated to cases by considering properties that may change, such as an object's attribute value. In this section, we discuss several correlation mechanisms, which can be used and combined as needed; each method is accompanied by a reference CPN formalization.

Singleton and Any Correlation. To illustrate the Singleton correlation mechanism, we consider a company with exactly one central budget data object, which is used for all expenses. As a result, correlation is trivial. The Any correlation is similar, since also multiple objects exist, but they can be correlated arbitrarily. In the CPN formalism, no adaptations are required for Singleton and Any correlation. If a case requires a cross-case data object, it can access any object available, as illustrated in Fig. 5.

Key and Context-based Correlation. If, however, an organization has different departments with respective budgets, more sophisticated correlation is required. Whenever they send a request for office supplies, they add a reference to the budget. Using CPNs, we can refine the colorsets to support key-based and context-based correlation. In key-based correlation, the colorset for the control flow contains explicit attributes, i.e., a reference to the budget. In context-based correlation, data objects contain this information, for instance `budgetID` in colorset request. In the Petri net, correlation is established by the respective arc inscriptions. This method requires the generation of appropriate values in initial markings as well as during case instantiation and data object creation.

To illustrate these correlation mechanisms, we consider a research group having a general budget and project specific budgets. The corresponding CPN model contains an attribute (in control flow tokens or request tokens) to match a case to the respective budget object, as shown in Fig. 6 for key-based correlation.

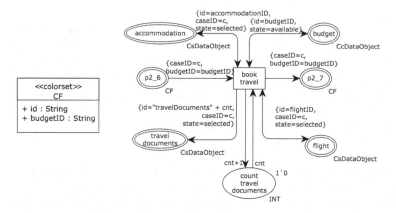

Fig. 6. The case references a budget using the attribute `budgetID` in the colorset on the left. Key-based correlation enforces that this reference is maintained. We use arc inscriptions to correlate cross-case object consistently to cases. (The variant writing budget in state *depleted* is modeled analogously)

Case-Based Correlation. If multiple objects of the same type correlate to a given case, consistent correlation is usually desirable: assuming that an object is required multiple times during a case, it has to be the same object. Consider a company with a travel booking process similar to the one depicted in Fig. 1. Assume furthermore that a flight can be booked right after choosing it, and accommodation can be payed immediately after selecting it. Consequently, there are two separate booking activities, each accessing a budget object. If a company guideline requires that one trip is paid from only one budget, we can enforce consistency by storing a reference to the cross-case object in the case upon first usage. Therefore, we add a new auxiliary place (e.g., `case to budget correlation` in Fig. 7) that stores tuples consisting of a caseID and an ID for the cross-case object (e.g., the budget). When a new case is started, a token with an empty cross-case data object ID is produced on that place. If an activity accesses a corresponding data object for the first time, its ID is set in the

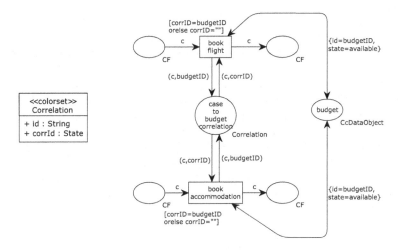

Fig. 7. CPN formalization of the activities *book flight* and *book accommodation* accessing a *budget* object. Case-based allocation asserts that the two activities access the same object. (The variants writing budget in state *depleted* are modeled analogously)

token for the case. Whenever the case requires an object of the same type, the data object ID must match the stored one (cf. Fig. 7).

5   Verifying *k*-Soundness with Cross-Case Data Objects

Formal semantics of processes have a variety of applications in business process management. They communicate behavioral details about a process, form the foundation for engine implementations, and are an essential requirement for verification, to detect flaws in the process prior to enactment. Verification includes checking soundness and compliance checking. However, such checks cover only isolated cases, neglecting dependencies between real-world processes.

In this paper, we have introduced cross-case data to process models, described their semantics formally, and investigated correlation mechanisms. This section introduces an approach to verifying multiple interconnected cases, which may detect flaws that remain unnoticed by traditional approaches.

The approach is based on the classical soundness criterion [3], which states that each process will terminate in a state with exactly one token in the final place, and that each transition takes part in at least one trace. By definition soundness is verified on a single workflow-net. This makes it impossible to verify properties of multiple connected processes sharing data objects. To address this deficiency, this section draws on the notion of *k*-soundness [11] supporting multiple cases and its variant for workflow-nets (resource-constraint workflow nets) with control flow places and resource places to handle data objects [5,12].

To verify a set of process models with cross-case data objects, we follow the steps depicted in Fig. 8. Given the respective BPMN models, we translate them to a CPN. To do so, we implemented a compiler that maps process models to a CPN using Access/CPN[25][1]. Once the CPN has been generated, it is transformed to a Coloured Resource-Constrained Workflow Net (RCWF-net)[11], with one dedicated input place and one dedicated output place. Notice that we use resource places to represent data objects, both case-specific and cross-case data objects. The resulting net is used to check the model for k-soundness.

To provide basis for our consideration, k-soundness is defined as follows. A Coloured RCFW-net is k-sound, if from an initial marking containing k tokens in i and a set of resource tokens, (i) all reachable final states (dead markings, in which no transition can fire) have k tokens in o, (ii) there are no other non-resource tokens, (iii) on every resource place there are at least as many tokens as in the initial marking, and (iv) every transition participates in at least one trace (the net contains no dead transitions).

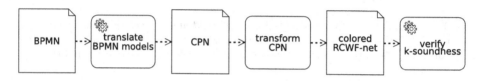

Fig. 8. Steps of verifying k-soundness on process models with cross-case data objects.

To transform the CPN to a Coloured RCWF-net, a new initial place i of type UNIT is added, and each transition representing a start event consumes a token from this place. Furthermore, we add an additional transition for each end-event and a place o of type UNIT. This transition consumes a token from a previously final place and produces a token on o. Furthermore, we need to add an initial marking, where there are k tokens on i. To reflect data objects, the net has initialized cross-case data objects. To represent two cases, two tokens are placed on i; a single budget object is created in state *available*. Once these steps have been taken, k-soundness can be verified. Based on the procedure shown in Fig. 8, CPNtools extracts the resulting state space. In the example (Fig. 1),

- the state space does not contain dead transitions, i.e., criterion (iv) is met;
- there are 36 dead markings, 14 of which violate k-soundness;
- a deadlock prevents proper completion of either process. It is reached if a case of either process depletes the budget. Any other case will get stuck upon booking the travel or paying the goods since no budget is left.

To detect dead markings that violate k-soundness, we implemented a search function in CPNTools (see Listing 1.1), as follows. The dead markings (line 3) are searched for states that contain a number of tokens in o unequal to k (criterion i) (l. 4) or containing control flow tokens (l. 5) (criterion ii). Note that the translation never deletes data object tokens (criterion iii).

---

[1] The source and binary of the tool as well as complete CPN formalizations are available online at https://bptlab.github.io/fcm2cpn/.

Listing 1.1. Standard ML function to find markings violating k-soundness

```
1  fun  FindNodesViolatingKSoundness  k =
2       SearchNodes(
3           ListDeadMarkings(),
4           fn  n => ((length (Mark.System'o 1 n)) <> k
5                   orelse (String.isSubstring "`\"" (NodeDescriptor n))),
6           NoLimit,
7           fn  n => n,
8           [],
9           op  ::);
```

The example in Sect. 3 with any number n of initial budget objects in state *available* violates k-soundness for any $k > n$. Each case must access a budget data object in state available once. However, if there are multiple cases, any case may deplete a budget. When all budgets have been depleted, subsequent cases will deadlock when they reach an activity requiring an available budget, i.e., book travel and pay goods. It is not trivial to avoid or resolve this error since multiple cases run concurrently. Adding a decision before *book travel* and *pay goods* does not solve the issue, because another case may deplete the budget between the decision and payment. The processes can, for example, first allocate the amount and then perform the payment, or it can use error boundary events.

6 Conclusion

Business process management supports organizations throughout the BPM life cycle from process elicitation to execution data analysis. Process models are the basis to design, communicate, analyze, and enact business processes. It is crucial that such models can capture the business operations accurately in order to be meaningful. However, it is generally assumed that processes do not share data, which contradicts real world scenarios.

This paper presents formal semantics for cross-case data using Coloured Petri nets, which distinguishes between case-specific and cross-case data objects. To relate cross-case data objects to cases, we present different correlation mechanisms: singleton/any correlation, key/context-based correlation and case-based correlation. These mechanisms allow defining the data access semantics precisely.

Furthermore, we discuss k-soundness as an application for the approach, which captures the fact that new dependencies among cases are added through cross-case data objects, which may lead to deadlocks which went unnoticed in existing verification approaches.

It is worth mentioning that the approach presented in this paper can be useful in further areas of our field, for instance process mining. Actually, in many process mining scenarios, different traces—reflecting cases—do access the same data objects. So far, there is no convincing approach that is capable of handing these situations. Consequently, cross-case data objects may spark new research in process mining, ranging from event log extraction to discovery of processes models.

Acknowledgments. We thank Leon Bein for major contributions to the compiler translating a set of BPMN models to CPNs.

References

1. van der Aalst, W.M.P., Artale, A., Montali, M., Tritini, S.: Object-centric behavioral constraints: Integrating data and declarative process modelling. In: Proceedings of the 30th International Workshop on Description Logics, Montpellier, France, 18–21 July 2017. http://ceur-ws.org/Vol-1879/paper51.pdf
2. van der Aalst, W.M.P., Barthelmess, P., Ellis, C.A., Wainer, J.: Workflow modeling using proclets. In: Scheuermann, P., Etzion, O. (eds.) CoopIS 2000. LNCS, vol. 1901, pp. 198–209. Springer, Heidelberg (2000). https://doi.org/10.1007/10722620_20
3. van der Aalst, W.M.P., et al.: Soundness of workflow nets: classification, decidability, and analysis. Formal Asp. Comput. **23**(3), 333–363 (2011). https://doi.org/10.1007/s00165-010-0161-4
4. Awad, A., Decker, G., Lohmann, N.: Diagnosing and repairing data anomalies in process models. In: Rinderle-Ma, S., Sadiq, S., Leymann, F. (eds.) BPM 2009. LNBIP, vol. 43, pp. 5–16. Springer, Heidelberg (2010). https://doi.org/10.1007/978-3-642-12186-9_2
5. Bashkin, V.A., Lomazova, I.A.: Decidability of k-soundness for workflow nets with an unbounded resource. In: Koutny, M., Haddad, S., Yakovlev, A. (eds.) Transactions on Petri Nets and Other Models of Concurrency IX. LNCS, vol. 8910, pp. 1–18. Springer, Heidelberg (2014). https://doi.org/10.1007/978-3-662-45730-6_1
6. Combi, C., Oliboni, B., Weske, M., Zerbato, F.: Conceptual modeling of interdependencies between processes and data. In: Proceedings of the 33rd Annual ACM Symposium on Applied Computing, SAC, Pau, France, pp. 110–119 (2018). https://doi.org/10.1145/3167132.3167141
7. Dijkman, R.M., Dumas, M., Ouyang, C.: Semantics and analysis of business process models in BPMN. Inf. Softw. Technol. **50**(12), 1281–1294 (2008). https://doi.org/10.1016/j.infsof.2008.02.006
8. Fahland, D.: Describing behavior of processes with many-to-many interactions. In: Donatelli, S., Haar, S. (eds.) PETRI NETS 2019. LNCS, vol. 11522, pp. 3–24. Springer, Cham (2019). https://doi.org/10.1007/978-3-030-21571-2_1
9. Fdhila, W., Gall, M., Rinderle-Ma, S., Mangler, J., Indiono, C.: Classification and formalization of instance-spanning constraints in process-driven applications. In: La Rosa, M., Loos, P., Pastor, O. (eds.) BPM 2016. LNCS, vol. 9850, pp. 348–364. Springer, Cham (2016). https://doi.org/10.1007/978-3-319-45348-4_20
10. Haarmann, S., Batoulis, K., Weske, M.: Compliance checking for decision-aware process models. In: Daniel, F., Sheng, Q.Z., Motahari, H. (eds.) BPM 2018. LNBIP, vol. 342, pp. 494–506. Springer, Cham (2019). https://doi.org/10.1007/978-3-030-11641-5_39
11. van Hee, K., Serebrenik, A., Sidorova, N., Voorhoeve, M.: Soundness of resource-constrained workflow nets. In: Ciardo, G., Darondeau, P. (eds.) ICATPN 2005. LNCS, vol. 3536, pp. 250–267. Springer, Heidelberg (2005). https://doi.org/10.1007/11494744_15
12. van Hee, K., Sidorova, N., Voorhoeve, M.: Soundness and separability of workflow nets in the stepwise refinement approach. In: van der Aalst, W.M.P., Best, E. (eds.) ICATPN 2003. LNCS, vol. 2679, pp. 337–356. Springer, Heidelberg (2003). https://doi.org/10.1007/3-540-44919-1_22

13. Indiono, C., Mangler, J., Fdhila, W., Rinderle-Ma, S.: Rule-Based Runtime Monitoring of Instance-Spanning Constraints in Process-Aware Information Systems. In: Debruyne, C., et al. (eds.) OTM 2016. LNCS, vol. 10033, pp. 381–399. Springer, Cham (2016). https://doi.org/10.1007/978-3-319-48472-3_22
14. Jensen, K., Kristensen, L.M.: Coloured Petri Nets. Modelling and Validation of Concurrent Systems. Springer, Heidelberg (2009). https://doi.org/10.1007/b95112
15. Künzle, V., Reichert, M.: Philharmonicflows: towards a framework for object-aware process management. J. Softw. Maintenance **23**(4), 205–244 (2011). https://doi.org/10.1002/smr.524
16. de Leoni, M., Felli, P., Montali, M.: A holistic approach for soundness verification of decision-aware process models. In: Trujillo, J.C., Davis, K.C., Du, X., Li, Z., Ling, T.W., Li, G., Lee, M.L. (eds.) ER 2018. LNCS, vol. 11157, pp. 219–235. Springer, Cham (2018). https://doi.org/10.1007/978-3-030-00847-5_17
17. Meyer, A., Pufahl, L., Fahland, D., Weske, M.: Modeling and enacting complex data dependencies in business processes. In: Daniel, F., Wang, J., Weber, B. (eds.) BPM 2013. LNCS, vol. 8094, pp. 171–186. Springer, Heidelberg (2013). https://doi.org/10.1007/978-3-642-40176-3_14
18. Montali, M., Rivkin, A.: DB-Nets: on the marriage of colored petri nets and relational databases. In: Koutny, M., Kleijn, J., Penczek, W. (eds.) Transactions on Petri Nets and Other Models of Concurrency XII. LNCS, vol. 10470, pp. 91–118. Springer, Heidelberg (2017). https://doi.org/10.1007/978-3-662-55862-1_5
19. OASIS: Webservice business process execution language (2007). https://docs.oasis-open.org/wsbpel/2.0/OS/wsbpel-v2.0-OS.html
20. Object Management Group (OMG): Business process model and notation (BPMN) version 2.0 (2014). https://www.omg.org/spec/BPMN/
21. Rinderle-Ma, S., Gall, M., Fdhila, W., Mangler, J., Indiono, C.: Collecting examples for instance-spanning constraints. CoRR (2016). http://arxiv.org/abs/1603.01523
22. Rosa-Velardo, F., de Frutos-Escrig, D.: Name creation vs. replication in petri net systems. Fundam. Inform. **88**(3), 329–356 (2008)
23. Steinau, S., Andrews, K., Reichert, M.: The relational process structure. In: Krogstie, J., Reijers, H.A. (eds.) CAiSE 2018. LNCS, vol. 10816, pp. 53–67. Springer, Cham (2018). https://doi.org/10.1007/978-3-319-91563-0_4
24. Steinau, S., Marrella, A., Andrews, K., Leotta, F., Mecella, M., Reichert, M.: DALEC: a framework for the systematic evaluation of data-centric approaches to process management software. Softw. Syst. Model. **18**(4), 2679–2716 (2018). https://doi.org/10.1007/s10270-018-0695-0
25. Westergaard, M.: Access/CPN 2.0: a high-level interface to coloured petri net models. In: Kristensen, L.M., Petrucci, L. (eds.) PETRI NETS 2011. LNCS, vol. 6709, pp. 328–337. Springer, Heidelberg (2011). https://doi.org/10.1007/978-3-642-21834-7_19
26. Winter, K., Stertz, F., Rinderle-Ma, S.: Discovering instance and process spanning constraints from process execution logs. Inf. Syst. **89**, 101484 (2020). https://doi.org/10.1016/j.is.2019.101484

Dynamic Process Synchronization Using BPMN 2.0 to Support Buffering and (Un)Bundling in Manufacturing

Konstantinos Traganos[1]($\boxtimes$), Dillan Spijkers[1], Paul Grefen[1], and Irene Vanderfeesten[1,2]

[1] Eindhoven University of Technology, Eindhoven, The Netherlands
{k.traganos,p.w.p.j.grefen,i.t.p.vanderfeesten}@tue.nl,
dillanspijkers@gmail.com
[2] Open University of the Netherlands, Heerlen, The Netherlands
irene.vanderfeesten@ou.nl

Abstract. The complexity of manufacturing processes is increasing due to the production variety implied by mass customization of products. In this context, manufacturers strive to achieve flexibility in their operational processes. Business Process Management (BPM) can help integration, orchestration and automation of these manufacturing operations to reach this flexibility. BPMN is a promising notation for modeling and supporting the enactment of manufacturing processes. However, processes in the manufacturing domain include the flow of physical objects (materials and products) apart from information flow. Buffering, bundling and unbundling of physical objects are three commonly encountered patterns in manufacturing processes, which require fine-grained synchronization in the enactment of multiple process instances. Unfortunately, BPMN lacks strong support for this kind of dynamic synchronization as process instances are modeled and executed from a single, isolated point of view. This paper presents a mechanism based on BPMN 2.0 that enables process modelers to define synchronization points by using the concept of recipes. The recipe system uses a dynamic correlation scheme to control many-to-many interactions among process instances to implement required inter-instance synchronizations. We formally describe the involved BPMN patterns, implement and evaluate them in a manufacturing scenario in the high-tech media printing domain.

Keywords: BPMN patterns · Dynamic process instances synchronization · Manufacturing · Buffering · (Un)bundling

1 Introduction

In discrete manufacturing, processes get more complex and organizations strive to manage and orchestrate their operations. Activities on a factory shop floor should also be integrated with business functions for a seamless, end-to-end process management [1]. Business Process Management (BPM) is a paradigm that is often employed to help with process orchestration and improve cross-functional integration. While BPM has proven

© Springer Nature Switzerland AG 2020
D. Fahland et al. (Eds.): BPM Forum 2020, LNBIP 392, pp. 18–34, 2020.
https://doi.org/10.1007/978-3-030-58638-6_2

its strength in business sectors where information processing is dominant, e.g. finance [2], it has also been extensively applied in healthcare [3] and transportation [4], where physical entities are included as well. Without surprise, the application of BPM in manufacturing has increasingly gained attention [5], especially in the Industry 4.0 era, where many advanced robots and automated guided vehicles (AGV) have been introduced. That is because the need for orchestration of the activities of all the versatile actors is more imperative.

Modelling and supporting the execution of processes, is a core part in applying BPM concepts. BPMN, as the de-facto standard for business process modelling [6], is widely used for business processes [7]. Its interdisciplinary understandability [8, 9] and the expressiveness with respect to integration to execution [10], together with the need of integration of business processes and manufacturing operations [1, 11, 12], make it a promising candidate for use in the discrete manufacturing domain. Various extensions of BPMN for manufacturing processes have already been proposed [13, 14] and the comparison to other languages [15] shows the language's strengths.

Despite its maturity and the recent interest of applying the notation in manufacturing, BPMN has inherent limitations. One of these is the fact that process models in BPMN are designed from a single, isolated process instance perspective, disregarding possible interactions among instances during execution [16, 17]. Often, process instances need to interact and collaborate based on information that is outside of the scope of one single instance. This collaboration is more important in manufacturing processes, where physical objects, and not only data information, are under consideration. Think of example of buffering points, where inventory is kept at an intermediate stage of a process, or the situation of bundling or batching products (multiple entities) for further processing as single entity (e.g. placing a number of items in a box for transporting). There should be hence, synchronization points where a process instance, representing the flow of activities of entities, waits or sends information regarding the state from or to other instances, commonly from different process definitions. BPMN provides basic synchronization with elements such as Signals or Messages. But the former is a broadcast message without any payload while the latter sends a payload message (with e.g. process instance identifiers or process definition keys) to only one instance. There is a lack of dynamic synchronization expressibility and functionality in the sense that the synchronization of the control flow of process instances cannot currently be decided based upon runtime state and content information of other process instances.

Buffering of entities and (un)bundling of entities and activities are constructs frequently encountered in the physical world of manufacturing processes. Using BPMN for manufacturing processes, entails explicit support for these constructs. Thus, we present in this paper an approach, called recipe system, to address the dynamic synchronization issue described in the previous paragraph. The approach uses standard BPMN 2.0 elements to form a dynamic controller that works as a correlation mechanism for synchronization points amongst independent process instances.

In Sect. 2 we discuss related work of the synchronization shortcoming of BPMN. In Sect. 3, we discuss the characteristics of the manufacturing concepts that we aim to support in terms of modeling and execution. In Sect. 4, the correlation mechanism is described and formalized. The implementation of the mechanism, its application and evaluation are discussed in Sect. 5. Finally, we conclude and reflect on the presented work in Sect. 6.

2 Background and Related Work

In the introduction, we briefly discussed the increasing interest of applying BPMN in manufacturing. More studies make it prominent [18–22]. The fact though that BPMN originates from the service industry, indicates that the notation cannot fully support concepts occurring in the physical, manufacturing world and consequently in the manufacturing sector. Various extensions have been proposed to capture the specific manufacturing characteristics, e.g. manufacturing activities, resource containers and material gateways [13], sensory event definitions, data sensor representations and specific smart manufacturing task types [23], sensing and actuating tasks [24], assets, properties and relationships among these entities [25].

However, all the aforementioned studies do not touch the synchronization problem that BPMN lacks to support. This shortcoming of the language has already been studied, but rather as a general problem, not targeting at the physical and manufacturing world. In general, we see two different paradigms; activity-centric ones (e.g. what BPMN follows) focusing on describing the ordering of activities, and artifact-centric ones focusing on describing the objects that are manipulated by activities [26–30]. From a BPMN perspective, artifact-centric modeling support is limited, though extension elements to support the artifact-centric paradigm have been defined [31]. Fahland et al. [32] approach the process synchronization from a dualistic point of view, both from the activity-centric and the artifact-centric paradigm perspectives. The study argues that processes are active elements that have agents, actors that execute activities. These actors drive the processes forward. Artifacts, on the other hands, are passive elements that are object to the activities. The activities are performed on these objects. While Petri nets are used as a means of process specification, Fahland argues that locality of transitions, which synchronize by "passing" tokens, are at the core of industrial process modeling languages, just like BPMN. Steinau et al. [33] also consider many-to-many process interactions in their study, proposing a relational process structure, realizing many-to-many relationship support in run-time and design-time. Earlier work on process interactions by van der Aalst et al. [34] (e.g. proclets), allowed for undesired behavior in many-to-many relations [35]. Pufahl et al. [36] put forward the notion of a "batch activity", which is an activity that is batched over multiple process instances of the same process definition. The batch is activated upon the triggering of an activation rule. The concept is similar to the approach presented in this paper, but our study includes a strong focus on the correlation of process instances of different process definitions, that typically contain different activities. Finally, Marengo et al. [37] study the interplay of process instances and propose a formal language, inspired by Declare [38], for process modeling in the construction domain.

The importance of BPMN and the still ongoing research on the issue of multi-instance synchronization, is the motivation of our work on supporting frequent manufacturing patterns with this specific language. Our presented approach enriches BPMN and allows practitioners to use the notation for modeling and enactment of their manufacturing processes. In the next section, we first discuss the characteristics of the manufacturing constructs that this paper aims to support with BPMN, namely buffering, bundling and unbundling.

3   Characteristics and Limitations of Manufacturing Constructs

This section introduces the buffering, bundling and unbundling constructs. Due to their similar but inverse relationship, the bundling and unbundling constructs are jointly discussed. Limitations of BPMN to express them is also discussed.

3.1   Manufacturing Constructs Characteristics

Buffering

From an operations management perspective, buffering is considered as maintaining excess resources to cover variation or fluctuation in supply or demand [39]. The concept is also referred to as decoupling inventory between process steps, as these can be performed independently from each other [40]. Buffering, as an operations type of storage, is covered by the manufacturing operations taxonomy, under the Inventory operations category [41]. From [42], we can define buffering as "a form of (temporary) storage with the intention to synchronize flow material between work centers or production steps that may have unequal throughput". From the five types of inventory from [43], we focus on the decoupling inventory/buffers in this study.

In the BPM field, van der Aalst [44] had already argued that places in Petri nets correlate to physical storage locations, in his effort to use high-level Petri nets to describe business processes. Thus, from a process management perspective the notion of a buffer can be explained as follows. An instance enters the buffer and is kept in a holding state. Once a condition is met (e.g. capacity becomes available in the downstream production step), one or more entities are released. The selection of which entity to be released can be based on multiple queuing policies, e.g. the First-In-First-Out (FIFO) policy. Once an entity is released, control flow continues as normal.

The above explanation though, considers the buffering from a single process instance perspective, leading to the process instance isolation issue we described in the previous sections. There is a need to approach the construct from a process control perspective, as such that buffer-level attributes and information from many process instances are captured and managed, as illustrated in Fig. 1.

Bundling and Unbundling

Manufacturing operations literature recognizes operations that bundle, merge, unitize and package entities, as well as their inverse counterparts, but to the best of our knowledge no literature exists that describes how these entities are selected during operations. This is assumed to be described by the modelers in another part of the models or in different models. In this paper we define bundling as "the synchronization of instances that are grouped in some way, either physically or virtually, whose control flow shall continue or terminate simultaneously as a group". Note this is a process-oriented definition and caution should be taken for generalization.

Examples of bundling are commonly encountered when physical entities need to be grouped into some sorts of a container. Imagine for instance products being produced and put in a packaging box. Once the capacity of the box is reached, the box can be transported as a single entity. Upon arrival of the box to a distribution center, entities are

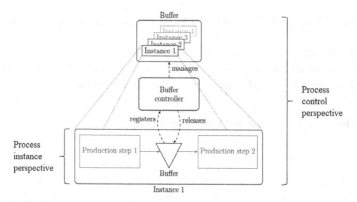

Fig. 1. Buffering construct from both process control and process instance perspective.

unbundled again. We use the term bundling, as a more generic term than batching, since the latter normally refers to putting together entities of the same type, while in bundling we can merge entities of different types. Bundling is often encountered together with buffering, as quite often, (sub-)entities are buffered before the bundling operation can take place, to ensure all (sub-)entities are present.

3.2 BPMN Limitations

A buffering point between two activities (or process fragments), as shown with the triangle element in Process Instance 1 in Fig. 1, could be naively modeled in BPMN 2.0 with the use of conditional or (intermediate) message catching events. These elements can offer the "holding" state of the control flow. However, none approach is suitable. Conditional events use local-instance variables, ignoring information of other process instances. Message events are targeted to a specific, pre-defined instance, missing dynamic correlation information.

Bundling and unbundling constructs can be probably modeled with AND-gateways. But these gateways (un)merge control flows that can be modelled on the same definition, which is not always possible. In many scenarios, different processes have to be correlated and gateways cannot perform this. Erasmus et al. [41] discuss also the use of multi-instance activities (among the other BPMN patterns proposed for modeling manufacturing processes) for unitizing, as they call it, entities. The spawning of repeated instances can serve (un)bundling functionality. However, the isolation problem appears here as well. Each child process instance is unaware of the information of the rest child instances.

4   Concept and Functionality of a Recipe System

In this section we first present the approach to overcome the synchronization issues. The approach is later formalized.

4.1 The Recipe Controller

Having provided the definitions of the constructs under consideration, along with their attributes, and discussed the BPMN limitations to support them (in Sect. 3), we present here the solution of a recipe control system. The work of Spijkers [45] discusses also a list of functional and non-functional requirements for designing such a system.

Recipe

The central notion of the system is the *recipe*. It corresponds to a synchronization (or integration) point, where (previously uncorrelated) control flows in independent process instances may be synchronized. It consists of a set of input rules and output rules. A recipe is fulfilled once all input rules are satisfied. We link two important concepts in a recipe. The *instance type* and the *selector attribute*. The first is used to group process instances of the same type in a *pool*. Think for example a car assembly process. It requires a number of wheels, a number of doors and a chassis. Each of these elements are produced independently according to their process models. Thus, we can have three pools, one with "CarWheel" instance type, one with "CarDoor" type and one with "CarChassis" type. The *selector attribute* is used for discriminating instances that are of the same type, yet of a different variant. For example, the "CarDoor" instance type can have the color (e.g. blue/red) as attribute. A pool is a virtual "container" to keep homogenous process instances; homogeneous from an instance type perspective, as these can have different attributes. All the concepts are illustrated in Fig. 2. Process instances are denoted as shape figures.

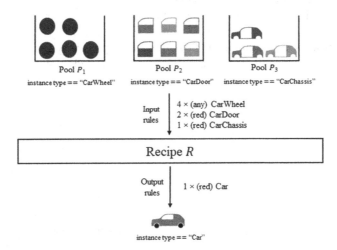

Fig. 2. Illustration of the recipe concepts with an example (Color figure online)

The configuration of each pool plays a crucial role for the fulfillment of a recipe. The following options are considered:

- **Genericity.** A pool can be either *generic* or *specific*. In the first case, the pool does not consider the selector attribute of the buffered process instances (e.g. in Pool $\mathcal{P}_1$

of Fig. 2). In the latter case, recipe fulfillment candidates are nominated based on the selector attribute (e.g. on the color in pools $\mathcal{P}_2$ and $\mathcal{P}_3$ of **Fig.** 2).

- **Availability mask.** Pools can represent physical buffers but as such should account for physical availability, i.e. how instances/objects are accessed. This paper considers three availability masks:

- ALL: All instances are available (e.g. in a virtual or physical pool that we do not care about the physical layout).
- FIRST: The instance that was first placed in the pool is considered as available. Subsequent instances are marked as available if and only if they share the selector attribute value of the first instance, in one sustained sequence.
- LAST: The instance that was placed last in the pool is considered as available. Subsequent instances are marked as available if and only if they share the selector attribute value of the last instance, in one sustained sequence.

- **Release policy.** The release policy ranks instances for recipe fulfillment (and thus "release" from the pool). This paper considers three policies:

- FIFO: instances that have been in the pool the longest are released first.
- LIFO: instances that have been in the pool the shortest are released first.
- ATTR: instances are released based on a selector attribute value.

Fulfillment Cardinality. The fulfillment cardinality determines how many instances of a pool are needed to lead to recipe fulfillment. It can be a single value, i.e. all instances are nominated for fulfillment or it can take a minimum (n) and a maximum (m) value, i.e. the pools needs at least n and less than m.

With the configuration options described above, the recipe can be specified with the following notation, shown in Fig. 3 (same example as in Fig. 2). Upon a recipe fulfilment, a process may continue its flow after the respective synchronization points or a new process instance (mainly from a different process definition) can start.

| Recipe name: | Final car assembly | | | | | | |
| Selector attribute: | ordernumber | | | | | | |
Input instance type		min	max	gen	relpol	mask	rel
CarWheel		4	4	•	FIFO	LAST	○
CarDoor		4	4	○	LIFO	ALL	○
CarChassis		1	1	○	LIFO	ALL	•
num	Start process definition key (output)						
1	Final_Car_Assembly_Process						

Fig. 3. Specification of a recipe (proposed notation)

4.2 Formalization

Recipes ($\mathcal{R}$) are treated as sequences that contain Pools ($\mathcal{P}$) that are treated as sequences that contain instances. The notation $|\mathcal{P}|$ is used to denote the number of instances currently in pool $\mathcal{P}$. The notation $\mathcal{P}(i)$, with $i \in \{1, \ldots, |\mathcal{P}|\}$, refers to the i-th instance in the pool. Not to be confused with the powerset notation $\mathcal{P}(A)$, referring to the powerset of set A. Note that this instance indexing is based on the time at which an instance was added to the pool. In other words, from a mathematical perspective, a pool is an array of instances that is sorted on arrival timestamp. In general, the symbol i is used to either denote an array index (like in the $\mathcal{P}(i)$ notation) or a process instance, like $i \in \mathcal{P}$. The latter should be read as *instance i in pool $\mathcal{P}$*. The mathematical model, which extends the content presented in the previous section, uses the following symbols:

$\mathcal{R}$ a recipe.
$\mathcal{P}$ a pool. Is a member of a recipe, i.e. $\mathcal{P} \in \mathcal{R}$.
$\mathcal{S}$ the (abstract) set of possible selector attributes.
$s_{\mathcal{P}}$ the selector attribute for pool $\mathcal{P}$.
$\mathcal{V}_s$ the (abstract) set of possible selector attribute values for selector attribute $s \in \mathcal{S}$.
v_i the selector attribute value for instance $i \in \mathcal{P}$.
$c_{\mathcal{P}}^-$ the minimum fulfillment cardinality for pool $\mathcal{P}$.
$c_{\mathcal{P}}^+$ the maximum fulfillment cardinality for pool $\mathcal{P}$.
$\alpha_{\mathcal{P}}(i)$ availability mask function for pool $\mathcal{P}$. $\alpha_{\mathcal{P}}(i) \in \{0, 1\}|\forall i \in \mathcal{P}$.
$\rho_{\mathcal{P}}(i)$ release policy ranking function for pool $\mathcal{P}$. $\rho_{\mathcal{P}}(i) \in \{1, \ldots, |\mathcal{P}|\}|\forall i \in \mathcal{P}$.
$g_{\mathcal{P}}$ boolean whether pool $\mathcal{P}$ is generic (1) or specific (0). $g_{\mathcal{P}} \in \{0, 1\}$.
$\mathcal{S}(\mathcal{P})$ the set of selector attribute values for which at least $c_{\overline{\mathcal{P}}}$ instances exist in pool $\mathcal{P}$.

Formally defined as

$$\mathcal{S}(\mathcal{P}) \equiv \left\{ v \in \left\{ v_p : p \in \mathcal{P} \right\} : \left| \left\{ v_p : p \in \mathcal{P} \wedge v_p = v \right\} \right| \geq c_{\overline{\mathcal{P}}} \right\} \tag{1}$$

Note that, by definition, $\mathcal{S}(\mathcal{P}) \subseteq \mathcal{V}_{s_{\mathcal{P}}}$ holds.

$m(\mathcal{P})$ a map that maps an attribute value to a sequence of fulfillment candidate instances (of the same attribute value) in pool $\mathcal{P}$.

$$m(\mathcal{P}) : v \rightarrow I \tag{2}$$

with $v \in \mathcal{S}(\mathcal{P})$ and set of instances $I \subseteq \mathcal{P}$.

Later in the discussion, Fig. 4 introduces an example of such a mapping.

Availability Mask Functions

Availability masking uses a boolean mask to indicate whether an instance is available for recipe fulfillment. The mask $\alpha_{\mathcal{P}}(i)$ equals to 1 if and only if the instance argument i is available for recipe fulfillment (otherwise 0). Consequently, an instance may only be nominated for a fulfillment if $\alpha_{\mathcal{P}}(i) = 1$ holds for instance $i \in \mathcal{P}$. There are three flavors of availability masks. First, there is the ALL mask, which means that all instances are available. Alternatively, there is the FIRST mask, which marks the first element as available. Subsequent instances are available if and only if they share the selector attribute value of the *first* instance, in one sustained sequence (as is often the case in physical stacks only accessible from the stacking direction). Somewhat inversely, there is the LAST mask. As the name suggests, this mask marks the last element as available. Preceding instances are available if and only if they share the selector attribute of the

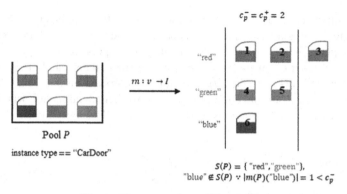

Fig. 4. Map generation $m(\mathcal{P})$ example

last instance, in one sustained sequence. All three masks are defined with the following equations:

$$\alpha_{\mathcal{P}}^{\text{ALL}}(\mathcal{P}(i)) \equiv 1, \quad \forall i \in \{1, \ldots, |\mathcal{P}|\} \tag{3}$$

$$\alpha_{\mathcal{P}}^{\text{FIRST}}(\mathcal{P}(i)) \equiv \begin{cases} 1 \text{ if } i = 1 \vee \left(v_{\mathcal{P}(i)} = v_{\mathcal{P}(i-1)} = \ldots = v_{\mathcal{P}(1)}\right) \\ 0 \qquad\qquad\qquad \text{otherwise} \end{cases} \quad \forall i \in \{1, \ldots, |\mathcal{P}|\} \tag{4}$$

$$\alpha_{\mathcal{P}}^{\text{LAST}}(\mathcal{P}(i)) \equiv \begin{cases} 1 \text{ if } i = |\mathcal{P}| \vee \left(v_{\mathcal{P}(i)} = v_{\mathcal{P}(i+1)} = \ldots = v_{\mathcal{P}(|\mathcal{P}|)}\right) \\ 0 \qquad\qquad\qquad \text{otherwise} \end{cases} \quad \forall i \in \{1, \ldots, |\mathcal{P}|\} \tag{5}$$

Release Policy Functions

Release policies use a ranking function to prioritize instances for fulfillment. A lower rank means the instance is preferred. First off, there is the First-In-First-Out (FIFO) release policy, which orders instances based on the timestamp t at which they were added to the recipe pool.

$$i_1 \prec i_2 \Leftrightarrow t_{i_1} \leq t_{i_2}, \quad \forall(i_1, i_2) \in \mathcal{P} \times \mathcal{P} \tag{6}$$

Instance i_1 is preferred over i_2 for release, if and only if the time added to the pool of i_1, t_{i_1} is smaller than or equal to that of i_2, t_{i_2}. In other words: the instances are ranked such that their timestamps are non-decreasing. The ranking function, $\rho_{\mathcal{P}}^{\text{FIFO}}$ is therefore defined simply as the instance index of the time-sorted sequence of instances in a pool:

$$\rho_{\mathcal{P}}^{\text{FIFO}}(\mathcal{P}(i)) \equiv i, \quad \forall i \in \{1, \ldots, |\mathcal{P}|\} \tag{7}$$

Secondly, there is the inverse of FIFO, Last-In-First-Out (LIFO), again based on timestamp t.

$$i_1 \prec i_2 \Leftrightarrow t_{i_1} \geq t_{i_2}, \quad \forall(i_1, i_2) \in \mathcal{P} \times \mathcal{P} \tag{8}$$

Notice that Eq. (8) results in the reverse ranking of Eq. (6). The resulting ranking function, $\rho_{\mathcal{P}}^{\text{LIFO}}$ is therefore the inverse ranking of Eq. (7):

$$\rho_{\mathcal{P}}^{\text{LIFO}}(\mathcal{P}(i)) \equiv 1 + |\mathcal{P}| - i, \quad \forall i \in \{1, \ldots, |\mathcal{P}|\} \tag{9}$$

Lastly, there is the attribute based policy (**ATTR**), which sorts instances based on some attribute, denoted by #. As an example instantiation of this policy, one could think of a priority based policy.

$$i_1 \prec i_2 \Leftrightarrow \#_{i_1} \geq \#_{i_2}, \ \forall (i_1, i_2) \in \mathcal{P} \times \mathcal{P} \tag{10}$$

To define the **ATTR** release policy ranking function, we first define the sequence $\text{sort}^{\downarrow}(A, \#) \subseteq A$ to be the result of sorting sequence A on some attribute # in descending order (i.e. the result is nonincreasing). Furthermore, we define $\text{index}(i, A) \in \{1, \ldots, |A|\}$ to return the index at which element i occurs in sequence A. Using these intermediate definitions, we can arrive at the final definition:

$$\rho_{\mathcal{P}}^{\text{ATTR}}(\mathcal{P}(i)) \equiv \text{index}\big(\mathcal{P}(i), \text{sort}^{\downarrow}(\{\mathcal{P}, \#\})\big), \ \forall i \in \{1, \ldots, |\mathcal{P}|\} \tag{11}$$

where # refers to the priority attribute to be sorted.

Given the properties of these functions, the discussion above can be generalized to

$$i_1 \prec i_2 \Leftrightarrow \rho_{\mathcal{P}}(i_1) \leq \rho_{\mathcal{P}}(i_2) \ \forall (i_1, i_2) \in \mathcal{P} \times \mathcal{P} \tag{12}$$

This generalized form is used in the subsequent implementation. The function definition denoted by $\rho_{\mathcal{P}}$ is to be replaced with an appropriate release policy function variant.

Note that, since output rules are released instantaneously once a recipe is fulfilled, the effect of these release policies is only observable if there is a choice which instances should remain in the pool. This choice is only there if there are more instances in the pool than the maximum fulfillment cardinality, i.e. $|m(v \in \mathcal{S}(\mathcal{P}))| > c_{\mathcal{P}}^{+}$. Otherwise, exactly $\min\big(c_{\mathcal{P}}^{+}, |\mathcal{P}|\big)$ instances are selected in the fulfillment and the ordering is irrelevant, as becomes apparent in the following algorithmic discussion.

The Pool Algorithm
As mentioned before, a pool can produce a mapping $m : v \in S(P) \rightarrow I \subseteq P$ upon request. This mapping maps an attribute value v to a sequence of fulfillment candidate instances I. A visual example that explains how that mapping works, can be found in Fig. 4. In this figure, the "CarDoor" pool from Fig. 2 is used as an example.

The pool's mapping algorithm is listed in Fig. 5.

The Recipe Algorithm
The recipe algorithm collects and analyzes pool maps to determine fulfillment feasibility. If a fulfillment can be achieved for a particular selector attribute value, the algorithm releases the appropriate instances from the pools and returns them in a list. The algorithm is listed in Fig. 6.

5   Prototype Implementation, Demonstration and Evaluation

This section discusses the technical implementation of the proposed synchronization approach, as it was prototyped, demonstrated and evaluated in a real use case.

Algorithm: Pool's mapping algorithm, i.e. $m(\mathcal{P})$.

Input: Pool $\mathcal{P}$.
Output: Mapping of attribute values to sequence of fulfillment candidate
    instances, $m : v \in \mathcal{S}(\mathcal{P}) \rightarrow I \subseteq \mathcal{P}$.

```
/* Map generation phase.                                                    */
m₁ ← ({} → {});                          /* Initialize empty map m₁. */
foreach i ∈ {i ∈ P : αₚ(i) = 1} do    /* For every available instance i in the pool. */
  if m₁(vᵢ) = ∅ then                     /* If value vᵢ not in map m₁ yet. */
    │ m₁(vᵢ) ← {};                       /* Add new value vᵢ to map m₁. */
  end
  │ m₁(vᵢ) ← m₁(vᵢ) ∪ {i};              /* Add instance i to map m₁. */
end
/* Map pruning phase.                                                       */
m₂ ← ({} → {});                          /* Initialize empty map m₂. */
foreach v ∈ m₁ do                        /* For every key value in map m₁. */
  if |m(v)| ≥ c⁻ₚ then                   /* At least c⁻ₚ instances exist for value v. */
    │ m₁(v) ← sort↑(m(v), ρₚ);          /* Rank instances based on release policy. */
    │ l ← {};                            /* Initialize empty candidate list. */
    │ x ← min(|P|, c⁺ₚ);                 /* Determine how many instances to nominate. */
    │ for (i ← 1; i ≤ x; i ← i + 1) do  /* For every nominated instance. */
    │   │ l ← l ∪ {m₁(v)(i)};           /* Add instance m₁(v)(i) to list of candidates. */
    │ end
    │ m₂(v) ← l;                         /* Place list of candidates in pruned map m₂. */
  end
end
return m₂;                               /* Return the pruned map m₂. */
```

Fig. 5. Pool's mapping algorithm

5.1 Technical Implementation

The aforementioned mathematical model and algorithms are implemented in a digital artifact using the Java programming language. The classes of the model are also represented by a technical data model. The Java code interacts with the process engine of a BPM System that executes the process models. The code is embedded in a typical process model definition, which offers the functionality of the recipe controller. The high-level internal implementation of the controller in BPMN 2.0 is shown in the bottom part of Fig. 7. It receives Submit (or Cancel) messages from specific synchronization points from the main process definitions (e.g. after Production task A1 of Production Process A and at the end of Production Process B), evaluates the recipes based on the messages' content and releases (via Release messages) the continuation of control flow once recipes are fulfilled.

5.2 Demonstration and Evaluation

The implemented recipe system was demonstrated in a real-world use case in the manufacturing printing domain, within the European EIT OEDIPUS[1] project. The scenario consisted of several printers, binding and trimming machines, and a robotic arm mounted on an AGV to grasp and transport paper and books between the devices and storage places. Activities performed by all these agents were modelled and enacted by a BPMS.

---

[1] https://www.eitdigital.eu/innovation-factory/digital-industry/oedipus/.

Algorithm: Recipe's fulfillment algorithm.

Output: Sequence of buffered instances that are part of the fulfillment. Empty
       sequence if recipe cannot be fulfilled.

```
/* Map analysis phase.                                                                    */
v ← ∅;                              /* Initialize sequence of potential fulfillment values. */
foreach p ∈ R do                                          /* For each pool in recipe. */
 │  m_p ← m(p);                                    /* Query and store the pool's map. */
 │  if |keys(m_p)| = 0 ∧ c_p^- ≠ 0 then              /* If this pool cannot be fulfilled. */
 │   │  v ← ∅;                              /* A global fulfillment is infeasible. */
 │   │  break;
 │  end
 │  if v = ∅ then                          /* If this is the first pool to analyze. */
 │   │  /* Take the first pool's potential fulfillment values as starting point.         */
 │   │  if c_p^- = 0 then
 │   │   │  v ← {∅};          /* Add generic null value as potential fulfillment value. */
 │   │  else
 │   │   │  v ← keys(m_p);      /* Add potential fulfillment values to sequence. */
 │   │  end
 │  end
 │  if g_p = 0 ∧ c_p^- ≠ 0 ∧ |p| ≠ 0 then              /* If this pool should be accounted for in
 │  fulfillment feasibility. */
 │   │  if v = {∅} then        /* If the previous pool was a generic pool (or was a satisfied
 │   │  pool with 0 candidates), but this pool is not. */
 │   │   │  v ← keys(m_p);                  /* Overwrite potential fulfillment values. */
 │   │  else
 │   │   │  v ← v ∩ keys(m_p);              /* Prune potential fulfillment values. */
 │   │  end
 │  end
end
/* Fulfillment feasibility analysis phase.                                                 */
if v = ∅ ∨ |v| = 0 then                          /* If no fulfillment is feasible. */
 │  return {};                              /* Return empty sequence. */
end
f ← v(1);                      /* Pick the or a fulfillment value and store it in f. */
r ← {};                          /* Initialize sequence of released instances. */
/* Note: f = ∅ can hold true by design, in case of a generic fulfillment.                 */
/* Data restructure phase.                                                                 */
foreach p ∈ R do                                              /* For each pool. */
 │  if ¬(c_p^- = 0 ∧ |m(p)| = 0) then                  /* Skip empty optional pools. */
 │   │  foreach i ∈ m(p)(f) do                  /* For every to-be-released instance. */
 │   │   │  release(p, i);                      /* Release instance from pool. */
 │   │   │  r ← r ∪ {i};                  /* Add instance i to sequence of released. */
 │   │  end
 │  end
end
return r;                          /* Return sequence of released instances. */
```

Fig. 6. Recipe's fulfillment algorithm

Various synchronization points existed in the scenario, mapping to the buffering and
un(bundling) constructs described in this paper. Recipes, using the notation of Fig. 3,
were described and configured for supporting these points. The points were modeled in
the process models as well. One such point was the output tray of a printer. There, sequentially produced books were placed and were ready for transportation to a binder. The corresponding recipe took care to synchronize the activities for bundling (and unbundling)
of the books from the tray, onto the AGV, and then into the binder. For the sake of brevity,
the complete process models are not presented here. Figure 8 shows a simpler example
of independent process models interacting with the Recipe controller.

The recipe controller system was primarily evaluated on its functionality to support
the modeling in BPMN 2.0 of physical manufacturing constructs (i.e., buffering and
(un)bundling). It was also evaluated in terms of usability by asking practitioners, through

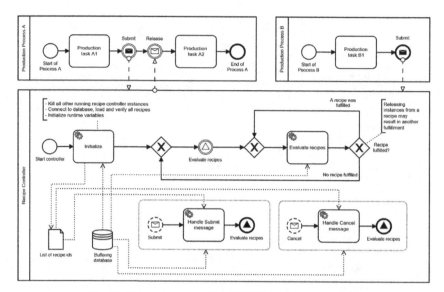

Fig. 7. The recipe controller (BPMN 2.0 process model)

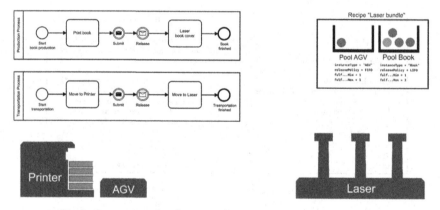

Fig. 8. Interaction of recipe controller with manufacturing processes

a structured survey, to define recipes and model synchronization points. In general, they find the approach useful.

6   Conclusion

This paper presents a solution for realizing dynamic process synchronization and correlation through a structured use of BPMN messages, with the goal to support manufacturing constructs, more specifically, buffering, bundling and unbundling. The solution is a novel approach to address the general dynamic synchronization issue, stemming from the process instance isolation, that the language fails to support. The recipe system that we propose is formally described, implemented and demonstrated in a real-world case study at a large, international firm in the printing industry. While this work does not claim complete suitability of the solution to all cases, the demonstration of the solution proves its feasibility. The usability was also evaluated by practitioners, engineers and researchers that modeled processes with the recipe system approach, who perceived it as useful.

However, there are limitations in the current work that are opportunities for further research. Assumptions in the definition of the models, such as that pools have infinite capacity or their cardinality is only expressed in units of process instances, may need to be relaxed. Workaround solutions exist, such as using an external knapsack problem solving engine, which passes group information to the recipe system in the form of selector attribute values, so that the recipe system can perform the appropriate bundling operations. Similarly, the assumption that a selector attribute is shared across all pools should be addressed by giving unique object identifiers to each pool. Furthermore, to make the system more dynamic and flexible, the recipes should be (re)configured during runtime and the fulfillment conditions should be variable instead of static. Not forget to mention that new BPMN elements can be crafted as extension to the notation for representing buffer and synchronization points.

References

1. Erasmus, J., Vanderfeesten, I., Traganos, K., Grefen, P.: The case for unified process management in smart manufacturing. In: IEEE Computer Society Digital Library (2018)
2. Brahe, S.: BPM on top of SOA: experiences from the financial industry. In: Alonso, G., Dadam, P., Rosemann, M. (eds.) BPM 2007. LNCS, vol. 4714, pp. 96–111. Springer, Heidelberg (2007). https://doi.org/10.1007/978-3-540-75183-0_8
3. Reichert, M.: What BPM technology can do for healthcare process support. In: Peleg, M., Lavrač, N., Combi, C. (eds.) AIME 2011. LNCS (LNAI), vol. 6747, pp. 2–13. Springer, Heidelberg (2011). https://doi.org/10.1007/978-3-642-22218-4_2
4. Baumgraß, A., Dijkman, R., Grefen, P., Pourmirza, S., Völzer, H., Weske, M.: A software architecture for transportation planning and monitoring in a collaborative network. In: C-Matos, L.M., Bénaben, F., Picard, W. (eds.) PRO-VE 2015. IAICT, vol. 463, pp. 277–284. Springer, Cham (2015). https://doi.org/10.1007/978-3-319-24141-8_25
5. Janiesch, C., et al.: The Internet-of-Things meets business process management: MutualBenefits and challenges. arXiv:1709.03628 (2017)
6. Decker, G., Barros, A.: Interaction modeling using BPMN. In: ter Hofstede, A., Benatallah, B., Paik, H.-Y. (eds.) BPM 2007. LNCS, vol. 4928, pp. 208–219. Springer, Heidelberg (2008). https://doi.org/10.1007/978-3-540-78238-4_22
7. Wohed, P., van der Aalst, W.M.P., Dumas, M., ter Hofstede, A.H.M., Russell, N.: On the Suitability of BPMN for Business Process Modelling. In: Dustdar, S., Fiadeiro, J.L., Sheth, A.P. (eds.) BPM 2006. LNCS, vol. 4102, pp. 161–176. Springer, Heidelberg (2006). https://doi.org/10.1007/11841760_12
8. Rosa, M., ter Hofstede, A., Wohed, P., Reijers, H., Mendling, J., van der Aalst, W.: Managing process model complexity via concrete syntax modifications. IEEE Trans. Ind. Inform. **7**(2), 255–265 (2011). https://doi.org/10.1109/TII.2011.2124467
9. Witsch, M., Vogel-Heuser, B.: Towards a formal specification framework for manufacturing execution systems. IEEE Trans. Ind. Inform. **8**(2), 311–320 (2012). https://doi.org/10.1109/TII.2012.2186585

10. Ko, R., Lee, S., Wah Lee, E.: Business process management (BPM) standards: a survey. Bus. Process Manag. J. **15**(5), 744–791 (2009)
11. Pauker, F., Mangler, J., Rinderle-Ma, S., Pollak, C.: Centurio.work - modular secure manufacturing orchestration. In: Proceedings of the Dissertation Award, Demonstration, and Industrial Track of the 16th International Conference on Business Process Management (BPM), CEUR-WS.org, Sydney, Australia (2018)
12. Prades, L., Romero, F., Estruch, A., García-Dominguez, A., Serrano, J.: Defining a methodology to design and implement business process models in BPMN according to the standard ANSI/ISA-95 in a manufacturing enterprise. Procedia Eng. **63**, 115–122 (2013). https://doi.org/10.1016/j.proeng.2013.08.283
13. Zor, S., Schumm, D., Leymann, F.: A proposal of BPMN extensions for the manufacturing domain. In: Proceedings of the 44th CIRP International Conference on Manufacturing Systems (2011)
14. Abouzid, I., Saidi, R.: Proposal of BPMN extensions for modelling manufacturing processes. In: 2019 5th International Conference on Optimization and Applications (ICOA), Kenitra, Morocco, pp. 1–6 (2019). https://doi.org/10.1109/icoa.2019.8727651
15. García-Domínguez, A., Marcos, M., Medina, I.: A comparison of BPMN 2.0 with other notations for manufacturing processes. In: AIP Conference Proceedings, Cadiz, vol. 1431, pp. 593–600 (2012). https://doi.org/10.1063/1.4707613
16. Van der Aalst, W., Artale, A., Montali, M., Tritini, S.: Object-centric behavioral constraints: integrating data and declarative process modelling. In: Description Logics (2017)
17. Leitner, M., Mangler, J., R-M, S.: Definition and enactment of instance-spanning process constraints. In: Wang, X.S., Cruz, I., Delis, A., Huang, G. (eds.) WISE 2012. LNCS, vol. 7651, pp. 652–658. Springer, Heidelberg (2012). https://doi.org/10.1007/978-3-642-35063-4_49
18. Kim, B.H., Park, S.B., Lee, G.B., Chung, S.Y.: Framework of integrated system for the innovation of mold manufacturing through process integration and collaboration. In: Gervasi, O., Gavrilova, M.L. (eds.) ICCSA 2007. LNCS, vol. 4707, pp. 1–10. Springer, Heidelberg (2007). https://doi.org/10.1007/978-3-540-74484-9_1
19. Cadavid, J., Alférez, M., Gérard, S., Tessier, P.: Conceiving the model-driven smart factory. In: ACM International Conference Proceeding Series, August 2015, vol. 24-26, pp. 72–76. Association for Computing Machinery (2015). https://doi.org/10.1145/2785592.2785602
20. Jasiulewicz-Kaczmarek, M., Waszkowski, R., Piechowski, M., Wyczółkowski, R.: Implementing BPMN in maintenance process modeling. In: Świątek, J., Borzemski, L., Wilimowska, Z. (eds.) ISAT 2017. AISC, vol. 656, pp. 300–309. Springer, Cham (2018). https://doi.org/10.1007/978-3-319-67229-8_27
21. Kavka, C., Campagna, D., Milleri, M., Segatto, A., Belouettar, S., Laurini, E.: Business decisions modelling in a multi-scale composite material selection framework. In: 4th IEEE International Symposium on Systems Engineering (2018). https://doi.org/10.1109/syseng.2018.8544386
22. Knoch, S., et al.: Enhancing process data in manual assembly workflows. In: Daniel, F., Sheng, Q.Z., Motahari, H. (eds.) BPM 2018. LNBIP, vol. 342, pp. 269–280. Springer, Cham (2019). https://doi.org/10.1007/978-3-030-11641-5_21
23. Yousfi, A., Bauer, C., Saidi, R., Dey, A.K.: uBPMN: A BPMN extension for modeling ubiquitous business processes. Inf. Softw. Technol. **74**, 55–68 (2016). https://doi.org/10.1016/j.infsof.2016.02.002
24. Petrasch, R., Hentschke, R.: Process modeling for industry 4.0 applications: towards an industry 4.0 process modeling language and method. In: 13th International Joint Conference on Computer Science and Software Engineering, JCSSE (2016)

25. Lindorfer, R., Froschauer, R., Schwarz, G.: ADAPT - a decision model-based approach for modeling collaborative assembly and manufacturing tasks. In: Proceedings of the IEEE 16th International Conference on Industrial Informatics, INDIN, pp. 559–564 (2018)

26. Cohn, D., Hull, R.: Business artifacts: a data-centric approach to modeling business operations and processes. IEEE Data Eng. Bull. **32**, 3–9 (2009)

27. Lohmann, N., Wolf, K.: Artifact-centric choreographies. In: Maglio, P.P., Weske, M., Yang, J., Fantinato, M. (eds.) ICSOC 2010. LNCS, vol. 6470, pp. 32–46. Springer, Heidelberg (2010). https://doi.org/10.1007/978-3-642-17358-5_3

28. Meyer, A., et al.: Data perspective in process choreographies: modeling and execution. Techn. Ber. BPM Center Report BPM-13-29. BPMcenter. org. (2013)

29. Meyer, A., et al.: Automating data exchange in process choreographies. In: Jarke, M., et al. (eds.) CAiSE 2014. LNCS, vol. 8484, pp. 316–331. Springer, Cham (2014). https://doi.org/10.1007/978-3-319-07881-6_22

30. Meyer, A., Weske, M.: Activity-centric and artifact-centric process model roundtrip. In: Lohmann, N., Song, M., Wohed, P. (eds.) BPM 2013. LNBIP, vol. 171, pp. 167–181. Springer, Cham (2014). https://doi.org/10.1007/978-3-319-06257-0_14

31. Lohmann, N., Nyolt, M.: Artifact-centric modeling using BPMN. In: Pallis, G., et al. (eds.) ICSOC 2011. LNCS, vol. 7221, pp. 54–65. Springer, Heidelberg (2012). https://doi.org/10.1007/978-3-642-31875-7_7

32. Fahland, D.: Describing behavior of processes with many-to-many interactions. In: Donatelli, S., Haar, S. (eds.) PETRI NETS 2019. LNCS, vol. 11522, pp. 3–24. Springer, Cham (2019). https://doi.org/10.1007/978-3-030-21571-2_1

33. Steinau, S., Andrews, K., Reichert, M.: The relational process structure. In: Krogstie, J., Reijers, H.A. (eds.) CAiSE 2018. LNCS, vol. 10816, pp. 53–67. Springer, Cham (2018). https://doi.org/10.1007/978-3-319-91563-0_4

34. Van der Aalst, W., Barthelmess, P., Ellis, C., Wainer, J.: Proclets: a framework for lightweight interacting workflow processes. Int. J. Coop. Inf. Syst. **10**, 443–481 (2001). https://doi.org/10.1142/S0218843001000412

35. Fahland, D., De Leoni, M., Van Dongen, B., Van der Aalst, W.: Many to-many: some observations on interactions in artifact choreographies. ZEUS **705**, 9–15 (2011)

36. Pufahl, L., Weske, M.: Batch activity: enhancing business process modeling and enactment with batch processing. Computing **101**(12), 1909–1933 (2019). https://doi.org/10.1007/s00607-019-00717-4

37. Marengo, E., Nutt, W., Perktold, M.: Construction process modeling: representing activities, items and their interplay. In: Weske, M., Montali, M., Weber, I., vom Brocke, J. (eds.) BPM 2018. LNCS, vol. 11080, pp. 48–65. Springer, Cham (2018). https://doi.org/10.1007/978-3-319-98648-7_4

38. Pesic, M., Schonenberg, H., Van der Aalst, W.: DECLARE: full support for loosely-structured processes. In: Proceedings of the 11th IEEE International Enterprise Distributed Object Computing Conference, pp. 287–300. IEEE (2007)

39. Nahmias, S., Olsen, T.: Production and Operations Analysis, 7th edn. Waveland Press, Long Grove, Ill (2015). (OCLC: 935795578)

40. Cachon, G., Terwiesch, C.: Matching Supply with Demand: An Introduction to Operations Management. McGraw-Hill/Irwin, Boston (2009). (OCLC: ocn191732546)

41. Erasmus, J., Vanderfeesten, I., Traganos, K., Grefen, P.: Using business process models for the specification of manufacturing operations. In: Computers in Industry (to appear)

42. Defense Acquisition University: Integrated Product Support (IPS) Element Guidebook. Defense Acquisition University, Fort Belvoir (2011)

43. De Groote, X.: Inventory theory: a road map. teaching note. Department of Decision Sciences, The Whanon School (1989)

44. Van der Aalst, W.: Putting high-level Petri nets to work in industry. Comput. Ind. **25**(1), 45–54 (1994). https://doi.org/10.1016/0166-3615(94)90031-0
45. Spijkers, D.: Expressing and supporting buffering and (un)bundling in the manufacturing domain using BPMN 2.0. Master's thesis, Eindhoven University of Technology, Eindhoven (2019)

Feature Development in BPMN-Based Process-Driven Applications

Konrad Schneid[1]([✉]) [ID], Sebastian Thöne[1], and Herbert Kuchen[2] [ID]

[1] Münster University of Applied Sciences, Münster, Germany
{konrad.schneid,sebastian.thoene}@fh-muenster.de
[2] University of Münster, Münster, Germany
kuchen@uni-muenster.de

Abstract. In the context of Continuous Software Engineering, it is acknowledged as best practice to develop new features on the mainline rather than on separate feature branches. Unfinished work is then usually prevented from going live by some kind of feature toggle. However, there is no concept of feature toggles for Process-Driven Applications (PDA) so far. PDAs are hybrid systems consisting not only of classical source code but also of a machine-interpretable business process model. This paper elaborates on a feature development approach that covers both the business process model and the accompanying source code artifacts of a PDA. The proposed solution, *Toggles for Process-Driven Applications (T4PDA)*, equipped with an easy to use modeling tool extension, enables the developer to safely commit unfinished work on model and source code to the project's mainline. It will be kept inactive during productive deployments unless the feature is finally released. During an AB/BA crossover design experiment, the T4PDA approach, including the provided tool support, showed higher software quality, a faster development process, and contented developers.

Keywords: Feature-driven development · Process-Driven Application · Continuous Software Engineering

1 Introduction

Process-Driven Applications are not only based on program code but also on executable process models. This type of process-aware information systems plays an increasingly important role, especially in enterprises living Business Process Management [18]. Executable process models can be modeled in the Business Process Model and Notation (BPMN) [7,12]. The language, published as ISO standard, is especially characterized by high comprehensibility without extensive prior (technical) knowledge. To further separate decision logic from process logic, decision models using Decision Model and Notation (DMN) can be added [13]. BPMN and DMN are both maintained by the Object Management Group (OMG) and can be ideally used in combination. Forming a common language, software developers and process designers can collaborate on these process

© Springer Nature Switzerland AG 2020
D. Fahland et al. (Eds.): BPM Forum 2020, LNBIP 392, pp. 35–50, 2020.
https://doi.org/10.1007/978-3-030-58638-6_3

models. To make a model executable and link certain model elements to units of source code, technical parameters can be configured directly on the model level. For this purpose, the meta-model of BPMN offers dedicated attributes, which do not influence the visualization and, thus, do not disturb any business stakeholder [6].

The concepts of Continuous Software Engineering intend to increase software quality through immediate integration (and sometimes also delivery) of new features. The approach requires the automatic building and testing of software artifacts in short cycles and is getting more and more popular in software development projects. Besides a strengthened software quality, the high degree of automation in testing and deployment enables shorter release intervals for new features (time-to-market) [4].

Techniques are necessary for the Continuous Integration (CI) of changes in order not to jeopardize the productive status by unfinished implementation parts. The widespread approach *Branch by Feature* employs separate development branches for features which are merged to the mainline after completion only. However, this approach is regarded as an anti-pattern for CI due to the missing integration of all changes to the common mainline. A CI-compliant alternative is *Develop on Mainline*. Here, all changes are continuously committed to the common mainline. But then, additional techniques such as *Feature Toggles* are necessary in order to exclude unfinished work from undesired release and to keep deployments production-ready [9].

Feature toggles are well established for traditional software consisting of program code. However, when it comes to the hybrid artifacts of PDAs, we are facing new challenges with the integration of process models into the Develop on Mainline approach: How can toggles be expressed in BPMN and DMN, how can they be activated depending on inputs from the environment, and how can they be integrated with source code artifacts? In this paper, we address these research questions and the problem of how to support both process designers and programmers in the definition of common feature toggles in a PDA. We aim at a solution which does not compromise over the comprehensibility of the process model. Tool support is provided so that manual effort is minimized and more business-focused process designers are not excluded from collaboration. This way, we want to enable a continuous engineering of PDAs based on CI and CD.

As methodology, we use Design Science Research (DSR) [14]. Roughly, the idea is that for an observed problem a prototypical artifact is developed to solve it. Then this artifact is evaluated and possibly improved in further iterations.

The rest of this paper is organized as follows. In Sect. 2, we explain the relevant background (BPMN, DMN, PDA, CI & CD, and feature development strategies). In Sect. 3, we discuss related work. As our main contributions, we propose toggles for PDA in Sect. 4 and an implementation as plugin for the Camunda Modeler[1] in Sect. 5. In Sect. 6, we evaluate our approach, before we conclude in Sect. 7.

---

[1] The plugin is available at https://git.fh-muenster.de/winfo/code-pro/t4pda.

2 Background

2.1 BPMN and DMN

BPMN is a well-established standard, maintained by the Object Management Group (OMG), for modeling business processes with a token-based execution semantics [7,12]. Its notation is intended to be comprehensible for all stakeholders involved. The main notation elements of BPMN can be divided into four categories: flow objects, data objects, connection objects, and collaboration elements (see Figs. 4 and 7 for examples). Flow objects play a key role and describe the process behavior. The most relevant ones are the different types of tasks, e.g., user tasks or automated service tasks, and gateways, which control branchings in the process flow. Flow objects are interconnected by directed sequence flows. Collaboration elements such as pools and message flows model the interaction between different processes, whereby data objects represent data flow aspects. The process model is stored in an XML file, which contains both the abstract representation and the layout information of a model [1].

In 2015, the OMG released their proposal *Decision Model and Notation (DMN)* for separation of decision logic from process schemes [13]. In a DMN model, the domain expert defines a set of related decision rules that a rule engine can automatically evaluate at runtime. As with BPMN, the approach intends to support transparency and usability. The central element is the decision table (see Fig. 6 for an example). In this table, results ("output") are derived based on modeled input data ("input"). Rules of the decision model are captured in the Friendly Enough Expression Language (FEEL). For more complex decision making, the Decision Requisitions Graph is used, which is visualized as a Decision Requisitions Diagram. This is used, especially if various decisions depend on each other [6].

2.2 Process-Driven Applications

Process-Driven Applications are a special kind of process-aware information systems which are used in enterprises and organizations to automate the execution of business processes. Their application domain covers human-centric workflows up to highly integrated service orchestrations [18,20]. Technically, they are hybrid systems, implemented not only by source code but centered around a machine-interpretable process model, typically in BPMN and DMN.

The deployment artifacts of a PDA include the process model and, optionally, decision models. To make the process model executable, its elements have to be equipped with technical parameters such as assignment logic for user tasks and references to classes or operations implementing service tasks. The source code is used to program the behavior of automated tasks (in BPMN called "service tasks"), accompanying event listeners, and adapters for accessing external applications and services. Eventually, the developer has to provide the implementation of view components that are related to human-centered user tasks.

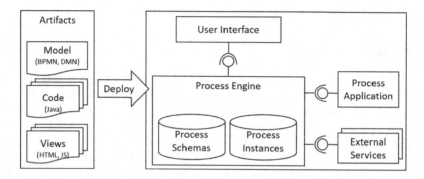

Fig. 1. Artifacts and runtime components of a Process-Driven Application.

To enact a process based on these artifacts, a process engine such as the open-source product Camunda BPM needs to be employed. Such a process engine can either be embedded into the application as a library or, as shown in Fig. 1, the engine is run as a separate service interacting with user interfaces (e.g. task lists), the implemented process applications, and external services.

2.3   Continuous Integration and Delivery

Continuous Integration (CI) is an established best practice of software development in teams, which has become increasingly popular since the article by Martin Fowler, but already was introduced in 1991 by Grady Booch [2,5]. All artifacts of an application, such as source code, should be kept in a single repository of a configuration management system. In the case of PDAs, this also includes the process model as a central element. A well-known configuration management system for such repositories is Git[2]. The main idea of the CI approach is a continuous integration of all changes. For this purpose, all developers (and process modelers) involved are required to commit their changes to a common mainline branch. It is claimed to run the integration process and corresponding integration tests at least on a daily basis, better after every single commit to the central repository [11].

The integration process is performed by a fully automated pipeline which triggers several sequential or parallel jobs (see Fig. 2). These jobs include building, analysis, and testing of the deployment artifacts. Afterwards, the developer receives immediate feedback on his committed changes. A successful run of all tests results in a build that can be deployed [11]. For the automated process, a CI-Server such as Travis CI or Jenkins is required [10].

The Continuous Delivery (CD) approach goes even one step further. While a CI run provides a deployable, packaged deliverable, the CD approach also includes the final step of deploying the application to a target environment, which could even mean releasing it to the production stage [21].

---

[2] https://git-scm.com/.

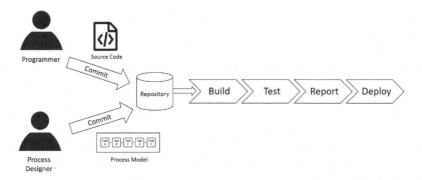

Fig. 2. A CD pipeline of a Process-Driven Application fed with source code and process model.

Benefits from CI and CD are increasing levels of software quality and developer productivity. Due to the high degree of automation, it is possible to keep feedback loops and release intervals short and at the same time test coverage and code quality high.

2.4 Feature Development Strategies in Context of CD

In ongoing projects, developers continuously have to implement new requirements and change requests, also referred to as features, in their program code. Having CI and CD in mind, the question arises on how to deal with implementations that are still under development and not yet ready for release. To cope with this question, the software engineering community mainly distinguishes between two alternative approaches: Develop on Mainline and Branch by Feature.

Following the Branch by Feature approach, a separate branch is created for committing the work on a new feature (see Fig. 3a). The mainline will not be disturbed by uncompleted changes and thus always keeps a release-ready status. As soon as the development in the dedicated feature branch is complete, the new feature will be merged into the mainline. However, this approach contradicts the core idea of CI: Due to the separation of branches, continuous integration of new feature code does not take place from the beginning on but only at a final stage. In the case of parallel developments, these branches can diverge strongly and end up in awful merging conflicts [21]. HUMBLE even states that "branching by feature is really the antithesis of continuous integration" [9].

Fig. 3. Branch by Feature (a) versus Develop on Mainline (b).

Feature development following the Develop on Mainline approach only takes place directly on the mainline; there are no other branches (see Fig. 3b). Changes are immediately available to all other developers, and merging conflicts are avoided. This smoothly accords with the CI approach. However, additional concepts such as feature toggles are necessary to ensure that uncompleted changes do not disturb the deployability and the productive status. Features in development can be activated and deactivated via such toggles. Thus, unfinished code has no negative impact on productive operations. In its simplest form, feature toggles are just if-else-statements [8]. More sophisticated approaches are provided by frameworks such as Togglz for Java [19] or by individually selectable profiles when using the popular Java framework Spring.

An alternative idea to feature toggles is called Branch by Abstraction. The term is a little bit misleading, since actually no branch in the code repository is created. Rather, an abstraction layer is introduced for the feature scope to be refined by different (old and new) concrete implementations. This approach is used especially for larger features, if small incremental changes are less appropriate [9].

All concepts discussed above consider programming artifacts only. PDA-compliant solutions that also cover process models still need to be designed. The difficulty with, e.g., feature toggles is how to smoothly integrate them into a process model and how to manage them across both model and code.

3   Related Work

RODRÍGUEZ ET AL. have observed in their study that in practice, both Develop on Mainline and Feature Branches are used in the CI context [16]. SCHNEID has evaluated, based on current literature-review, the development strategies in connection with Continuous Delivery for traditional software based on program code. He recommends the strategy Develop on Mainline combined with the techniques feature toggles and branch by abstraction as the most suitable solution for CD [17]. In a study by RAHMAN ET AL., the promises and perils of feature toggles were analyzed [15]. MAHDAVI-HEZAVEH ET AL. performed a qualitative analysis of the use of feature toggles in practice. They have identified various practices in the areas of management, initialization, implementation, and cleanup [15]. According to CHEN, there is further need for research in the field of software development within the Continuous Delivery approach [4].

All of the listed publications develop strategies for source code only. To the best of our knowledge, there is no approach yet which also includes process models. In the following sections, we will close this gap and propose an extended kind of toggles that enable feature development in continuously engineered PDA projects.

4   Toggles for Process-Driven Applications (T4PDA)

Developing new features for a PDA on the mainline, i.e. with feature toggles rather than feature branches, requires a solution which meets the following objectives:

- Must allow to define features across both source code and process model.
- Unfinished parts of a feature must not disturb productive operation. The old version has to be kept productive unless, after completion of modeling and implementation, the feature toggle is released.
- Activation of toggles must not be hard-coded but should be controllable by environment configurations.
- Toggles should be releasable gradually, e.g., depending on deployment stage, user group, or user quantity.
- Should not impair the comprehensibility of the process model.
- Should be applicable by process designers with low technical expertise, ideally through appropriate tool support.
- Should make easy to find out which features are currently active and why.
- Should support a clean removal of all toggle-related elements at completion of a feature.

With *Toggles for Process-Driven Applications (T4PDA)*, we propose a solution that extends classical feature toggles by suitable constructs of BPMN and DMN. Developing a feature this way comprises non-invasive model transformations, a sophisticated toggle evaluation, and useful code generation.

4.1   Model Transformations

The first question to answer is how toggles can be inserted into an existing BPMN model. One important element of our approach is to create a business rule task that refers to a decision model in DMN notation. It decides which toggles are currently active. The second transformation step encapsulates the model section that is to be changed in a new subprocess. This subprocess is then cloned, and both original and clone are placed as alternative paths subsequent to a newly inserted XOR-split. The original subprocess can be run for the time the feature is under development, and the cloned subprocess forms the work space for the feature developer. The XOR gateway is controlled by the toggle activation rules defined in the first mentioned DMN table.

Let's look at the example shown in Fig. 4. The two parallel sections form the scope of the planned feature. This scope has to be identified by the modeler. It might, in principle, contain any connected fraction of the current model. In the first transformation step, the selected feature scope is encapsulated behind a (collapsed) subprocess. The collapsed notation allows to keep the process clear and understandable. In the next step, the subprocess is copied and both versions are surrounded by exclusive gateways. Now, the designer can expand the copied subprocess and start working on it.

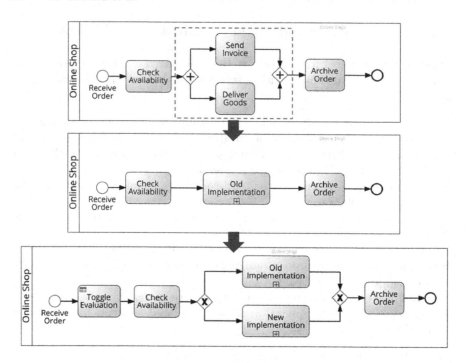

Fig. 4. Adding a new feature toggle - step by step illustration.

If the feature scope is spread over several lanes or unconnected model regions, the procedure has to be repeated for each occurrence. However, the same toggle (by id) can be used and has to be evaluated only once.

After the feature is completed and the toggle has been finally released, it needs to be removed from the model again in order to keep the model clean - unless it is supposed to be a permanent business toggle. The removal comprises the decision rules, the two inserted gateways, one of the two subprocesses, and any dangling sequence flow edges. As discussed in Sect. 5, these steps can easily be automated.

4.2   Toggle Evaluation

The new business rule task preceding the XOR split refers to a DMN model for toggle evaluation. Its outcome can be used as branching condition of the subsequent gateway and, thus, determines whether the control flow proceeds with the old or new version of this model part. As long as a toggle is deactivated, the old path is executed. Hence, the productive operation is not affected during the feature development phase, even in projects running Continuous Delivery.

The DMN model can either be used locally or, if certain functionalities affect several process models, globally across the affected process landscape. Due to its simple notation, the decision model is easily comprehensible and manageable by process designers and software developers. It even helps operators to understand the reason behind a process flow.

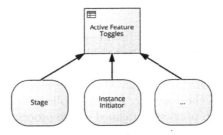

Fig. 5. A simplified version of a decision model for evaluating toggles.

Figure 5 shows possible evaluation criteria. One considerable option is to include the current deployment stage of the PDA. At the beginning, the new feature could be active at the development stage, only. Once development is complete, the feature can be released to the test stage for functional tests before it is finally released to the production stage. Another meaningful decision input might be the process instance initiator or a certain user group. For instance, the developer could define a rule, which binds the activation of a toggle to his user id, and test his feature without disturbing other developers or users.

At design time, the modeler can exploit the available input variables in order to define as many toggle activation rules as desired. At runtime, these rules are automatically evaluated by a process or rule engine executing the generated business rule task. Every evaluation run results in a list of activated toggles, which can be accessed onwards as data object or process variable.

Decision_Table_Feature_Toggles			
C	Input +		Output +	
	Stage	InstanceInitiator	Toggle	
	string	string	string	Annotation
1	"Dev"	Developer01	"feature_A"	Active for User Developer01 at Dev-Stage
2	"Dev", "Test"	-	"feature_B"	Active for all Users at Dev-, and Test-Stage
3	"Dev", "Test", "Prod"	-	"feature_C"	Active for all Users at all Stages

Fig. 6. Basic structure of the decision table, which can be extended according to individual needs.

Figure 6 provides an example of such a decision table. It shows that Feature A is supposed to be active only for user "Developer01" and on stage "Dev". Feature B is active for all users on the development and test stage, while Feature C is active on all systems regardless of the user.

Another option is the application of formulas and expressions. This could be applied to realize, e.g., A/B-testing or lower-risk canary deployments, which first release the new feature to a small percentage of users only and then systematically cover more and more of them.

4.3 Code Generation

So far, the T4PDA concept mainly addresses the process modeler. However, software developers working on the technical level of the model and on accompanying source code artifacts such as the implementation of service tasks have to be considered, too. Obviously, they should be enabled to make use of the same feature toggles as the modeler. If they applied the same toggles in their source code as have been previously defined at the process model, then it would be possible to manage these toggles synchronously.

In order to achieve this goal, the programmer can access the list of active toggles stored in a data object after execution of the DMN decision task. For this purpose, we can assume a suitable API provided by the process engine in use. The available toggle IDs defined at the process level can be provided to the programmer by a small code generator that updates a declaration of suitable constants in the desired target programming language every time a feature toggle is created or removed.

5 Camunda Modeler Plugin to Support T4PDA

Tool support to follow the T4PDA approach is essential for both software developers and process designers, as otherwise manual effort would be high. Besides, using a supporting tool is less error-prone and, therefore, safer to use. As technical platform, we decided to build on *Camunda Modeler* due to the fact that the tool is popular and open source [3].

Figure 7 demonstrates how a feature development is started using the plugin in Camunda Modeler: The user selects the feature scope and starts the automated process with the button "Start Feature". In the subsequent dialog, the user provides a name of the feature or toggle, respectively, and decides if he wants the tool to copy the elements of his selection into the new subprocess or to generate an empty subprocess as starting point of the feature development. Afterwards, the tool automatically generates all required model transformations as defined in Sect. 4.1.

If not yet existing, a business rule task for evaluating the toggles is automatically inserted and the corresponding decision table is created (cf. Sect. 4.2). The generator also inserts a first toggle-evaluation rule into the DMN decision table, which leaves the new toggle inactive by default to avoid an accidental activation. The generated DMN table can be executed by the Camunda process engine and stores the resulting list of active toggles in a process variable. This process variable is accessed in the branching condition which is automatically assigned to the subsequent XOR gateway in order to determine the feature execution.

In addition, the tool supports software developers as explained in Sect. 4.3: Whenever the modeler starts a new feature, the tool uses the assigned name to

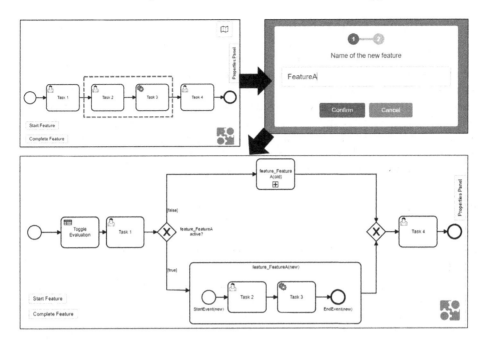

Fig. 7. Screenshots of the extended Camunda Modeler while adding a new feature.

generate the declaration of a constant in Java code and inserts it into a small but useful helper class. The programmer is supposed to use these constants for any toggle-dependent code. Using the same toggle names in both model and code, allows to also generate Java code for checking whether a toggle is active or not, which requires querying the Camunda process engine. Using the generated constants helps to avoid unnoticed typing errors and makes future renaming of a toggle easier.

Eventually, the tool enables an automated completion and removal of a toggle. For this purpose, the process section to be kept must be selected. Then the toggle and the path that is no longer required are removed from the process model by clicking the "Complete Feature" button. The entry in the decision model and the constant in the helper class are deleted, too. Optionally, the business-rule task is also removed, if there are no further toggles in the PDA. The time-consuming removal of toggles, which is listed as a disadvantage of feature toggles in literature [15], is not necessary. Only in the source code, any temporary toggle checks must be removed manually. If the constants of the helper class have been used in the program code, all affected lines can easily be detected by compile errors caused by the missing declaration of constants.

6   Empirical Evaluation

The proposed T4PDA approach and the corresponding tool are intended to increase software quality and developer productivity. In order to evaluate the success, we conducted an empirical experiment whose focus was on comparison of two groups of feature developers, working either with or without T4PDA. More precisely, a two-stage AB/BA crossover design experiment has been performed and certain key figures indicating productivity, software quality, usability, and user acceptance have been measured. The following sections report on the settings and findings.

6.1   Setting

Since T4PDA is a new proposal whose contribution cannot empirically be observed in the field, we decided for a practical experiment with selected feature development tasks. In order to be able to compare solving those tasks with and without our approach, we required two comparison groups with similar skills. That's why we chose participants of a bachelor degree program in business informatics at our university. The attendees of class "Business Engineering" were especially appropriate for our purpose because they are already experienced in traditional software development (after about 50 credits in software engineering according to the European credit transfer system ECTS) and have intensively learned how to model business processes in BPMN and how to implement them as PDAs using the Camunda tools. The experiment took place at the end of the course so that extensive knowledge in the field could be assumed.

Eighteen volunteer students registered for the experiment. At the beginning, all participants received a brief recap on Continuous Integration and feature development strategies such as Feature Branches and Develop on Mainline, which had already been taught in a previous course. Then, the experimental group was randomly divided into two subgroups of equal size and spatially separated. One and a half hours were planned for the subsequent experiment. This time span was scheduled into two rounds working on two different sample PDAs (a simple pizza delivery process and a more complex claim settlement process) and a final questionnaire. The difference between the two groups was that group I applied T4PDA in the first round only, and group II in the second round only. This typical AB/BA setting allows to uncover effects resulting from a certain sequence of tasks only (cf. Table 1).

Table 1. AB/BA experimental design.

Round:	Pizza delivery PDA	Claim settlement PDA
Group I:	T4PDA	Traditional
Group II:	Traditional	T4PDA

Before a group was asked to apply the new tool, they got a short introduction to and handout on T4PDA and a five minutes training with the Modeler plugin only. For the second group, this means that they did not know our approach at all when they followed the traditional approach in the first round.

In the first round of the experiment, tasks had to be performed on a lightweight PDA for a pizza delivery system. Its BPMN model consists of only 15 elements[3]. For this application, both groups received an identical set of tasks such as starting new features and completing or discarding them later on. The changes affected both, the process model and the program code.

Every time having fulfilled the next task and additionally after some random intervals, the participants had to commit their (intermediate) results to Git repositories provided beforehand. They were asked not to run any local tests but to push their changes blindly. In order to measure the quality of their work, every commit triggered a small CI pipeline, which validated the artifacts and the correct solution of the task by appropriate unit tests. The frequency of commits and the testing outcomes could be used as indicators for productivity and software quality. In order to avoid impact on the participants, the results were investigated by the research team only and not returned as direct feedback.

The PDA used in the second round is more complex than the first one. Provided by an industry partner, it deals with the settlement of claims in the event of vehicle glass breakage. The corresponding process model includes a total of 39 BPMN elements[4]. The higher complexity of the sample application also increased the difficulty of the feature development tasks to be solved in that round. This structure was reasonable because of the growing familiarization during the experiment.

After the two rounds, the participants had to fill in a short questionnaire about their experience during the experiment. A Likert scale was used to find out, among others, how far T4PDA supported Develop on Mainline, how helpful the modeler plugin was to solve the tasks, and how intuitive it was to learn. Eventually, a nominal-scale question asked the participants which approach they would prefer for feature development in PDAs.

Documents related to the experiment, such as the survey and the results, are available online[5].

6.2   Result and Discussion

The figures generated by the experiment give evidence on increased productivity, software quality, usability, and user acceptance. In the following discussion, the term "traditional" refers to those rounds of the experiment where a group was asked not to use the T4PDA approach. Then, the group members had to come up with own ad-hoc (and sometimes not CI-compliant) heuristics such as leaving new feature elements completely unconnected to the existing flow unless the feature is released.

---

[3] Six sequence flows, five tasks, three events, and one gateway.

[4] Twenty-one sequence flows, eight tasks, seven gateways, and three events.

[5] https://git.fh-muenster.de/winfo/code-pro/t4pda/-/tree/master/experiment.

Developer Productivity: The two subgroups worked on the same tasks in parallel, one with and the other one without T4PDA (altered after the break). Hence, an indicator for developer productivity is the number of accomplished tasks grouped by the two development modes. Since the participants have been asked to commit their results after every task completion, the total number of commits directly correlates to the number of accomplished tasks and serves as a measure of the developers' performance. In Table 2, column "Total Commits" reveals that when applying T4PDA the aggregated number of commits grew by 12%. It seems that T4PDA improves the productivity of feature-driven development.

Software Quality: The intention of Continuous Software Engineering is to always have production-ready executables. That T4PDA supports this objective is stressed by the other columns in Table 2. They reveal the ratio of successful builds with all unit tests passed: it is twice as high when T4PDA is used. Here, software quality certainly benefits from the fact that T4PDA ensures that the old implementation is still working while a new feature is under development. The apparently disappointing success rates for both approaches are justified by the given instruction not to perform any local checks before committing. This way, the success rate could be measured without effects by manual interferences.

Table 2. Number of commits and associated builds per approach.

Approach	Total commits	Successful builds	Success rate
Traditional	49	14	28,57%
T4PDA	55	27	49,01%

Usability: The survey conducted at the end of the experiment was mainly intended to aggregate the users' subjective assessments of the T4PDA approach and tool. The diagrams in Fig. 8 show the outcome on usability-related questions. In summary, strong majorities of the participants agree or even completely agree that T4PDA enables Develop on Mainline (a), was helpful in solving the feature development tasks (b), and offers an intuitively to use tool (c) - despite the short preparation time.

User Acceptance: When finally asked which feature development approach for PDAs they personally prefer, 88.9% of the participants voted for T4PDA (see Fig. 8d), after having tested both techniques during the experiment. This figure indicates a high user acceptance rate.

In summary, the various results of the experiment evidence the advantage of T4PDA over the traditional approach. That trend is considerable enough, even if one takes the lack of practical experience of the participating students compared to more experienced professionals into account. However, the evidence should next be strengthened by further observations in practical fields, ideally with a larger number of test persons.

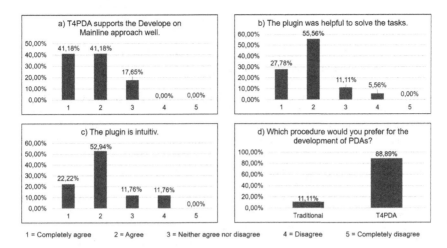

Fig. 8. Extract of the results of the survey after the experiment.

7   Conclusion and Further Work

We have presented T4PDA, the first approach which enables feature development on mainline for PDAs. The concept combines the strengths of process modeling using BPMN and DMN with proven software engineering methods such as feature toggles and transfers them to the artifacts of a PDA. This solution closes a significant research gap and enables the development of PDAs according to Continuous Integration and Delivery.

By outsourcing the toggle logic to a decision model (potentially global for the entire process landscape), both, software developers and process engineers, can easily set and control toggles at one central place. Merging conflicts caused by diverging versions in collaborative teams are avoided. Process operators and even PDA users benefit from the concept, too. For them, the process flow is always transparent and comprehensible. Moreover, T4PDA offers possibilities such as canary releasing to reduce the risk of introducing new features as well as techniques such as A/B testing. For the technical support of the concept, a plugin for the Camunda Modeler was developed. This extension automates the manual creation and removal of toggles following our T4PDA approach. This saves time and also reduces the error rate.

The evaluation of the T4PDA approach and the provided tool support in an AB/BA crossover design experiment proved its applicability. The results reflect a higher software quality, a faster development process, and satisfied developers. The maturity of the tool remains to be evolved such that it can be applied and evaluated in large-scale projects.

The future aim is to improve the T4PDA plugin for productive use. Especially the positioning of the BPMN elements during the automatic model transformation must be improved. A further goal is a stronger automatic synchronization of the toggles with the source code.

References

1. Allweyer, T.: BPMN 2.0: Introduction to the Standard for Business Process Modeling. Books on Demand, Norderstedt (2016)
2. Booch, G., Maksimchuk, R.A., Engle, M.W., Young, B.J., Connallen, J., Houston, K.A.: Object-oriented analysis and design with applications. ACM SIGSOFT Softw. Eng. Notes **33**(5), 29 (2008)
3. Camunda: Camunda modeler (2020). https://camunda.com/products/modeler/
4. Chen, L.: Continuous delivery: huge benefits, but challenges too. IEEE Softw. **32**(2), 50–54 (2015)
5. Fowler, M.: Continuous integration (2006). https://martinfowler.com/articles/continuousIntegration.html
6. Freund, J., Rücker, B.: Real-Life BPMN: with introductions to CMMN and DMN. CreateSpace Independent Publishing Platform (2006)
7. Harmon, P., Wolf, C.: Business process modeling survey. BPtrends report (2011)
8. Hodgson, P.: Feature toggles (aka feature flags) (2017). https://martinfowler.com/articles/feature-toggles.html
9. Humble, J., Farley, D.: Continuous Delivery: Reliable Software Releases through Build, Test, and Deployment Automation. Pearson Education, London (2010)
10. Larrea, V.G.V., Joubert, W., Fuson, C.: Use of continuous integration tools for application performance monitoring. Concurrency and Computation Practice and Experience on the Cray User Group (2015)
11. Meyer, M.: Continuous integration and its tools. IEEE Softw. **31**(3), 14–16 (2014)
12. OMG: About the business process model and notation specification version 2.0 (2011). https://www.omg.org/spec/BPMN/2.0/About-BPMN/
13. OMG: About the decision model and notation specification version 1.2 (2019). https://www.omg.org/spec/DMN/About-DMN/
14. Peffers, K., Rothenberger, M., Tuunanen, T., Vaezi, R.: Design science research evaluation. In: Peffers, K., Rothenberger, M., Kuechler, B. (eds.) DESRIST 2012. LNCS, vol. 7286, pp. 398–410. Springer, Heidelberg (2012). https://doi.org/10.1007/978-3-642-29863-9_29
15. Rahman, M.T., Querel, L.P., Rigby, P.C., Adams, B.: Feature toggles: practitioner practices and a case study. In: Proceedings of the 13th International Conference on Mining Software Repositories, pp. 201–211 (2016)
16. Rodríguez, P., et al.: Continuous deployment of software intensive products and services: a systematic mapping study. J. Syst. Softw. **123**, 263–291 (2017)
17. Schneid, K.: Branching strategies for developing new features within the context of continuous delivery. In: Proceedings of the 2nd Workshop on Continuous Software Engineering Co-located with Software Engineering (SE 2017), Hannover, Germany, 20 February 2017. CEUR Workshop Proceedings, vol. 1806, pp. 28–35. CEUR-WS.org (2017)
18. Stiehl, V.: Process-Driven Applications with BPMN. Springer, Cham (2014). https://doi.org/10.1007/978-3-319-07218-0
19. Togglz: Togglz - feature flags for the java platform (2020). https://www.togglz.org/
20. Weske, M.: Business process management architectures. Business Process Management, pp. 333–371. Springer, Heidelberg (2012). https://doi.org/10.1007/978-3-642-28616-2_7
21. Wol, E.: Continuous Delivery: Der pragmatische Einstieg. dpunkt.verlag, Heidelberg (2016)

Say It in Your Own Words: Defining Declarative Process Models Using Speech Recognition

Han van der Aa$^{1(\boxtimes)}$, Karl Johannes Balder^2, Fabrizio Maria Maggi^3,
and Alexander Nolte^{2,4}

^{1} University of Mannheim, Mannheim, Germany
han@informatik.uni-mannheim.de
^{2} University of Tartu, Tartu, Estonia
Karl.Johannes.Balder@tudeng.ut.ee, alexander.nolte@ut.ee
^{3} Free University of Bozen-Bolzano, Bolzano, Italy
maggi@inf.unibz.it
^{4} Carnegie Mellon University, Pittsburgh, PA, USA

Abstract. Declarative, constraint-based approaches have been proposed to model loosely-structured business processes, mediating between support and flexibility. A notable example is the DECLARE framework, equipped with a graphical declarative language whose semantics can be characterized with several logic-based formalisms. Up to now, the main problem hampering the use of DECLARE constraints in practice has been the difficulty of modeling them: DECLARE's formal notation is difficult to understand for users without a background in temporal logic, whereas its graphical notation has been shown to be unintuitive. Therefore, in this work, we present and evaluate an analysis toolkit that aims at bypassing this issue by providing users with the possibility to model DECLARE constraints using their own way of expressing them. The toolkit contains a DECLARE modeler equipped with a speech recognition mechanism. It takes as input a vocal statement from the user and converts it into the closest (set of) DECLARE constraint(s). The constraints that can be modeled with the tool cover the entire Multi-Perspective extension of DECLARE (MP-DECLARE), which complements control-flow constraints with data and temporal perspectives. Although we focus on DECLARE, the work presented in this paper represents the first attempt to test the feasibility of speech recognition in business process modeling as a whole.

Keywords: Declarative process modeling · Speech recognition · Natural language processing · Text mining

1 Introduction

Process models are an important means to capture information on organizational processes, serving as a basis for communication and often as the start-

Work supported by the Estonian Research Council (project PRG887).

D. Fahland et al. (Eds.): BPM Forum 2020, LNBIP 392, pp. 51–67, 2020.
https://doi.org/10.1007/978-3-030-58638-6_4

ing point for analysis and improvement [10]. For processes that are relatively structured, *imperative* process modeling notations, such as the Business Process Model and Notation (BPMN), are most commonly employed. However, other processes, in particular knowledge-intensive ones, are more flexible and, therefore, less structured. An important characteristic of such processes is that it is typically infeasible to specify the entire spectrum of allowed execution orders in advance [9], which severely limits the applicability of the imperative process modeling paradigm. Instead, such processes are better captured using *declarative* process models defined in process modeling languages, like DECLARE, whose semantics can be characterized using temporal logic properties. These models do not require an explicit definition of all allowed execution orders, but rather use constraints to define the boundaries of the permissible process behavior [5].

Although their benefits are apparent, establishing declarative process models is known to be difficult, especially for domain experts that generally lack expertise in temporal logics, and in most of the cases find the graphical notation of DECLARE constraints unintuitive [12]. Due to these barriers to declarative model creation, a preliminary approach has been presented in [1] that automatically extracts declarative process models from natural language texts, similar to others works that have investigated the generation of imperative process models from natural language descriptions (cf., [3,11,18,22]).

In this work, we go beyond this state-of-the-art by presenting and evaluating an interactive approach, which takes vocal statements from the user as input, and employs speech recognition to convert them into the closest (set of) DECLARE constraint(s). The approach has been integrated into a declarative modeling and analysis toolkit. With this tool, the user is not required to have any experience in temporal logics nor to be familiar with the graphical notation of DECLARE constraints, but can express temporal properties using her/his own words. The temporal properties that can be modeled with the tool cover the entire Multi-Perspective Declare (MP-DECLARE) language that cannot only express control-flow properties, but also conditions on the data, time, and resource perspectives. A further important contribution of the paper is that the user evaluation presented in this paper can be considered the first test of the usability of speech recognition in business process modeling. This evaluation revealed that participants found that the tool would improve their efficiency, particularly in mobile settings. However, it also pointed towards a learning curve, especially for less experienced users. Based on the obtained feedback, we were able to make various improvements to the user interface.

The remainder of the paper is structured as follows. Section 2 introduces MP-DECLARE and the transformation of natural language into DECLARE. Section 3 presents our conceptual contributions to the generation of multi-perspective constraints based on natural language input, while Sect. 4 presents the implemented toolkit. Section 5 discusses the evaluation conducted to assess the usability of our tool. Finally, Sect. 6 considers related work, before concluding in Sect. 7.

2   Background

This section introduces MP-DECLARE followed by the main challenges associated with the translation of natural language into declarative constraints.

2.1   Declarative Process Modeling

DECLARE is a declarative process modeling language originally introduced by Pesic and van der Aalst in [19]. Instead of explicitly specifying the flow of the interactions among process activities, DECLARE describes a set of constraints that must be satisfied throughout the process execution. The possible orderings of activities are implicitly specified by constraints and anything that does not violate them is possible during execution. MP-DECLARE is the Multi-Perspective extension of DECLARE that was first introduced in [8] and can express constraints over perspectives of a process like data, time, and resources.

To explain the semantics of DECLARE and MP-DECLARE, we have to introduce some preliminary notions. In particular, we call a *case* an ordered sequence of events representing a single "run" of a process (often referred to as a *trace* of events). Each event in a trace refers to an *activity*, has a *timestamp* indicating when the event occurred, and can have additional *data attributes* as payload. Consider, e.g., the occurrence of an event *ship order* (O) and suppose that, after the occurrence of O at timestamp τ_O, the attributes *customer type* and *amount* have values *gold* and 155€. In this case, we say that, when O occurs, two special relations are valid *event*(O) and $p_O(gold,155€)$. In the following, we identify *event*(O) with the event itself O and we call (*gold*,155€), the *payload* of O.

Table 1. Semantics for declare templates

Template	LTL semantics	Activation
Participation(A)	$\mathbf{F}A$	A
Init(A)	A	A
Absence(A)	$\neg\mathbf{F}A$	A
AtMostOne(A)	$\neg\mathbf{F}(A \wedge \mathbf{X}(\mathbf{F}A))$	A
Responded Existence(A,B)	$\mathbf{F}A \rightarrow \mathbf{F}B$	A
Coexistence(A,B)	$\mathbf{F}A \leftrightarrow \mathbf{F}B$	A, B
Response(A,B)	$\mathbf{G}(A \rightarrow \mathbf{F}B)$	A
Chain Response(A,B)	$\mathbf{G}(A \rightarrow \mathbf{X}B)$	A
Precedence(A,B)	$\mathbf{G}(B \rightarrow \mathbf{O}A)$	B
Chain Precedence(A,B)	$\mathbf{G}(B \rightarrow \mathbf{Y}A)$	B
Not Coexistence(A,B)	$\mathbf{F}A \rightarrow \neg\mathbf{F}B$	A, B
Not Succession(A,B)	$\mathbf{G}(A \rightarrow \neg\mathbf{F}B)$	A, B

Declare. A DECLARE model consists of a set of constraints applied to activities. Constraints, in turn, are based on templates. Templates are patterns that define parameterized classes of properties, and constraints are their concrete instantiations (we indicate template parameters with capital letters and concrete activities in their instantiations with lower case letters). Templates have a graphical representation and their semantics can be formalized using different logics [16], the main one being LTL over finite traces, making them verifiable and executable. Each constraint inherits the graphical representation and semantics from its template. Table 1 summarizes some Declare templates (the reader can refer to [5] for a full description of the language). Here, the **F**, **X**, **G**, and **U** LTL (future) operators have the following intuitive meaning: formula $\mathbf{F}\phi_1$ means that ϕ_1 holds sometime in the future, $\mathbf{X}\phi_1$ means that ϕ_1 holds in the next position, $\mathbf{G}\phi_1$ says that ϕ_1 holds forever in the future, and, lastly, $\phi_1\mathbf{U}\phi_2$ means that sometime in the future ϕ_2 will hold and until that moment ϕ_1 holds (with ϕ_1 and ϕ_2 LTL formulas). The **O**, **Y**, and **S** LTL (past) operators have the following meaning: $\mathbf{O}\phi_1$ means that ϕ_1 holds sometime in the past, $\mathbf{Y}\phi_1$ means that ϕ_1 holds in the previous position, and, lastly, $\phi_1\mathbf{S}\phi_2$ means that sometime in the past ϕ_2 holds and since that moment ϕ_1 holds.

Consider, for example, constraint Response(a,b). This constraint indicates that if a *occurs*, b must eventually *follow*. Therefore, this constraint is satisfied for traces such as $\mathbf{t}_1 = \langle a, a, b, c\rangle$, $\mathbf{t}_2 = \langle b, b, c, d\rangle$, and $\mathbf{t}_3 = \langle a, b, c, b\rangle$, but not for $\mathbf{t}_4 = \langle a, b, a, c\rangle$ because, in this case, the second instance of a is not followed by a b. Note that, in $\mathbf{t}_2$, the considered response constraint is satisfied in a trivial way because a never occurs. An *activation* of a constraint in a trace is an event whose occurrence imposes, because of that constraint, some obligations on other events (targets) in the same trace. For example, a is an activation for Response(a,b) and b is a target because the execution of a forces b to be executed, eventually. In Table 1, for each template, the corresponding activations are specified.

Multi-Perspective Declare. MP-DECLARE extends DECLARE with additional perspectives. We here describe its semantics informally and refer the interested reader to [8] for more details.

The standard semantics of DECLARE is extended by requiring additional conditions on data, i.e., the *activation condition*, the *correlation condition*, and a *time condition*. As an example, we consider constraint Response(ship order, send invoice), with *ship order* as activation and *send invoice* as target. The activation condition φ_a is a relation that must be valid when the activation occurs. If the activation condition does not hold the constraint is not activated. The activation condition has the form $p_A(x) \wedge r_a(x)$, meaning that when A occurs with payload x, the relation r_a over x must hold. For example, we can say that whenever *ship order* occurs, the order amount is higher than 100€, and the customer is of type *gold*, eventually an invoice must be sent. In case *ship order* occurs, but these conditions are not satisfied, the constraint is not activated.

The correlation condition φ_c is a relation that must be valid when the target occurs. It has the form $p_B(y) \wedge r_c(x, y)$, meaning that when B occurs with payload

y, the relation r_c involving the payload x of A and the payload y of B must hold. For example, we can say that whenever *ship order* occurs with order amount higher than 100€, and customer type *gold*, eventually an invoice must be sent with the same order amount. Finally, a time condition can be specified through an interval $(I = [\tau_0, \tau_1))$ indicating the minimum and the maximum temporal distance allowed between the occurrence of the activation and the occurrence of the corresponding target.

2.2   From Natural Language to Declarative Models

A crucial component of our work involves the extraction of declarative constraints from natural language. This extraction step involves the identification of the described actions (activities), as well as the identification of the constraint that applies to these actions. Due to the inherent flexibility of natural language, this extraction step can be highly challenging. Its difficulty manifests itself in the sense that, on the one hand, the same declarative constraint can be expressed in a wide variety of manners, whereas, on the other hand, subtle textual differences can completely change the meaning of the described constraint.

Table 2. Different descriptions of Precedence (create invoice, approve invoice)

ID	Description
s_1	An invoice must be created before the invoice can be approved
s_2	A bill shall be created prior to it being approved
s_3	Invoice creation must precede its approval
s_4	Approval of an invoice must be preceded by its creation
s_5	Before an invoice is approved, it must be created

Variability of Textual Descriptions. As shown in Table 2, the same declarative constraint can be described in a broad range of manners. Key types of differences occur due to: the use of synonyms (e.g., *create invoice* in s_1 and *create bill* in s_2) and due to different grammatical structures (e.g., s_1 uses verbs to denote activities, whereas s_3 uses nouns, like *"invoice creation"*). Finally, constraint descriptions can differ in the order in which they describe the different components of binary constraints, i.e., whether they describe the constraint in a chronological fashion, e.g., s_1 to s_3, or in the reverse order, such as s_4 and s_5. To support users in the elicitation of declarative process models, an approach must not limit users too much in terms of the input that they can provide. Rather, an approach must accommodate different manners in which users may describe constraints. However, a successful approach must be able to do this while also being able to recognize subtle distinctions among constraint types.

Subtle Differences Leading to Different Constraints. Small textual differences can have a considerable impact on the semantics of constraint descriptions and, thus, on the constraints that should be extracted from them. To illustrate this, consider the descriptions in Table 3. In comparison to description s_6, the three other descriptions each differ by only a single word. However, as shown in the right-hand column, the described constraints vary greatly. For instance, the difference between the Response constraint of s_6 and the Precedence constraint described by s_7 lies in the obligation associated with the *send invoice* action. The former specifies that this *must* occur, whereas the latter specifies that it *can* occur. Further, the direction in which a constraint is described is often signaled through small textual elements, typically through the use of temporal prepositions. For instance, in s_8, the use of *first* completely reverses the meaning of the described constraint. Finally, the presence of a negation also drastically changes the meaning of a constraint, as seen for s_9. The addition of *not* to description s_6 changes Response into Not Succession.

Table 3. Subtle textual differences (A as *ship order*, B as *send invoice*)

ID	Description	Constraint
s_6	If an order is shipped, an invoice must be sent	Response(A,B)
s_7	If an order is shipped, an invoice can be sent	Precedence(A,B)
s_8	If an order is shipped, an invoice must be sent first	Precedence(B,A)
s_9	If an order is shipped, an invoice must not be sent	Not Succ.(A,B)

State of the Art. So far, only one approach has been developed to automatically extract declarative constraints from natural language text. This approach, by van der Aa et al. [1], is able to extract five types of Declare templates, Init, End, Precedence, Response, Succession, as well as their negated forms. Its evaluation results show that it is able to handle a reasonable variety of inputs. Recognizing its potential as well as its limitations, we extend this approach as follows: (1) we generalize the pattern recognition mechanisms in order to handle more flexible inputs, (2) we cover eight additional constraint templates, and (3) we add support for augmentation with data and time conditions.

3   Conceptual Approach

Figure 1 provides an overview of the main components of our Speech2RuM approach. As shown, the user provides inputs through speech as well as the interaction with the Graphical User Interface (GUI). In this way, users are able to construct a declarative process model through three main functions: (1) using speech to describe constraints in natural language, (2) augmenting constraints with data and time conditions, and (3) editing and connecting the constraints.

In this section, we outline our conceptual contributions with respect to the first two functions. We cover the implementation of the approach in Sect. 4. There, we also show how a model can be edited after its creation.

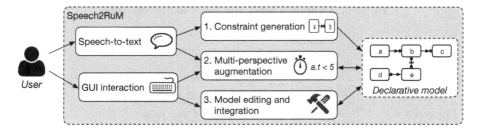

Fig. 1. Overview of the Speech2RuM approach

3.1 Constraint Generation

In this component, our approach turns a constraint description, recorded using speech recognition, into one or more constraints. For this task, we take the result of the parsing step of the state-of-the-art approach [1] as a starting point. Given a sentence S, parsing yields a list A_S of actions described in the sentence and the interrelations that exist between the actions, i.e., a mapping $rel_S : A_S \times A_S \rightarrow relationType$, with $relationType \in \{xor, and, dep\}$. As depicted in Fig. 2, an action $a \in A_S$ consists of a verb and optional subjects and objects.

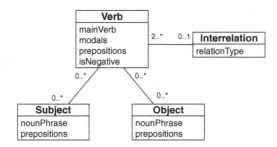

Fig. 2. Semantic components returned in the parsing step of [1]

Based on this parsing step, we have added support to handle eight additional constraint types, presented in Table 4. Given a set of activities A_S and a relation rel_S extracted for a sentence S, these types are identified as follows:

- **Participation and Absence.** If A_S contains only one action, either a Participation or an Absence constraint is established for the action, depending on whether it is negative or not.
- **AtMostOne.** If a Participation constraint is originally recognized, we subsequently check if S specifies a bound on its number of executions, i.e., by checking for phrases such as *at most once, not more than once, one time.*
- **Coexistence.** Coexistence relations are identified for sentences with two actions in an *and* relation, typically extracted from a coordinating conjunction like *shipped and paid.* Furthermore, some notion of obligation should be

Table 4. Additional constraint types in Speech2RuM

Constraint	Example
Participation	*An invoice must be created*
Absence	*Dogs are <u>not</u> allowed in the restaurant*
AtMostOne	*An invoice should be paid <u>at most once</u>*
Coexistence	*An order should be <u>shipped and paid</u>*
Responded Existence	*If a product is produced, it <u>must be tested</u>*
Not Coexistence	*If an application is accepted, it cannot be rejected*
Chain Precedence	*After an order is received, it may <u>immediately</u> be refused*
Chain Response	*After an order is received, it must <u>directly</u> be checked*

present, in order to distinguish between actions that *can* happen together and ones that *should*.

- **Responded Existence.** Responded Existence constraints are extracted when two actions are in a dependency relation, i.e., a_1 *dep* a_2, for which it holds that the target action, a_2, is indicated to be mandatory, e.g., using *must*. The key difference between Response and Responded Existence is that the former includes a notion of order, i.e., a_1 precedes a_2, whereas no such order is specified for Responded Existence constraints.
- **Not Coexistence.** This constraint type is identified as the negated form of both Coexistence and Responded Existence. Semantically, in both cases, the description states that two actions should not appear in the same process instance. This is, e.g., seen in Table 4, where *"If an application is accepted, it cannot be rejected"* is actually the negative form of a Responded Existence description.
- **Chain Precedence and Chain Response.** These constraint types are specializations of Precedence and Response constraints. Chain constraints are recognized by the presence of a temporal preposition that indicates immediacy. Generally, such a preposition is associated with the verb of either action a_1 or a_2 in a relation a_1 *dep* a_2. For this, we consider the preposition *immediately* and several of its synonyms, i.e., *instantly*, *directly*, and *promptly*.

3.2 Multi-perspective Augmentation

Our approach supports the augmentation of constraints with conditions on the data and time perspectives, turning DECLARE constraints into MP-DECLARE ones. Our approach allows users to express three types of conditions through speech recognition, i.e., *activation*, *correlation*, and *time* conditions (see Sect. 2.1).

We note that descriptions of conditions likely reflect textual fragments rather than full sentences. Furthermore, given that these are short statements, the expected variance is considerably lower than for descriptions of declarative constraints. For these reasons, our approach to extract conditions is based on pattern matching as opposed to the grammatical parsing used in Sect. 3.1.

Table 5. Supported patterns for activation conditions

Condition	Pattern	Example
> or ≥	[greater|higher|more] than [or equal to]	*Amount higher than 500*
< or ≤	[smaller|lower|less] than [or equal to]	*Quantity is less than or equal to 12*
= or ≠	is [not] [equal to]	*The color is not red*
∈ or ∉	is [not] in [*list*]	*Size is not in small, medium, large*

Activation Conditions. Activation conditions denote specific requirements that must be met in order for a constraint to be applicable, e.g., stating that Response(ship order, send invoice) should only apply when the amount associated with *ship order* activity is above 500. Our approach allows users to express complex conditions, i.e., conditions that concatenate multiple logical statements, such as *"The amount is higher than 500 and the color is not red"*. Therefore, given an input string S, our approach first splits S into sub-strings, denoted by $split(S)$. This is done by recognizing the presence of coordinating conjunctions (*and* and *or*) and dividing S accordingly. Each sub-string $s \in split(S)$ is expected to correspond to an individual expression, which together are joined using logical $\wedge$ and $\vee$ operators. On each $s \in split(S)$, we use pattern matching based on the patterns depicted in Table 5. In this manner, we are able to handle conditions related to both numerical, (e.g., *amount* and *length*) and categorical attributes (e.g., *color* and *customer type*).

Correlation Conditions. Correlation conditions must be valid when the *target* of a constraint is executed. These express relations that must exist between attributes associated with the activation and the target of the constraint, e.g., the employee *receiving* a loan application should not be the same as the employee that *checks* it. Our approach handles two patterns here, either allowing a user to express that an attribute should be equal for both activities, e.g., *"The amount is the same"* or that they not equal, e.g., *"The applicant is different"*.

Table 6. Supported patterns for time conditions

Condition	Pattern	Example
time ≤ *x*	[in|at most|no later than] [x]	*At most 3 h*
x ≤ *time* ≤ *y*	between [x] and [y]	*Between 3 and 5 days*
x ≤ *time* ≤ *y*	not [before|earlier than] [x] and [within|not later than|not after] [y]	*Not before 3 h and within 12 h*

Time Conditions. A time condition bounds the time between the occurrence of the activation and target of a constraint, e.g., stating that the target of a Response constraint should occur within 5 days after its activation. As shown in Table 6, we allow users to specify time conditions according to three general patterns. The first pattern only specifies an upper bound on the duration, whereas the other two patterns specify ranges. Note that we support time conditions specified using seconds, minutes, hours, and days.

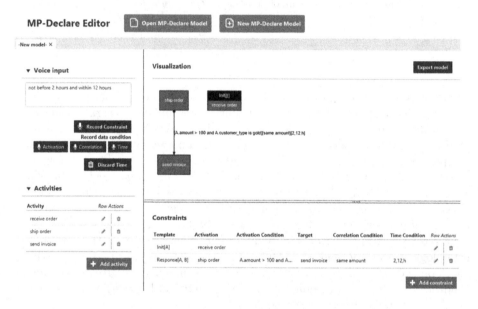

Fig. 3. Screenshot of the Speech2RuM implementation

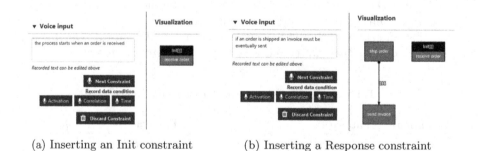

(a) Inserting an Init constraint   (b) Inserting a Response constraint

Fig. 4. Inserting control-flow properties

4   The Tool

We integrated the proposed Speech2RuM approach into RuM, a modeling and analysis toolkit for Rule Mining.[1] The tool is implemented in Java 11 and uses the speech recognition API provided by *Google Cloud Speech*.[2] In Fig. 3, a screenshot of the latest version of the tool is shown. In the top-left area of the screen, the recognized vocal input is shown. It is possible to record a constraint (control-flow perspective) as shown in Fig. 4 where an Init and a Response constraint are modeled. After having inserted a DECLARE constraint, the user can record an activation, a correlation, and a time condition as shown in Fig. 3 where an activation (*amount* is greater than 100 and *customer type* is *gold*) and a correlation (activation and target share the same value for attribute *amount*) condition have already been recorded and a time condition is extracted from the vocal input *"not before 2 h and within 12 h"*. The constraints expressed with their graphical notation are shown in the top-right area of the screen, and modifiable lists of activities and constraints are shown in the bottom part of the screen.

5   User Evaluation

To improve the current implementation and to assess the feasibility of the developed Speech2RuM approach based on user feedback, we conducted a qualitative user study. Our aim was to (1) identify means for improving the interface, (2) identify issues and demands of individuals with different backgrounds, and (3) identify usage scenarios. In the following, we will describe the study before discussing our findings and potential threats to validity.

5.1   Study

Participants. For our study, we selected eight participants (P1 to P8) with differing characteristics. We selected two participants that had experience related to Business Process Management (BPM, tier 1, P6 and P7), two participants with formal background on temporal logics (tier 2, P3 and P8), two participants that had used DECLARE to model temporal constraints (tier 3, P2 and P4), and two participants that had used different tools to model and analyze temporal constraints with DECLARE (tier 4, P1 and P5). We chose this differentiation to study how individuals with different levels of expertise perceive the tool and identify potential issues and demands.

Study Setup. The study was conducted via Skype by a team consisting of a *facilitator* and an *observer*, with the facilitator guiding the participant and the observer serving in a supporting role. We used a manufacturing process in a bicycle factory[3] to provide a scenario that is likely to be familiar for all participants.

---

[1] The tool can be found at https://rulemining.org.
[2] https://cloud.google.com/speech-to-text/.
[3] The scenario can be found at https://git.io/JfHbl.

We opted to use a common scenario rather than asking participants to describe a process of their choice to ensure the comparability of our findings. We also created a help document[4] to outline the tool's capabilities to the participants.

Prior to the study, the facilitator sent the participants a link to the scenario, the help document, and the tool itself suggesting to read the documents and install the tool. At the beginning of the study, the facilitator introduced the study procedure and the tool. When ready, the participant shared her/his screen, whereas the facilitator started the video recording and began guiding the participant through the prepared scenario. The scenario required the participants to construct a declarative process model. First, they added constraints by reading prepared sentences, before being asked to establish their own constraint descriptions. During this time, the facilitator and the observer noted down any issues the participant mentioned or that appeared to come up.

At the end of the study, the facilitator conducted a follow-up interview, focusing on clarifying questions about issues the participant had encountered. He also asked the participants what s/he *"liked about the tool"*, *"wished would be different"*, and *"under which circumstances s/he would use it"*. After the study, the participants were asked to complete a short post-questionnaire. The questionnaires consisted of multi-point Likert scales including the System Usability Scale (SUS) [7] and scales covering satisfaction, expectation confirmation, continuation intention, and usefulness which were adapted from the ones presented by Bhattacherjee in [6] and used to assess continued use of information systems.[5] The individual studies lasted between 14 and 35 min each.

Result Analysis. To analyze the collected data, the research team first built an affinity diagram [13] based on the video recordings, observations, and follow-up interviews, by creating a single paper note for each issue that was reported. The team then clustered the notes based on emerging themes, which resulted in 139 notes divided over 21 distinct clusters. The results of the questionnaires served as an additional qualitative data point during the analysis.

5.2 Findings

In general, our study revealed that the tested interface was reasonably usable as evidenced by an average SUS score of 73.13, which can be considered to be a good result compared to the commonly reported average of 68 [7]. However, our study also revealed usability issues, which served as a basis to improve the tool (Fig. 3). The main changes include the possibility to undo changes to the model based on the last recording, editing the recorded text directly rather than having to re-record it, and adding data conditions through speech recognition (instead of specifying those conditions manually).

Common Observations. First, we found that all participants **wrote sentences down** before saying them aloud. They either wrote them *"on paper"*

---

[4] The help document can be found at https://git.io/JfHbc.
[5] The complete questionnaire can be found at https://git.io/JfHb8.

(P1), *"typed them on [their] computer"* (P2), or noted down key parts they wanted to say (P7). Most participants also used somewhat **unnatural sentences** when entering constraints. This is evidenced by them, e.g., using activity names instead of natural sentences (*"after [activity1], [activity2] is executed."*, P7). Some even tried to identify keywords that the algorithm might pick up (*"is activity like a keyword?"*, P8). Only P1 used natural sentences from the start. Taking these findings together indicates that users might have felt insecure to just use natural sentences and rather tried to create sentences that they felt the system would be able to understand, but that were difficult for them to formulate without writing them down first.

Trying to identify keywords and adapting the input statements for the tool might have **increased the complexity of entering constraints** while being counterproductive when using speech recognition since our tool was designed to understand natural expressions. These issues indicate that **individuals need to learn how to effectively use speech as a means of modeling constraints** regardless of their previous experience related to BPM or DECLARE. Therefore, it would be desirable for users to go through a training phase to better understand how to interact with the tool and also appreciate the variety of vocal inputs the tool can deal with.

Difference Between Participants with Different Backgrounds. There were also noticeable differences between individuals from different backgrounds. For example, participants from tiers 1 and 2 (having low confidence with DECLARE) **mainly relied on the visual process model** to assess whether the tool had understood them correctly, while participants from tiers 3 and 4 (more familiar with DECLARE and analysis tools based on DECLARE) **mainly relied on the list providing a semistructured representation** of the entered constraints (bottom-right area in Fig. 3). This was evidenced by how they tried to fix potential errors. Participants from tiers 1 and 2 initially tried to edit the visual process model (observation of P6, *"the activity label should be editable"*, P2) – which was not possible in the version of the tool we tested – while participants from tiers 3 and 4 directly started editing the constraint list (observation of P1 and P5). Only P2 attempted both.

Findings from our study also indicated that the way the **recognized text** was displayed (text field below *Voice input* in the top-left area of the screenshot in Fig. 3) might not have been ideal. Participants checked the input correctness after they entered the first constraint, but did not continue to do so afterward (observation of P3 and P6) despite few sentences being wrongly translated from speech into text during the tests. This led to confusion, particularly among less experienced participants, who thought that they did not formulate a constraint correctly, while the issue was that their vocal input was not recorded properly. One way to address this issue is to highlight the text field containing the recognized text directly after a new text has been recorded to indicate to the user that s/he should pay attention to the recognized text.

These findings indicate that, even if using speech recognition users can somehow bypass the interaction with the graphical notation of DECLARE constraints,

it still appears difficult for less experienced users to assess the correctness of their entered constraints. One approach to mitigate this issue might be to provide the option to mark constraints that a user cannot verify and subsequently ask an experienced user to assess their correctness. In alternative, traces representing examples and counterexamples of the recorded behaviors together with a textual description of the constraint could be shown.

The aforementioned issues might also have led to **less experienced participants perceiving the tool as less useful** (m = 2.63 for tier 1 and m = 3.50 for tier 4) and being **less inclined to use the tool in the future** (m = 2.33 for tier 1 and m = 3.65 for tier 4) than experienced participants. Related to this finding, it thus appears counter-intuitive that experienced participants were slightly less satisfied with the tool than less experienced participants (m = 3.00 for tier 3, m = 3.50 for tier 4, and m = 3.67 for tiers 1 and 2). This can however potentially be explained by experienced participants having higher expectations regarding the functionality of the tool (*"what about recording data conditions as well?"*, P5).

Finally, it should be noted that participants from different backgrounds were *"excited"* (P5) to use the tool. Some even continued to record constraints after the actual evaluation was over (*"let me see what happens when I [...]"*, P8). It thus appears reasonable to assume that addressing the identified shortcomings can positively influence the perception of the tool.

Potential Usage Scenarios. The participants mentioned multiple scenarios during which they would consider speech as useful for entering constraints. Most of them stated that using speech would **improve their efficiency** because they could *"quickly input stuff"* (P4) especially when reading *"from a textual description"* (P1) with P5 pointing out that it would be particularly useful for *"people that are not familiar with the editor"* (P5). On the other hand, P8 had a different opinion stating that *"manually is quicker"* (P8). S/he did however acknowledge that speech might be useful *"when I forget the constraint name"* (P8). Moreover, participants also suggested that vocal input might be useful *"for designing"* (P3) which points towards its usefulness for specific activities at the early stages of modeling a process. Finally, P1 mentioned that speech input would be useful for **mobile scenarios** such as *"using a tablet"* (P1), because entering text in a scenario like that is usually much more time consuming than when using a mechanical keyboard.

5.3   Threats to Validity

The goal of our study was to collect feedback for improving the tool, identify potential usage scenarios, and pinpoint any issues for individuals with different backgrounds. It thus appeared reasonable to conduct an in-depth qualitative study with selected participants from a diverse range of backgrounds related to their knowledge and experience with BPM in general and temporal properties and DECLARE in particular. Conducting a study with a small sample of participants is common because research has shown that the number of additional insights gained deteriorates drastically per participant [17]. There are, however, some threats to validity associated with this particular study design. A first threat is related to the fact that we developed a specific tool and studied its use by specific people in a specific setting over a limited period of time. Despite carefully selecting participants and creating a setting that would be close to how we envision the tool would be commonly used, it is not possible to generalize our findings beyond our study context since conducting the same study with different participants using a different setup might yield different results. In addition, the study results were synthesized by a specific team of researchers which poses a threat to validity since different researchers might potentially interpret findings differently. We attempted to mitigate this threat by ensuring that observations, interviews, and the analysis of the obtained data were collaboratively conducted by two researchers. We also abstained from making causal claims providing instead a rich description of the behavior and reporting perceptions of participants.

6   Related Work

Organizations recognize the benefit of using textual documents to capture process specifications [4, 23], given that these can be created and understood by virtually everyone [11]. To allow these documents to be used for automated process analysis, such as conformance checking, a variety of techniques have been developed to extract process models from texts [11, 18, 22]. Other works exploit textual process specifications for model verification [2, 20] or directly for process analysis [3, 21]. In the context of declarative process models, some recent works also provide support for the extraction of DCR graphs from textual descriptions [14, 15], whereas the preliminary work on the extraction of DECLARE constraints from texts presented in [1] represents the foundation of our work.

7   Conclusion

In this work, we presented an interactive approach that takes vocal statements from the user as input and employs speech recognition to convert them into multi-perspective, declarative process models. Our Speech2RuM approach goes beyond the state-of-the-art in text-to-constraint transformation by covering a broader range of DECLARE templates and supporting their augmentation with data and time conditions. The integration of our approach into the RuM toolkit enables users to visualize and edit the obtained models in a GUI. Furthermore, it also allows these models to directly serve as a basis for the toolkit's analysis techniques, such as conformance checking and log generation. Finally, we note that the conducted user evaluation represents the first study into the feasibility of using speech recognition for business process modeling. The results demonstrated its promising nature, although modeling purely based on speech recognition is especially suitable for mobile environments.

In future work, we aim at further developing Speech2RuM based on the obtained user feedback. In particular, we are going to integrate the speech recognition tool with a chatbot which represents a valid alternative in the context of desktop applications. In addition, the text-to-constraint component shall be improved to support less natural descriptions, e.g., those that explicitly mention the term *activity* to denote a process step. Finally, it will be highly interesting to investigate how speech recognition can be lifted to also support the elicitation of imperative process models.

References

1. van der Aa, H., Di Ciccio, C., Leopold, H., Reijers, H.A.: Extracting declarative process models from natural language. In: Giorgini, P., Weber, B. (eds.) CAiSE 2019. LNCS, vol. 11483, pp. 365–382. Springer, Cham (2019). https://doi.org/10.1007/978-3-030-21290-2_23
2. van der Aa, H., Leopold, H., Reijers, H.A.: Comparing textual descriptions to process models: the automatic detection of inconsistencies. Inf. Syst. **64**, 447–460 (2017). https://doi.org/10.1016/j.is.2016.07.010
3. van der Aa, H., Leopold, H., Reijers, H.A.: Checking process compliance against natural language specifications using behavioral spaces. Inf. Syst. **78**, 83–95 (2018). https://doi.org/10.1016/j.is.2018.01.007
4. van der Aa, H., Leopold, H., van de Weerd, I., Reijers, H.A.: Causes and consequences of fragmented process information: insights from a case study. In: 23rd Americas Conference on Information Systems, AMCIS 2017, Boston, MA, USA, 10–12 August 2017. Association for Information Systems (2017). http://aisel.aisnet.org/amcis2017/OrganizationalIS/Presentations/6
5. van der Aalst, W.M.P., Pesic, M., Schonenberg, H.: Declarative workflows: balancing between flexibility and support. Comput. Sci. Res. Dev. **23**(2), 99–113 (2009). https://doi.org/10.1007/s00450-009-0057-9
6. Bhattacherjee, A.: Understanding information systems continuance: an expectation-confirmation model. MIS Q. **25**(3), 351–370 (2001). http://misq.org/understanding-information-systems-continuance-an-expectation-confirmation-model.html
7. Brooke, J.: SUS-a quick and dirty usability scale. In: Jordan, P.W., Thomas, B., McClelland, I.L., Weerdmeester, B. (eds.) Usability Evaluation in Industry. CRC Press (1996). https://www.crcpress.com/product/isbn/9780748404605
8. Burattin, A., Maggi, F.M., Sperduti, A.: Conformance checking based on multi-perspective declarative process models. Expert Syst. Appl. **65**, 194–211 (2016). https://doi.org/10.1016/j.eswa.2016.08.040
9. Di Ciccio, C., Marrella, A., Russo, A.: Knowledge-intensive processes: characteristics, requirements and analysis of contemporary approaches. J. Data Semant. **4**(1), 29–57 (2015). https://doi.org/10.1007/s13740-014-0038-4
10. Dumas, M., La Rosa, M., Mendling, J., Reijers, H.A., et al.: Fundamentals of business process management, vol. 1. Springer, Heidelberg (2013). https://doi.org/10.1007/978-3-662-56509-4
11. Friedrich, F., Mendling, J., Puhlmann, F.: Process model generation from natural language text. In: Mouratidis, H., Rolland, C. (eds.) CAiSE 2011. LNCS, vol. 6741, pp. 482–496. Springer, Heidelberg (2011). https://doi.org/10.1007/978-3-642-21640-4_36

12. Haisjackl, C., et al.: Understanding declare models: strategies, pitfalls, empirical results. Softw. Syst. Model. **15**(2), 325–352 (2016). https://doi.org/10.1007/s10270-014-0435-z
13. Holtzblatt, K., Wendell, J.B., Wood, S.: Rapid contextual design: a how-to guide to key techniques for user-centered design. Ubiquity **2005**, 3 (2005). https://doi.org/10.1145/1066322.1066325
14. López, H.A., Debois, S., Hildebrandt, T.T., Marquard, M.: The process highlighter: from texts to declarative processes and back. In: Proceedings of the Dissertation Award, Demonstration, and Industrial Track at BPM 2018 co-located with 16th International Conference on Business Process Management (BPM 2018), Sydney, Australia, 9–14 September 2018, pp. 66–70 (2018). http://ceur-ws.org/Vol-2196/BPM_2018_paper_14.pdf
15. López, H.A., Marquard, M., Muttenthaler, L., Strømsted, R.: Assisted declarative process creation from natural language descriptions. In: 23rd IEEE International Enterprise Distributed Object Computing Workshop, EDOC Workshops 2019, Paris, France, 28–31 October 2019, pp. 96–99 (2019). https://doi.org/10.1109/EDOCW.2019.00027
16. Montali, M., Pesic, M., van der Aalst, W.M.P., Chesani, F., Mello, P., Storari, S.: Declarative specification and verification of service choreographiess. ACM Trans. Web **4**(1), 3:1–3:62 (2010). https://doi.org/10.1145/1658373.1658376
17. Nielsen, J., Landauer, T.K.: A mathematical model of the finding of usability problems. In: Human-Computer Interaction, INTERACT 1993, IFIP TC13 International Conference on Human-Computer Interaction, 24–29 April 1993, Amsterdam, The Netherlands, jointly organised with ACM Conference on Human Aspects in Computing Systems CHI 1993, pp. 206–213 (1993). https://doi.org/10.1145/169059.169166
18. Ferreira, R.C.B., Thom, L.H., de Oliveira, J.P.M., Avila, D.T., dos Santos, R.I., Fantinato, M.: Assisting process modeling by identifying business process elements in natural language texts. In: de Cesare, S., Frank, U. (eds.) ER 2017. LNCS, vol. 10651, pp. 154–163. Springer, Cham (2017). https://doi.org/10.1007/978-3-319-70625-2_15
19. Pesic, M., Schonenberg, H., van der Aalst, W.M.P.: Declare: full support for loosely-structured processes. In: 11th IEEE International Enterprise Distributed Object Computing Conference (EDOC 2007), 15-19 October 2007, Annapolis, Maryland, USA, pp. 287–300. IEEE Computer Society (2007). https://doi.org/10.1109/EDOC.2007.14
20. Sànchez-Ferreres, J., van der Aa, H., Carmona, J., Padró, L.: Aligning textual and model-based process descriptions. Data Knowl. Eng. **118**, 25–40 (2018). https://doi.org/10.1016/j.datak.2018.09.001
21. Sànchez-Ferreres, J., Burattin, A., Carmona, J., Montali, M., Padró, L.: Formal reasoning on natural language descriptions of processes. In: Hildebrandt, T., van Dongen, B.F., Röglinger, M., Mendling, J. (eds.) BPM 2019. LNCS, vol. 11675, pp. 86–101. Springer, Cham (2019). https://doi.org/10.1007/978-3-030-26619-6_8
22. Schumacher, P., Minor, M., Schulte-Zurhausen, E.: Extracting and enriching workflows from text. In: IEEE 14th International Conference on Information Reuse & Integration, IRI 2013, San Francisco, CA, USA, 14–16 August 2013, pp. 285–292 (2013). https://doi.org/10.1109/IRI.2013.6642484
23. Selway, M., Grossmann, G., Mayer, W., Stumptner, M.: Formalising natural language specifications using a cognitive linguistic/configuration based approach. Inf. Syst. **54**, 191–208 (2015). https://doi.org/10.1016/j.is.2015.04.003

Process Mining

IoT-Based Activity Recognition for Process Assistance in Human-Robot Disaster Response

Adrian Rebmann[1,2]($\boxtimes$), Jana-Rebecca Rehse[1,3]($\boxtimes$), Mira Pinter[1,2],
Marius Schnaubelt[4], Kevin Daun[4], and Peter Fettke[1,2]

[1] German Research Center for Artificial Intelligence (DFKI), Saarbrücken, Germany
{adrian.rebmann,jana-rebecca.rehse,mira.pinter,peter.fettke}@dfki.de
[2] Saarland University, Saarbrücken, Germany
[3] University of Mannheim, Mannheim, Germany
rehse@uni-mannheim.de
[4] Technical University of Darmstadt, Darmstadt, Germany
{schnaubelt,daun}@sim.tu-darmstadt.de

Abstract. Mobile robots like drones or ground vehicles can be a valuable addition to emergency response teams, because they reduce the risk and the burden for human team members. However, the need to manage and coordinate human-robot team operations during ongoing missions adds an additional dimension to an already complex and stressful situation. BPM approaches can help to visualize and document the disaster response processes underlying a mission. In this paper, we show how data from a ground robot's reconnaissance run can be used to provide process assistance to the officers. By automatically recognizing executed activities and structuring them as an ad-hoc process instance, we are able to document the executed process and provide real-time information about the mission status. The resulting mission progress process model can be used for additional services, such as officer training or mission documentation. Our approach is implemented as a prototype and demonstrated using data from an ongoing research project on rescue robotics.

Keywords: Rescue robotics · Process assistance · Internet of Things · Activity recognition · Emergency process management

1 Introduction

In any natural or man-made disaster, first response teams have to save and protect humans, animals, assets, and the environment. These responsibilities, along with a lack of real-time information, public interest, the need to act quickly, and an often hectic and confusing situation, put officers under a high amount of pressure [3]. To reduce both the mental burden and the physical danger to the officers, rescue robots are increasingly used to support disaster response operations [17]. Ground vehicles, for example, can transport heavy equipment

© Springer Nature Switzerland AG 2020
D. Fahland et al. (Eds.): BPM Forum 2020, LNBIP 392, pp. 71–87, 2020.
https://doi.org/10.1007/978-3-030-58638-6_5

or search for victims in areas that would be too dangerous for human officers to enter, while drones may take aerial pictures of a disaster site, which can help in identifying victims or finding access routes in unsafe territory [18].

However, integrating rescue robots into first response teams adds an additional dimension to an already complex and stressful situation, increasing the cognitive load of the officers in charge [16]. One way to support the officers in this situation is to provide them with real-time information about the status of the mission [29]. For this purpose, Business Process Management (BPM) approaches are particularly interesting, because disaster response missions are organized in a process-oriented way [7]. The actions of both officers and robots during an emergency mission follow predefined processes that are well documented and frequently practiced. Therefore, BPM methods can be used to capture and visualize ongoing processes for live-mission monitoring and to document them for after-mission reporting and debriefing. However, since the execution of disaster response processes depends on the concrete circumstances of a mission, any useful BPM solution requires real-time data about the executed activities.

For human first responders, we designed a system that interprets verbal communications, recognizes activities, structures them into processes, and visualizes the processes to provide assistance to the officers [30]. Since radio communication does not provide a complete status of the mission, there is a need for additional data sources. (Teleoperated) Rescue robots are equipped with sensors that record data about themselves and their surroundings, such as the robot's position, acceleration, or direction of movement. This sensor data can be used to recognize the activities that the robot operator executes during a mission.

In this paper, we design, develop, and evaluate an approach for the real-time recognition of robot activities that can be used for process assistance during robot-assisted disaster response. We define an ad-hoc process with a set of activities, which are executed as the mission status requires. Using a machine learning approach, we automatically recognize the ongoing activity from sensor data collected during execution. These activities are mapped to an instance of the ad-hoc process and stored in an event log, which is visualized to provide real-time information on the mission status to the officer in charge. In addition, this system supports the robot operator, as it documents all their activities during a mission. The collected data can afterwards be used for automatically generating a mission documentation report, as illustrative material for training new robot operators, or to improve the processes by means of post-mission process mining.

Our research is objective-centered, motivated by the opportunity to use available sensor data to provide process assistance. In this paper, we focus on detecting activities by a teleoperated ground robot in a specific application scenario described in Sect. 2. Sections 3 and 4 describe the design and development of our activity recognition approach, which is demonstrated and evaluated in Sect. 5. We report on related work in Sect. 6, before concluding the paper in Sect. 7.

Name	Search for unconscious human and take gas probe after industrial fire
Description	After an industrial fire, the robot UGV is sent into the factory to inspect a gas tank. Once the tank is found, the robot takes a gas probe to find out whether it was damaged by the fire and might therefore be leaking dangerous gases. After taking the probe, the robot searches for a missing human that could be unconscious due to the fire or any dangerous fumes. Once the human is detected, the robot returns to the base. Based on the result of the probe, first responders know if they can safely enter the building to rescue the unconscious human or if any additional steps must be taken beforehand.
Actors	UGV, Unconscious victim
Preconditions	A factory has burned down. A potentially damaged gas tank is located inside a building and a human is lying unconsciously on the floor of the same building.
Postconditions	The human is detected, a gas probe was taken, and UGV has arrived back at the base.
Main Success Scenario	1. UGV is navigated across the factory site to approach the entrance in a direct path. Once there, it enters the building. 2. UGV is navigated inside the room to search for the gas tank. 3. The operator detects the gas tank by classifying it on the camera picture. 4. UGV is navigated to the gas tank. 5. A gas probe is taken next to the tank, by raising UGV's sensor arm and holding it close to the tank. After taking the probe, the arm is lowered into the standard position. 6. UGV is navigated through the building to search for an unconscious human. 7. The operator detects an unconscious human by classifying it on the camera picture. 8. UGV is navigated towards the human so the operator can determine his/her health status. 9. UGV is navigated back to the base.

Fig. 1. Application scenario of an industrial fire formulated as a use case

2 Application Scenario

Automatic activity recognition depends on the capabilities of the robot and the situation in which it is used. In this paper, we focus on the application scenario of a medium-sized unmanned ground vehicle (UGV) in an industrial fire, where using a robot is particularly beneficial. Since production processes often require highly explosive or flammable materials, they are stored within the plants, making industrial fires both dangerous and frequent. The American National Fire Protection Association reports that between 2011 and 2015 municipal fire departments in the U.S. responded to an average of 37,910 industrial fires every year. Annual losses from these fires are estimated at 16 civilian deaths, 273 civilian injuries, and $1.2 billion direct property damage [2]. At the same time, those fires can be challenging for first responders. For instance, gas tanks stored within a plant might be damaged by a fire, making it dangerous to enter the building without proper safety measures. In this case, a robot can be sent into the building to locate the gas tank of interest and take a gas probe to determine if any gas has leaked. Furthermore, the robot can search for any victims that might be trapped or unconscious to communicate their position to a rescue team.

To illustrate the application possibilities of the concept and implementation of the artifact presented in this paper, we propose the use case depicted in Fig. 1 in the form of a modified version of the Use Case Template according to [4].

3   Conceptual Design

3.1   Outline

Our approach for automatic activity recognition consists of several steps to overcome an abstraction gap from low-level sensor data to a top-level process. These sensor data are obtained from sensing devices, which can communicate over communication networks, including the Internet, i.e. the Internet of Things (IoT)[8]. Given the set of available activities, we can use machine learning to automatically recognize the ongoing activity from fine-grained sensor data acquired from UGV's on-board sensor unit. Therefore, we train a classifier to assign a sequence of sensor data to one of the activity types, as described in Sect. 3.3.

Conceptually, the set of relevant robot activities constitutes an ad-hoc process, i.e., a process whose control flow cannot be modeled in advance, but instead is defined at run-time [5]. This is because the robot operator has to adapt his actions to the terrain and the needs in a new and unforeseeable emergency situation. This ad-hoc process is the basis for our process assistance tool, as described in Sect. 3.4, where we log and visualize each process execution as a sequential process instance. As a post-processing step, process discovery methods can be applied to the recorded event log, allowing insights into how the robot operators are performing their tasks and enable process improvements based on that.

3.2   Activity Type Definition

For our approach, we assume activity types that are defined a priori. Those activities are derived from the robot's capabilities within the application scenario and based on the human operators' understanding of high-level activities (e.g. "Search") instead of the low-level capabilities of the robot (e.g. "Drive Forward"), which are components of the higher-level activities. We aim at directly recognizing the activities that are relevant to the process. In this way, a direct mapping between activity and task in a process model is achieved. A different approach would be recognizing the robot's capabilities and deriving higher-level activities, which are only subsequently mapped onto a process model. Both approaches however, require prior knowledge of higher-level activities.

3.3   Recognizing Activities from IoT Data

Sensor data, like those acquired from the UGV, are frequently recorded representations of some kind of state. A state could be a temperature, a velocity, a level of humidity, a position, or a distance to a specific point in a space. The main challenge is to assign those fine-grained state representations to activities, more precisely to instances of activity types on a process-relevant abstraction level. Those activities commonly occupy larger time frames than the ones between two sensor captures. Therefore, time frames of sensor data have to be assigned to activity instances. This can be modeled as

$$\langle s_1, \ldots, s_n \rangle \to a \tag{1}$$

where $s_{1,...,n}$ are n data points captured by an individual sensor s and a is one of the given activity types. This solves an individual assignment problem between sensor data and activity type. In order to train a machine learning model, we need a sufficient amount of labeled input data, which is considered as a great challenge in detecting activities, due to its high effort [11]. In our case, we manually label the sensor data that belong to certain activity instances with the corresponding activity name. In the described application scenario, it was sufficient to rely on a single sensor as a data source to distinguish all relevant activity types. Limiting ourselves to one sensor with low payload per measurement, can also prevent problems such as transfer speeds from the mobile robot to other processing components, especially since data transmission can become difficult in an emergency situation. A scenario, where multiple sensors of the same type or even multiple types of sensors are necessary to fully capture all relevant activities, would require a mechanism for synchronizing, aggregating, filtering, and merging sensor data, making a reliable process assistance much more challenging.

3.4   Providing Ad-Hoc Process Assistance

After training a machine learning classifier for activity recognition, we can use it to recognize, document, and visualize the robot's actions in real time. Therefore, we model UGV's behavior as an ad-hoc process, as shown in Fig. 2. Within an instance of this sub-process, all tasks can be executed arbitrarily many times and in any order. Readers should note that the application scenario and the activity names indicate a partial ordering, which is not represented in the process model. For example, *Navigate Back to Base* appears to be the last activity in a trace. A prior activity filtering based on a more structured process model is also interesting and will be addressed in future work.

The process model is the basis for our process assistance system. We connect the activity recognition to a process model repository and process execution engine, where the model is instantiated at the beginning of a robot mission. The process execution engine is controlled by the activity recognition, such that the process status is continuously captured and automatically recorded in an event log. For example, whenever the recognition approach detects an activity instance that differs from the currently active activity instance, it triggers the completion of the active one and starts the newly detected one. With a continuous pipeline from the robotic sensors to the activity recognition and process execution engine to an event log, we are now able to provide process assistance to the first responders. The active process instance is visualized, providing process assistance to the

Fig. 2. Ad-hoc process for the behavior of UGV, modelled as a BPMN diagram

operator, as well as real-time information on the mission status to the officer in charge. After completing a mission, the logged event data captures all actions of the robot in sequence and can be used for post-mission process mining, debriefing the mission participants, improving the mission processes and protocols, as well as training new officers [30].

4   Implementation

In order to implement the conceptual design, we created an overall system architecture that is easily extendable by new components, shown in Fig. 3.

Data Acquisition. The UGV collects its IoT data using the Robot Operating System (ROS)[1], which uses a proprietary data format and is organized in topics, similar to Kafka[2]. Both systems implement a publisher subscriber pattern. *Topics* or *Channels* are categories or feed names to which records are published and subscribed. We implemented a bridge application that consumes sensor data from topics relevant for task recognition, translates them into a format, which is simpler to process by our system, and re-publishes them to an MQTT[3] broker. From there, other components of the system, in particular recognition and matching, can subscribe to the data channels relevant to them.

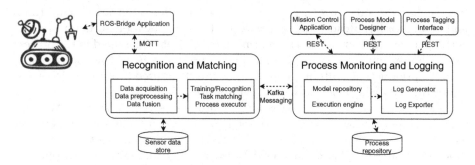

Fig. 3. System architecture and relevant connected components.

The robot collects a variety of different sensor data. For our goal of recognizing tasks on the process-level of an operator, we assume motion data to be most discriminative. Therefore, we base our implementation on the data collected by the Inertial Measurement Unit (IMU) sensor, attached to the base of the robot. Using a combination of accelerometer and gyroscope, the IMU measures linear accelerations, angular velocities, and the orientation for each 3D-axis at high frequency. Each of the three quantities has three dimensions yielding a

---

[1] https://www.ros.org.
[2] https://kafka.apache.org.
[3] http://mqtt.org.

9D-measurement in total. Even though the IMU is located at the base of the robot, operating actuators induce small motions, which are recognized by the IMU and allow for conclusions about the state of the manipulator arm.

Data Pre-processing. The recognition approach essentially consists of a continuous multi-class classification using sliding windows of pre-processed IoT data streams. These classification results need to be mapped to pre-defined activity types. Therefore, we define a fixed window size and compute feature vectors based on segments of sensor time series. Then, we classify each segment individually. To ensure that no boundaries of activity instances are missed, we use segments overlapping by 50%, as proposed by [26]. Thus, given an individual sensor s and segments consisting of t points in time when sensor data is captured (segment size), the sensor time series data are segmented as follows:

$$
\begin{aligned}
segment_1 &: s_1 \ \ ... \ \ s_t \\
segment_2 &: \quad\quad s_{t/2} \ ... \ s_{3t/2} \\
segment_3 &: \quad\quad\quad\quad s_t \ \ ... \ \ s_{2t} \\
&\ \ \vdots \quad\quad\quad\quad\quad \ddots \ \ \ddots \ \ \ddots
\end{aligned}
\tag{2}
$$

In order to achieve accurate classification results, we compute representative features of the raw sensor data and use them as input for the classification. We consider each dimension of the IMU measurements as a separate time series and compute the mean, standard deviation, root mean squares, kurtosis, skewness, interquartile range, and mean absolute deviation [1,10]. Besides those statistical features, frequency domain features are used as additional input because they provide additional insights. Specifically, we include the energy, entropy, and signal magnitude area for each time series as features. Moreover, for all time series within each quantity, we include the pairwise correlation.

Training. After the sensor data is pre-processed, segmented, and labeled with the activities from Fig. 2, it can be used to train the activity recognition. Therefore, we need a machine learning approach that allows for multi-class classification. We selected a Random Forest, as it produced promising results in past applications [21] and does not require vast amounts of training data, as opposed to, e.g., neural networks. The Random Forest classifier we finally deployed was trained as an ensemble of 200 trees with a maximum tree depth of 16. As entropy measure, we used the Gini impurity. The sensor data used for training is stored in a relational database as shown in the system architecture.

Recognition. While a disaster response process is executed, the trained model is queried continuously. Whenever a segment of the defined size is completed, we apply the same pre-processing as for the training data. Since segments are overlapping by 50%, the last half of the previous segment becomes the first half of the current segment for classification. To complete the activity recognition, the classification outcome needs to be mapped to the process model tasks, as stored in the process model repository. The available activity instances as published by the process execution engine are matched with the classification results of the activity classifier. This is done on the basis of the activity label, which in our case is identical to the respective activity label in the process model. Based on the activity label, life-cycle transitions of activity instances of the ad-hoc process can be controlled. For example, if an activity instance is currently active but the classification yields a different activity, the active activity instance is completed and the detected activity instance is started. This enables the process assistance.

Process Monitoring and Assistance. The process assistance is based on the deployed process execution engine, which instantiates processes at the beginning of a mission and monitors their execution depending on the results of the activity recognition. Therefore, the process engine bridges the gap between the activity recognition and the ad-hoc process execution. We used the open source process engine from jBPM[4] for this purpose. Accessing the results of all previous processing steps, we implemented a status monitoring visualization as a web-based application, which can be shown on any display. The app depicts an integrated view of the current mission status with all active and completed disaster response processes. Aggregating information about each process gives the officers in charge a better overview of the overall situation, as well as information about the progress in critical processes, such as reconnaissance.

5   Evaluation

5.1   Evaluation Approach

To ensure repeatable experiments and efficient data capturing, we train and evaluate our approach using data generated by a physics-based simulation of the UGV. The simulation is based on the open-source framework Gazebo[5], which includes accurate simulation of multi-body dynamics and realistic sensor data generation for common robotic sensors. We use a digital twin of the Telerob Telemax Hybrid UGV, which is deployed by the German Rescue Robotics Center.[6] The simulation exactly mirrors the real robot with respect to the software interface and the available sensors.

Based on this, we performed a two-stage evaluation. For the first step, we gave the use case description from Fig. 1 to ten case study participants and asked them to execute the described reconnaissance process using the simulation application. During the execution of each individual process instance, sensor data of the simulated robot were produced and consumed by our system. Simultaneously, the performed activities were manually labeled with the respective activity type. Based on this data, we trained the machine learning model as introduced in Sect. 4 and derived per-class accuracies as well as overall accuracies of the recognition. Using the trained and evaluated model, we performed a second case study to validate the process assistance system in real-time. Five study participants were asked to execute a reconnaissance sub-process in the

---

[4] https://jbpm.org.
[5] http://gazebosim.org/.
[6] https://rettungsrobotik.de.

same setting as the first case study. In order to measure the validation accuracy, we compared the actually executed activities to the ones the system detected and visualized in the developed mission control application. The use case executed in the simulation application with the running mission control app can be viewed in [22].

5.2 Case Study Implementation

While the study participants controlled the robot, we used a web-based tool (also depicted in Fig. 3) to tag the start and completion of process and activity instances. Available instances were determined according to the respective process model and the process instance's current state. This way, labelling was done implicitly via timestamps. Every recorded sensor data point was saved with the timestamp of its creation. Then, the activity time frames were used to label the data. We acquired a total of 45 process instances for training and testing including different variants of activity orders. Next, the data was prepared according to Sect. 3. The recorded and labeled data were used to train a random forest classifier, which is capable of performing multi-class classification as well as producing probability distributions over all possible classes for each unlabeled feature vector. The model was trained using different segment sizes (4-8-16). Segment size 16 yielded slightly better results, but since larger segments take longer to complete, this resulted in longer intervals between classifications and therefore a longer latency in the process assistance. We therefore chose to use a rather small segment size of 4, which performed comparably well and has the advantage of generating more training data. We initially evaluated each trained model using 20% of the collected data as test data. This classifier was then used for the overall evaluation in the second case study. Participants controlled the robot, performing both the same use case as for the training data acquisition and a different variant to show that the classifier can handle different orders of activities. For that second variant, we asked them to execute the original scenario, but since the human could be detected immediately, participants skipped the search step and directly approached the human. Afterwards, they searched for the gas tank, took a probe, and approached the tank to examine it visually. This was followed by another search activity to detect any other possible victims. 14 evaluation instances were captured, while the system was running. The efficacy of the approach can be shown by comparing the process instances produced by the system to the actual process execution.

5.3 Results

Activity Classifier Performance. Figure 4 shows the results of the individual performance evaluation of the activity classifier, which was trained on the sensor data of the previously recorded training process instances. We split the data into 80% training data and 20% test data. Additionally, Fig. 4 shows the evaluation metrics (precision, recall and f1-score) for the classifier evaluation. The results

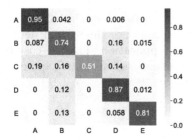

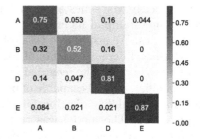

Fig. 4. Normalized confusion matrix (left) and classification metrics (right) for the activity classifier per class (A = Take probe; B = Approach Destination; C = Stand by; D = Search; E = Navigate back to base)

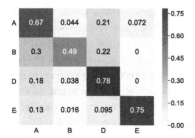

Fig. 5. Normalized confusion matrices showing all classified segments, including (left) and excluding (right) activity transitions (A = Take probe; B = Approach Destination; D = Search; E = Navigate back to base)

indicate that the classifier is capable of differentiating the robot's activities very accurately. A close look at the individual mis-classification revealed that confusions mainly occur because *Approach destination* is often classified mistakenly and therefore produces false positives. Also, the classifier mostly fails to correctly classify *Standby*, however, there are no false positives. Even in this case, the performance is sufficient to accurately determine the activities given the selected segment size and overlap. It should also be considered that standby times hardly occurred in the process executions of our case study, which is why this activity is underrepresented in training and test data. The activity *Detect target* is not considered in the classifier evaluation, as it is a point-based activity, which can be recognized using image data rather than motion data.

Overall System Performance. To evaluate the overall efficacy of the approach and the implemented artifact, we compare instances produced by the system in the form of an event log with the actual activities the robot performed, while being controlled by the case study participants. Figure 5 shows normalized confusion matrices of the classifications of the overall evaluation. The left matrix shows the performance of all classifications throughout all use case instances, whereas the right one displays all classifications that were executed as long as no activity transition took place. Label C (Stand by) is not depicted, as this activity was never executed during the overall system evaluation.

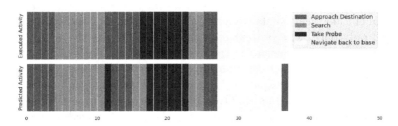

Fig. 6. Single overall execution instance of the scenario, with actual labels in the top row and predicted labels in the bottom row

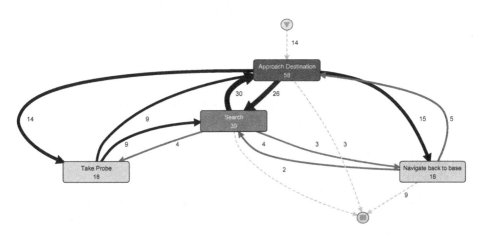

Fig. 7. Process map generated from the process instances that were recorded by the execution engine during the overall system evaluation (100% activities; 100% paths)

5.4   Discussion

The evaluation shows that our approach can monitor and capture process execution. The individual classifier performs well for all activities in our use case, despite limited training data, using a simple random forest classifier, and a single sensor as data source. Based on this classifier, we can map the results acquired in an online mode to tasks of our reconnaissance sub-process and control the process execution. As expected, the recognition accuracy is better for segments of the sensor data that are completely contained in an activity time frame. To further support this observation, Fig. 6 shows a single execution instance of our scenario comparing actual labels with predicted ones, illustrating that misclassifications mainly occur at activity transitions. To demonstrate post-mission process mining, we exported the event log produced during the overall system evaluation and used Disco to generate a process map, as shown in Fig. 7. For this log, subsequent identical events were merged, such that each recognized activity was logged only once. The process map shows that despite the ad-hoc process model, there is a clear structure to the executed process. For example, the first activity was always *Approach Destination* and *Take Probe* is always an intermediate activity. Other than expected, *Navigate back to base* was not always the last activity in the recorded cases. Overall, this discovered process model might give first hints to first responders regarding the recommended execution of a reconnaissance process.

Limitations. The conducted evaluation is limited due to rather few cases available for training. However, regarding the number of training samples, the availability of the simulation application allowed for obtaining far more data than we could have obtained using the actual robot. Moreover, in this case study, only two scenarios have been carried out. Future work will address these limitations.

Reproducibility. Our data repository[7] contains an export of the PostgreSQL database, which stores the IMU data as well as the process instance and respective activity instance data that we collected in the case study and used for the evaluation. This data can be used to reproduce our classification approach. Unfortunately, the simulation environment and our process assistance implementation cannot be made available due to licensing problems.

6   Related Work

Human Activity Recognition. The robot used in this paper could operate at least semi-automated, but currently, all actions are remotely controlled by a human operator, because disaster response missions require absolute dependability. Hence, our approach is related to Human Activity Recognition (HAR) in disaster response, as for example Lieser et al. present [12]. They describe an approach for situation detection based on activity recognition using wearable devices. Specifically, they detect people in disaster related situations by recognizing their physical activities using two smartwatches and three smartphones. Although the approach uses machine learning for HAR, an integration with a process-aware assistance system is not provided. The authors of [20] apply HAR based on data from wearable sensors with the goal of providing workers with ergonomic recommendations during the execution of manual commissioning processes. The approach allows for real-time monitoring, however, it uses predefined process sequences and does not consider ad-hoc behaviour. In their literature analysis, Mannhardt et al. give a comprehensive overview of approaches that

---

[7] https://github.com/a-rebmann/iot-processassistance.

employ activity recognition on sensor data, with the goal of enabling process discovery [13]. However, the focus is specifically on industrial settings.

Event Abstraction. Our approach aims at abstracting low-level events in the form of sensor data to tasks in a disaster response process. Therefore, it is related to a line of work in process mining that aims to identify high-level activities from low-level event logs. For example, Tax et al. present an approach for the abstraction of events in an XES event log based on supervised learning [27]. It consists of a feature representation of an XES event and a learning step based on a conditional random field. The authors evaluated their approach using, among others, a data set of sensor data from a smart home. Compared to our approach, however, no online monitoring is performed. In [28], a machine learning framework for log abstraction is proposed. It consists of two phases, log segmentation and classification. The approach is evaluated using a low-level SAP log and focuses on existing event logs in a narrower sense. Therefore, the applicability in IoT scenarios needs to be investigated. Mannhardt et al. present a method for event abstraction based on behavioral patterns [14]. Their goal is to create an abstracted log by matching low-level event logs with activity patterns. The user has to create a model for each high-level activity to be able to use the approach. Futhermore, analyses are conducted on historical data only. Compared to this approach, we focus on real-time monitoring of ad-hoc processes, with the aim of giving timely process assistance to relevant stakeholders. An approach that transforms sensor data, in particular real-time location system data (RTLS), into event logs is presented by the authors of [24]. Interactions are used as an additional knowledge layer to close the gap between sensor data and process instances. The approach was implemented and evaluated using simulated event logs and is focused on RTLS. Process assistance is not targeted.

Process Management in Rescue Robotics. From a process management view, disaster response missions are an interesting but challenging field. On the one hand, both the team structure and the processes, including communications and actions, are predefined and strictly followed. In many practice sessions, team members internalize all rules such that they are the most efficient during a mission. Therefore, there have been many attempts at applying business process management (BPM) methods in disaster response, including, e.g., [6,7,9]. On the other hand, each disaster situation is unique on its own. No one can foresee which processes have to be executed in which kind of environment. These decisions have to be made on-site and in real time, often without knowing the entire situation. Researchers have argued that this unforeseeable nature of disaster missions made conventional process management methods ineffective [19]. Our approach deviates from previous research, as we do not attempt to plan disaster response processes, but instead take a data-driven bottom-up approach for managing them in real-time [30].

BPM and IoT. The authors of [8] postulate a series of challenges for integrating BPM and IoT. The paper at hand addresses the specific challenge of *Dealing with unstructured environments*, in addition to the challenges *Detecting new pro-*

cesses from data and *Monitoring of manual activities*. Related work exists in the discovery of models of daily routines, where activity recognition is embedded into a BPM context. The authors of [25] use an existing sensor log from a smart space to mine processes of human habits, which they depict using process modeling techniques. Presence Infrared sensors were used as a single data source. In [21] the development and evaluation of an approach which recognizes and logs manually performed assembly and commissioning activities is presented. The goal is to enable the application of process discovery methods. The presented system uses a body area network, image data of the process environment and feedback from the executing workers in case of uncertainties. In [23] an architecture is proposed that integrates IoT and BPM. The authors present a provenance framework for IoT data and consider IoT objects in the process model. The system architecture is implemented and demonstrated, e.g. in a use cases in the production industry using wearable devices. The authors of [15] present an autonomous monitoring service for inter-organizational processes. Using this approach, human operators e.g. from service providers do not need to notify about the execution of their tasks. An extension of the Guard–Stage–Milestone (GSM) makes it possible to monitor the process even if the control flow is not strictly followed.

7   Conclusion and Outlook

In this paper, we presented an approach for process assistance based on IoT data collected during a robot's reconnaissance run during a disaster response mission in the application scenario of an industrial fire. We presented the design, implementation, and evaluation of a system for the real-time recognition of robot activities. The robots' set of capabilities defines a set of activities, which can be mapped to tasks in an ad-hoc process models. Based on this, the process assistance system captures process instances of the disaster response operation.

As our evaluation shows, process behavior can be monitored accurately using a single activity classifier. The classifier itself is trained on sensor data of rather few process instances, captured from a single sensor mounted on the robot. It can therefore be said that we presented a light-weight approach with respect to training effort as well as to the required capacity for the transmission of sensor data. We showed that the approach allows for the IoT-based capturing of unstructured processes, which can otherwise not be captured by information technology. Yet, there are many possible starting points for future research. First, the machine learning model used for activity recognition can be improved either by further optimizing segmentation and overlapping ratios of the sliding windows or by training different model types. In order to preserve an efficient data transmission, edge computing scenarios could be investigated. All data pre-processing or even model training and querying could be done on a machine deployed on the robot. Then, only the results need to be sent to a server to enable process assistance. Second, our system has been tested and evaluated using a simulation application to mimic real-life reconnaissance processes. Although the application is an accurate virtual representation, the results need to be validated in the field. This would also allow to test the approach with respect to environmental factors and phenomena such as lossy data streams happening due to connection problems between the robot and the other components.

Furthermore, the evaluation as it was conducted here, aimed at showing the efficacy of the presented IoT-based process monitoring and logging approach. We did not evaluate the acceptance of target users, which is required before introducing the system to real disaster response missions. In addition, incorporating further application scenarios would allow to show the generalizability of the approach. This should not only be limited to including other types of robots into the scenario, but different types of missions should be considered as well.

Acknowledgements. The research results presented here were partially developed within the research project A-DRZ (Grant No.: 13N14856), funded by the German Ministry for Education and Research (BMBF).

References

1. Bulling, A., Blanke, U., Schiele, B.: A tutorial on human activity recognition using body-worn inertial sensors. ACM Comput. Surv. **46**(3), 1–33 (2014)
2. Campbell, R.: Fire in Industrial or Manufacturing Properties. Technical report, National Fire Protection Association (2018). https://www.nfpa.org/News-and-Research/Data-research-and-tools/Building-and-Life-Safety/Fires-in-US-Industrial-and-Manufacturing-Facilities
3. Carver, L., Turoff, M.: Human-computer interaction: the human and computer as a team in emergency management information systems. Commun. ACM **50**(3), 33–38 (2007)
4. Cockburn, A.: Basic use case template. Humans and Technology, Technical report 96 (1998)
5. Dorn, C., Burkhart, T., Werth, D., Dustdar, S.: Self-adjusting recommendations for people-driven Ad-Hoc processes. In: Hull, R., Mendling, J., Tai, S. (eds.) BPM 2010. LNCS, vol. 6336, pp. 327–342. Springer, Heidelberg (2010). https://doi.org/10.1007/978-3-642-15618-2_23
6. Gašparín, M.: Identification and description of processes at the operational and information centre of the fire and rescue service of the Czech republic. Qual. Innov. Prosperity **19**(1), 1–12 (2015)
7. Hofmann, M., Betke, H., Sackmann, S.: Process-oriented disaster response management: a structured literature review. Busi. Process Manag. J. **21**(5), 966–987 (2015)
8. Janiesch, C., et al.: The Internet-of-Things meets business process management: mutual benefits and challenges. arXiv preprint arXiv:1709.03628 (2017)
9. Kittel, K., Sackmann, S., Betke, H., Hofmann, M.: Achieving flexible and compliant processes in disaster management. In: Hawaii International Conference on System Sciences, pp. 4687–4696. IEEE (2013)
10. Lara, O.D., Labrador, M.A.: A survey on human activity recognition using wearable sensors. IEEE Commun. Surv. Tutor. **15**(3), 1192–1209 (2013)
11. Lavania, C., Thulasidasan, S., LaMarca, A., Scofield, J., Bilmes, J.: A weakly supervised activity recognition framework for real-time synthetic biology laboratory assistance. In: International Joint Conference on Pervasive and Ubiquitous Computing, pp. 37–48. ACM (2016)

12. Lieser, P., et al.: Situation detection based on activity recognition in disaster scenarios. In: International Conference on Information Systems for Crisis Response and Management (2018)

13. Mannhardt, F., Bovo, R., Oliveira, M.F., Julier, S.: A taxonomy for combining activity recognition and process discovery in industrial environments. In: Yin, H., Camacho, D., Novais, P., Tallón-Ballesteros, A.J. (eds.) IDEAL 2018. LNCS, vol. 11315, pp. 84–93. Springer, Cham (2018). https://doi.org/10.1007/978-3-030-03496-2_10

14. Mannhardt, F., de Leoni, M., Reijers, H.A., van der Aalst, W.M.P., Toussaint, P.J.: From low-level events to activities - a pattern-based approach. In: La Rosa, M., Loos, P., Pastor, O. (eds.) BPM 2016. LNCS, vol. 9850, pp. 125–141. Springer, Cham (2016). https://doi.org/10.1007/978-3-319-45348-4_8

15. Meroni, G., Di Ciccio, C., Mendling, J.: An artifact-driven approach to monitor business processes through real-world objects. In: Maximilien, M., Vallecillo, A., Wang, J., Oriol, M. (eds.) ICSOC 2017. LNCS, vol. 10601, pp. 297–313. Springer, Cham (2017). https://doi.org/10.1007/978-3-319-69035-3_21

16. Mirbabaie, M., Fromm, J.: Reducing the cognitive load of decision-makers in emergency management through augmented reality. In: European Conference on Information Systems (2019)

17. Murphy, R.R.: Trial by fire [rescue robots]. IEEE Robot. Autom. Mag. 11(3), 50–61 (2004)

18. Murphy, R.R., et al.: Search and rescue robotics. In: Siciliano, B., Khatib, O. (eds.) Springer Handbook of Robotics. Springer, Heidelberg (2008). https://doi.org/10.1007/978-3-540-30301-5_51

19. Peinel, G., Rose, T., Wollert, A.: The myth of business process modelling for emergency management planning. In: International Conference on Information Systems for Crisis Response and Management (2012)

20. Raso, R., et al.: Activity monitoring using wearable sensors in manual production processes - an application of CPS for automated ergonomic assessments. In: Multikonferenz Wirtschaftsinformatik, pp. 231–242. Leuphana Universität Lüneburg (2018)

21. Rebmann, A., Emrich, A., Fettke, P.: Enabling the discovery of manual processes using a multi-modal activity recognition approach. In: Di Francescomarino, C., Dijkman, R., Zdun, U. (eds.) BPM 2019. LNBIP, vol. 362, pp. 130–141. Springer, Cham (2019). https://doi.org/10.1007/978-3-030-37453-2_12

22. Rebmann, A., Rehse, J.R., Pinter, M., Schnaubelt, M., Daun, K., Fettke, P.: IoT-Based Activity Recognition for Process Assistance in Human-Robot Disaster Response (Video) (June 2020). https://doi.org/10.6084/m9.figshare.12409577.v1

23. Schönig, S., Ackermann, L., Jablonski, S., Ermer, A.: IoT meets BPM: a bidirectional communication architecture for IoT-aware process execution. Softw. Syst. Model. 1–17 (2020)

24. Senderovich, A., Rogge-Solti, A., Gal, A., Mendling, J., Mandelbaum, A.: The ROAD sensor data to process instances via interaction mining. In: Nurcan, S., Soffer, P., Bajec, M., Eder, J. (eds.) CAiSE 2016. LNCS, vol. 9694, pp. 257–273. Springer, Cham (2016). https://doi.org/10.1007/978-3-319-39696-5_16

25. Sora, D., Leotta, F., Mecella, M.: An habit is a process: a BPM-based approach for smart spaces. In: Teniente, E., Weidlich, M. (eds.) BPM 2017. LNBIP, vol. 308, pp. 298–309. Springer, Cham (2018). https://doi.org/10.1007/978-3-319-74030-0_22

26. Tapia, E.M., et al.: Real-time recognition of physical activities and their intensities using wireless accelerometers and a heart rate monitor. In: International Symposium on Wearable Computers. IEEE (2007)
27. Tax, N., Sidorova, N., Haakma, R., van der Aalst, W.M.P.: Event abstraction for process mining using supervised learning techniques. In: Bi, Y., Kapoor, S., Bhatia, R. (eds.) IntelliSys 2016. LNNS, vol. 15, pp. 251–269. Springer, Cham (2018). https://doi.org/10.1007/978-3-319-56994-9_18
28. Tello, G., Gianini, G., Mizouni, R., Damiani, E.: Machine learning-based framework for log-lifting in business process mining applications. In: Hildebrandt, T., van Dongen, B.F., Röglinger, M., Mendling, J. (eds.) BPM 2019. LNCS, vol. 11675, pp. 232–249. Springer, Cham (2019). https://doi.org/10.1007/978-3-030-26619-6_16
29. Weidinger, J., Schlauderer, S., Overhage, S.: Analyzing the potential of graphical building information for fire emergency responses: findings from a controlled experiment. In: Internationale Tagung Wirtschaftsinformatik (2019)
30. Willms, C., Houy, C., Rehse, J.R., Fettke, P., Kruijff-Korbayová, I.: Team communication processing and process analytics for supporting robot-assisted emergency response. In: International Symposium on Safety, Security, and Rescue Robotics (2019)

Discovering Activities from Emails Based on Pattern Discovery Approach

Marwa Elleuch[1,2]([⊠]), Oumaima Alaoui Ismaili[1], Nassim Laga[1],
Walid Gaaloul[2], and Boualem Benatallah[3,4]

[1] Orange Labs, Paris, France
{marwa.elleuch,oumaima.alaouiismaili,nassim.laga}@orange.com
[2] Telecom SudParis, Institut Polytechnique de Paris, Paris, France
{marwa.elleuch,walid.gaaloul}@telecom-sudparis.eu
[3] UNSW Sydney, Sydney, Australia
boualem.benatallah@gmail.com
[4] LIRIS, Ecully, France

Abstract. Significant research work has been conducted in the area of process mining leading to mature solutions for discovering knowledge from structured process event logs analysis. Recently, there have been several initiatives to extend the scope of these analysis to consider heterogeneous and unstructured data sources. More precisely, email analysis has attracted much attention as emailing system is considered as one of the principal channel to support the execution of business processes (BP). However, given the unstructured nature of email logs data, process mining techniques could not be applied directly; thus it is necessary to generate structured event logs. Activities form a cornerstone component of BP that must be identified to obtain such structured logs. In this paper we propose to discover frequent activities from email logs. Existing approaches are usually supervised or require human intervention. In addition, they do not take into consideration activity business data (BD) perspective. In this paper, we introduce a pattern discovery based approach to tackle these limitations; we suggest mainly to discover frequent activities and their BD with unsupervised way. Additionally, our approach allows the detection of multiple activities per email and the automatic generation of their names while reducing human intervention. We validate our work using a public email dataset. We publicly provide our results to be a first step towards ensuring reproducibility in the studied area, which allows more practical analysis for further research.

Keywords: Business process · Activities · Emails · Business data.

1 Introduction

Process mining consists of discovering models of actual BP from structured event logs. However, some BP or BP parts are not necessary supported by a BP management system that would produce structured events logs. Therefore, applying

© Springer Nature Switzerland AG 2020
D. Fahland et al. (Eds.): BPM Forum 2020, LNBIP 392, pp. 88–104, 2020.
https://doi.org/10.1007/978-3-030-58638-6_6

traditional process mining techniques would generate at best partial view of such BP. Emails are widely used as a collaborative tool to support BP execution. However, given the unstructured nature of email traces, process mining techniques could not be directly applied. Thus, we need to generate structured event logs that identify BP-related components from email traces.

Activities form a BP cornerstone component that must be discovered from emails to build a structured BP event logs. In this paper, we focus on discovering frequent activities form emails. To achieve this goal, different approaches are proposed [1–8]. However, most of them rely on supervised learning techniques [1,4,6] or manual intervention to select relevant information [7]. This needs knowing activities in advance which is not always feasible. Additionally, it involves human intervention (e.g., for labeling training dataset). Moreover, approaches using unsupervised techniques as [8], considered a sentence as the lowest structure that could express an activity. Actually, in the case of non-controlled systems as emails, the expression of activities would not be constrained by emails' punctuation; employees could express more than one activity in the same sentence or email. Finally, the same approaches focused on discovering activities without considering associated BD. BD refers to the set of information that executed activities (e.g., trade energy) use (e.g., traded quantity) and produce (e.g., trade ID). Such information are mandatory for example to allow the discovery of data-centric BP [9], which is an area of growing interest nowadays.

Pattern discovery from textual data is one text mining technique which could be used for discovering activities from emails. It consists of uncovering relevant frequent substructures (e.g., set of words) that co-occur frequently. If one activity is discovered in the form of the set of patterns that are frequently used to express it, it would be detected more flexibly without being constrained by emails' punctuation. Based on recent surveys in such context [10], Sequential Patterns Mining (SPM) techniques present the state of the art today (e.g; n-gram generators, PrefixSpan). They identify words while paying attention to the relationships between them so that their semantic meanings can be preserved. However, they are based on the sequencing of words during the construction of patterns. For this reason, they are likely to miss the detection of important activities since in email bodies; words expressing the same activity (1) may not be identical, (2) may not follow same order, and (3) are not necessarily adjacent in sentences.

Based on pattern discovery, we propose in this work, a completely unsupervised approach for discovering activities from emails. We define an activity as the composition of: (1) a name that reflects its main goal, and (2) BD that reflect its consumed and produced data. We introduce an approach based on discovering frequent patterns. Such approach allows the discovery of common substructures of activities while tolerating using words that could be (1) different but sharing the same meaning, (2) discontinuous, and (3) not sequential. Our approach consists of two main steps: The first one analyzes per employee emails to reduce expression variance. It aims to capture the set of frequent patterns used by employees in order to express recurrent activities. The second step regroups similar activities of different employees. To regroup patterns belonging to the same activity (first step) and to regroup similar activities of different employees (second step), two similarity measures are used on the basis of: (1) words' synonyms and (2) activity/pattern business context (defined by their related BD).

The rest of the paper is organized as follows. Section 2 gives an overview of the state of the art related to the studied area. Then, in Sect. 3, we present email-activity model to define the main abstractions in our approach. In Sect. 4, we present our proposed approach and the algorithmic solution of some of its key steps. Finally and before concluding, we present, in Sect. 5, our evaluation results carried out using the public Enron dataset[1].

2   State of the Art

Activity discovery usually refers to the identification of activities occurring through event logs of BP management system. Such event logs are generally structured in the way that for each event, we dispose these information: activity name, process name and process instance. As consequence, activities can be easily inferred. However, starting from textual data to recognize activities, present a more challenging task. In fact, they are generally expressed in natural language, in a non-standard way. Some existing works started form textual descriptions (e.g. story telling) of BP to discover their activities [11,12]. They usually suppose that BP are described while respecting some writing rules (e.g. grammatical, semantic) enabling activities' extraction. Nevertheless, in the case of emails written by different employees, there will be more variance in the expression of activities.

Activity discovery approaches exploiting emails are divided into two categories: (1) those that allow the detection of one activity per email as [1,2]. However they rely on unrealistic assumption; one email can contain multiple activities, and (2) those that allow the detection of multiple activities per email [3–8]. Such approaches suppose that one activity is expressed at sentence level. They start by splitting each email into sentences. Then, each sentence is assigned to an activity using supervised learning techniques [3–6], rules based techniques [7] or clustering techniques [8] . This final step is, in some works, preceded by another one, carried out manually [7] or with supervised learning methods [8] , and which aims to identify relevant sentences.

Supervised approaches of the second category [3–6] classify email sentences according to the sender speech act (e.g.; request, propose). Obviously, this needs labeled data for training speech acts recognizers which requires human intervention. Furthermore, BP tasks differ from one BP to another. Thus, setting a unique list in advance disable the generation of the right BP models.

As for [7], BD discovery was studied in more general context by discovering activities' resources; emails were considered as storytelling narrative text

---

[1] https://www.cs.cmu.edu/~enron/.

where resources are supposed to appear in the same sentence (expressing an activity) having the same grammatical function (noun phrase object). Nevertheless, such suppositions would be not realistic and would generate non-significant and non-precise results; noun phrase objects in sentences do not always reflect business information. The same work requires manual tasks during its execution for selecting relevant activities or BP related emails, which is nor always feasible. As for Diana et al. [8], they discover activities without prior knowledge by using hierarchical clustering and proposing an automatic annotation technique. To ensure clustering task, similarity measures between relevant sentence pairs are calculated using the cosine similarity between the word2vec vectors of their verb-nouns. The limitation of such proposal is that human intervention is still required for experimentally tuning some clustering parameters (e.g; level cut) or for training dataset towards the selection of relevant sentences. Moreover, while integrating word2vect in this work presents an advantage for recognizing activities expressed differently, it seems to generate not accurate results. In fact, word2vect groups terms having the same context without differentiating between their meanings, which may create some confusion in the context of activity discovery. Taking the example of two antonyms: buy and sell, they would be grouped together when using their word2vect representations while they refer to two different activities (e.g., buy energy and sell energy).

3   Email Activity: Model

In this section, we present our model to define the main abstractions regarding email and activities.

Email body: An email is the composition of the following parts: (1) sender, (2) recipients, (3) subject, (4) timestamp and (5) body. We consider the body and the sender as our approach input.

Frequent Activities: A frequent activity is a task frequently discussed in emails and it is likely to belong to a BP. It is a composition of (1) Activity Name which reflects the goal of the activity (e.g; create deal), and (2) BD which represents the data used/generated in the execution of the activity (e.g; deal price).

Activities expression in emails: As illustrated in Fig. 1, we suppose that activity components (name & BD) are expressed in emails through coherent

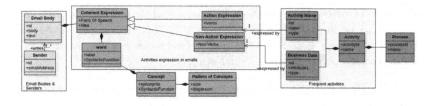

Fig. 1. Email activity: model

expressions. We define a coherent expression as a low dispersed set of words that contribute in precising the same idea (e.g; 'create ticket'). Taking the example of two emails' extracts (A & B) of enron dataset;

 - A = 'I have created deal tickets 241558 and 241560 for July 99..'
 - B = 'I have created a spreadsheet to assist in the tracking and booking of the gas...I will then update the demand fee on the deal ticket..'

A and B share the set of words {'create', 'ticket'} which is low dispersed in A and highly dispersed in B. Both words in A contribute to the expression of the idea 'ticket's creation', while in B, each word contributes to the expression of a different idea ('spreadsheet creation' and 'ticket update'). This shows that low dispersity of words, as in A, is essential to express activity objective.

Coherent expressions are of two types: (1) action expression containing at least one verb and that would be used to express activity names, and (2) non action expression that does not contain verb and that would be used to express BD.

Beside, one activity can be expressed differently through emails using different coherent expressions (e.g., 'buy energy' and 'purchase energy') sharing synonyms (Definition 2) . Therefore, we introduce the notion of concept (Definition 3) to regroup synonyms. These synonyms: (1) support one potential meaning, (2) have the same syntactic function (Definition 1) and (3) share the same context of use in natural language (e.g; the set {'purchase', 'buy'} forms one concept). If one concept appears in an email at a defined position, it will be expressed using a single word belonging to its synonyms. A pattern of concepts will be a set of concepts that coexist in the same email (they will not necessary appear continuously or in defined order). Such pattern will regroup the combination of words that would express the same thing (e.g., 'buy energy' and 'purchase energy' will be regrouped in the pattern of concept {{purchase, buy}, {energy}}). As for its size, it will depend on the size of expressions that it would reflect.

Definition 1 *(Syntactic Function).* *Let W be the set of words and T be the set of part of speech tags (e.g; NOUN, VERB) that indicates how a word functions grammatically in an email e. The syntactic function $ScFc_e : W \to T$ returns for each word $w \in W$, the corresponded part of speech tags $t \in T$.*

Definition 2 *(Synonymy Function).* *Let W be a set of words and T a set of possible syntactic function tags. A synonymy function $SynFc: W \times T \to W^*$ returns the synonyms of $w \in W$ in respect to its syntactic function $t \in T$.*

Definition 3 *(Concept).* *Let $S = \{w_0, w_1, ..w_p\}$ be a set of words and let $E = \{e_0, e_1, ..e_p\}$ be the set of emails $| \forall i, w_i \in e_i$. Let $C \subset S$ (C a subset of S) a concept $\Leftrightarrow \forall(w_i, w_j) \in C \times C$; (1) $ScFc_{e_i}(w_i) = ScFc_{e_j}(w_j) = t$ AND (2) $w_i \in SynFc(w_j, t)$, AND (3) $w_j \in SynFc(w_i, t)$.*

Each words' pair belonging to a concept must have the same syntactic function and reciprocal relations of synonymy. We consider the synonymy reciprocity criterion so that one concept supports the expression of a single meaning.

This fact would probably guarantee that two expressions sharing the same concepts will express the same idea. Taking the example of the words' pairs ('set', 'arrange') and ('set', 'define'), their elements are reciprocally synonyms, however, 'arrange' and 'define' are not reciprocally synonyms. If the words 'define', 'set' and 'arrange' belong to the same concept, this latter would express two potential meanings referring to two different activities (organizing or defining something).

4   Proposed Approach

4.1   Approach Overview

In this section, we will present the main steps of our approach which is based on the discovery of frequent patterns of concepts. In the following, we rely on these two emails (e_1 and e_2) as illustrative examples:

- e_1 = '..We shall arrange a preliminary interview for trader position with this person... I have some availability next week; can you contact him to define a time slot?'
- e_2 = '..another interview should be set with him... Last week a student forum was held and I probably found other potential candidates... I've set up some time slots with them for this week and I will send you my feedback concerning my preliminary interviews with them'.

As illustrated in Fig. 2, our approach is composed of 4 main steps:

Step 1: Preprocessing: In this step, the conversation histories are removed from email bodies to avoid redundancy in addition to some useless words (e.g.; thanking, salutaions, signatures). We detect then some numeric values and named entities, and we replace them by tags (e.g., pricenumeric, personname). We keep such information as they often reflect BD values in emails. Finally, we return for each email, a list of lemmatized words concatenated with their syntactic functions tags. We consider only two types of syntactic functions: verb (v) and non-verb (n) as they are important in defining activity components. For example, e_1 = ['arrange_v', 'preliminary_n', 'interview_n', 'trader_n', 'position_n', 'person_n', 'availability_n', 'contact_v', 'define_v', 'time_n', 'slot_n'] and e_2 = ['interview_n', 'set_v', 'student_n', 'forum_n', 'hold_v', 'probably_n', 'find_v',

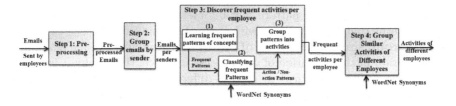

Fig. 2. Main steps

'potential_n', 'candidate_n', 'set_v', 'time_n', 'slot_n', 'send_v', 'feedback_v', 'concern_v', 'preliminary_n', 'interview_v'].

Step 2: Group emails by sender: This step regroups emails by their senders, which will be useful to take into account employees' writing style and deduce their recurring activities (step 3). We suppose that one employee, who is frequently implied in the execution of one or multiple activities, likely use close wordings to express the same frequent activity. Such close wordings would bring a kind of regularity at the level of coherent expressions that he uses for expressing activities components. This regularity will be reflected by the set of patterns shared by coherent expressions, which are used to express the same activity.

In other hand, this step would be useful to minimize the variance degree of emails' writing styles. It reduces the degree of noise per employee. Additionally, it can potentially enhance the interpretability and the coherency of the obtained patterns while decreasing the appearance of non-relevant combinations of words.

Step 3: Discover frequent activities per employee: This step generates a list of frequently discussed activities. It consists of three elementary sub-steps:

> **Sub-step 3.1: Learning frequent patterns of concepts:** It analyzes per employee emails' outbox to learn his/her frequent patterns of concepts forming coherent expressions. Such goal requires matching synonymous words. We rely then on the integration of a dictionary of synonyms (e.g. WordNet).

> **Sub-Step 3.2: Classifying frequent patterns of Concepts:** This sub-step aims to obtain potential activity names and BD. Therefore, it classifies frequent patterns into action pattern (of action expressions), if they contain at least one verb. Otherwise, they will be considered as non-action patterns (reflecting non action expressions). Patterns composed only of verbs are excluded because coherent expressions would not be only formed by verbs. At this level, each obtained action pattern will be formed by at least one non-verb concept that may refers to: (1) useful information contributing in specifying activity names, or (2) BD.

> **Sub-Step 3.3: Group patterns into activities** It aims to regroup action and non-action patterns into activities. An activity is associated to multiple action patterns since one employee can express differently activities using words sharing the same context (not necessary synonyms) (e.g.; 'I sold 50 mw's' expresses the activity of selling electricity power which was not explicitly mentioned). To ensure such regrouping, we suppose that patterns of the same activity must share: (1) similar verb concepts that identify its main action, and (2) similar business context. The business context of each pattern will be inferred from the set of non action patterns that highly coexist with it (in the same email).

Step 4: Group similar activities of different employees: Different employees collaborate in the execution of one process instance, that's why it is important to group similar activities executed by different employees. Each activity at this level is characterized by: (1) a set of action patterns sharing the same verb concepts (2) one verb concepts used to group such action patterns in sub-step 3.3 and (3) non-action patterns that represent activity BD. In this step, we proceed as in sub-step 3.3 by considering that similar activities must share similar: (1) verb concepts and (2) BD characterizing their business context.

In what follows, we will detail only the key steps of our proposal, namely sub-step 3.1 (*learning frequent patterns of concepts*) and sub-step 3.3 (*Group patterns into activities*).

4.2 Learning Frequent Patterns of Concepts

Algorithm 1   Discover Frequent Patterns Per Employee
1: **procedure** DISCOVERFREQUENTACTIVITIESPEREMPLOYEE($employeeEmails, thresh, minFreq$)
2:     $dic\_conceptsPatterns = \{\}$, $dic\_Affect = \{\}$, $dic\_ids = \{\}$
3:     **for** $(e1, e2)$ in $employeeEmails.Pairs()$ **do**
4:         $PatternsOfConcepts = $ **Correlation**$(e1, e2, thresh)$
5:         $dic_conceptsPatterns, dic_ids, dic_Aff = $ **UpdateDic**$(dic\_conceptsPatterns, dic\_ids,$
6:             $dic\_Aff, PatternsOfConcepts, e1.ID, e2.ID)$
7:     $dic_conceptsPatterns = $ **FilterByFrequency**$(dic\_conceptsPatterns, dic\_ids, minFreq)$
8:     **return** $dic\_conceptsPatterns$
end

This sub-step aims to learn the frequent patterns of concepts shared by the overall of emails sent by the same employee. It is mainly composed of three additional steps: (a) Correlating emails pairs (line 4, Algorithm 1) (b) Updating dictionary of patterns of concepts (line 5 & 6, Algorithm 1) and (c) Filtering patterns of concepts by frequency (line 7, Algorithm 1) after defining a threshold of frequency. In what follows, we will focus on explaining steps (a) and (b).

Correlating Emails' Pairs: This step disposes as input two preprocessed emails. The goal is to find their intersection in terms of patterns while verifying coherent expressions criteria (Definition 8) and tolerating the existence of synonyms. Given that two emails can share multiple activities (n) and given that one activity corresponds to one pattern, our solution for this step must allow obtaining n patterns per correlation. To this end, we propose the Algorithm 2 which is composed of three main sub-steps: (1) Find Fictive Intersection (Defintion 5) of the two emails (*FC* of line 2 in Algorithm 2) where each fictive concept (Definition 4) forming this intersection can support more than one meaning, (2) Search common Sub-Patterns of fictive concepts shared by the two emails under dispersion constraints (line 3 → 8, Algorithm 2, and (3) Build the common pattern of concepts where each concept support unique meaning (line 9 & 10 Algorithm 2).

Algorithm 2   Correlation of two emails

```
 1: procedure CORRELATION(e₁, e₂, thresh, dic_conceptsPatterns)
 2:     FC = FindFictiveIntersection(e₁, e₂, WordNet)
 3:     newe₁, newe₂ = Replace(e₁, e₂, FictiveConceptsDictionary)
 4:     FCpos_e₁ = FindOrderedPositions(newe₁, FC)
 5:     FCpos_e₂ = FindOrderedPositions(newe₂, FC)
 6:     SPpos1 = SearchSubPatterns(newe1, thresh, FCpos_e₁)
 7:     SPpos2 = SearchSubPatterns(newe2, thresh, FCpos_e₂)
 8:     intersections = FindIntersection(SPpos1, SPpos2, thresh, newe₁, newe₂)
 9:     MP_Words1, MP_Words2 = RealCorrespondance(intersection, e₁, e₂)
10:     PattensOfConcepts = UnionPatterns(MP_Words1, MP_Words2)
11:     return PattensOfConcepts
end
```

Sub-Step1: Find Fictive Intersection (line 2, Algorithm 2): The goal is to find the intersection of the two emails in the form of fictive intersection while tolerating the existence of synonyms. In fact, it is insufficient to rely solely on identical words shared by both emails. For instance, $e_1 \cap e_2 = \{$'preliminary_n', 'interview_n', 'candidate_n', 'time_n', 'slot_n'$\}$ does not include common concepts present in e_1 and e_2 but expressed differently (the concept $\{$'set_v', 'arrange_v'$\}$ that reflects organizing an interview and the concept $\{$'define_v', 'set_v'$\}$ that reflects fixing a time slot). Our fictive intersection is composed of a set of n fictive concepts FCi, $i \in [0...n]$ shared by the two emails. Each fictive concept (Definition 4) is the union of a set of concepts where each one (supporting unique meaning) must necessary share at least one synonym with another existing concept (which means that a fictive concept can support multiple meanings).

Definition 4 (Fictive Concept). *Let $FC = \{w_0, w_1, ..w_{p-1}\}$ be a set of p words, t be a syntactic function tag and $C = \{c_0, c_2, .., c_{m-1}\}$ be a set of m concepts where: (1) $m \leq p$, (2) $\forall c_j \in C, \forall w \in c_j, w \in FC$, and (3) $\bigcup_{i=0}^{m-1} c_i = FC$. FC is a Fictive Concept $\Leftrightarrow \forall i \in [0, m-1], \exists j \in [0, m-1] \mid c_i \cap c_j \neq \emptyset$.*

Definition 5 (Fictive Intersection). *Let $FI = \{FC_0, FC_1, ..FC_{d-1}\}$ be a set of fictive concepts and e_1 and e_2 two emails. FI is a fictive intersection of e_1 and $e_2 \Leftrightarrow \forall i \in [0, d-1], \exists w_1 \in e_1$ and $w_2 \in e_2 \mid w_1 \in FC_i$ and $w_2 \in FC_i$.*

We introduce fictive concept notion since it is ambiguous at this level to determine the set of shared concepts supporting unique meanings, especially if one email contains a word that supports two different meanings existing in the other email. Going back to our example, according to WordNet synonyms and the possible meanings appearing in the two emails; (1) 'set' of e_2 belongs to two meanings (m1 = {'set', 'arrange'}, m2 = {'set', 'define'}) and (2) the two meanings are not equivalent because 'arrange' $\notin$ SynFc('define', v). Therefore, we cannot replace the word 'set' by a tag that reflects unique meaning (because we will neglect the other possible meaning and we may prevent the appearance of some patterns). Now, if we apply our first step of calculating fictive intersection on e_1 and e_2, we obtain five fictive concepts: $FC_1 = \{$'arrange_v', 'set_v', 'define_v'$\}$, $FC_2 = \{$'interview_n'$\}$, $FC_3 = \{$'preliminary_n'$\}$, $FC_4 = \{$'time_n'$\}$, $FC_5 = \{$'slot_n'$\}$.

Sub-Step2: Find Common Sub-patterns of Fictive Concepts (line 3 → 8, Algorithm 2): The goal of this sub-step is to obtain, from the set of fictive concepts, multiple significant sub-patterns that reflect the common existing coherent expressions in a pair of emails. In the case of our example (e_1 and e_2), we must obtain the following set of multiple sub-patterns MSP = $\{SP_1, SP_2, SP_3\}$ = $\{\{FC_1, FC_2\}, \{FC_3, FC_2\}, \{FC_1, FC_4, FC_5\}\}$ that reflects respectively the following ideas: organizing an interview, preliminary interview and defining a time slot. For this purpose, our actual sub-step consists first of replacing words, in the preprocessed emails, by tags reflecting their fictive concepts ($newe_1$ and $newe_2$ in line 3 of Algorithm 2). Then, each email will be reduced to the list of actual appearance positions of these tags (where elements appear in increasing order) and without removing redundancy ($FCpos\_e_1$ and $FCpos\_e_2$ in lines 4 & 6 of Algorithm 2). After that, by relying on the dispersion constraints characterizing coherent expressions (Definition 8), we search the sub-patterns (Definition 6) positions that can be deduced from each email and reflecting coherent expressions ($SPpos1$ and $SPpos2$ in lines 6 & 7 of Algorithm 2). Finally, we search the intersection between each sub-patterns pairs belonging to different emails (*intersections*, line 8 of Algorithm 2). Such intersections will contain: (1) the shared sub-patterns by the two emails if they satisfy coherent expressions criteria, and (2) the actual appearance positions of the elements composing them.

Definition 6 *(Sub-Pattern).* *Let $FC = \{FC_0, FC_2, .., FC_{m-1}\}$ be a set of fictive concepts that are not necessary unique ($\exists i \neq j \mid FC_i = FC_j$), E be a set of emails, $e \in E \mid e \cap FC_j \neq \emptyset \; \forall i \in [0, m-1]$ and Pos : $FC \times E \to IN$ a function that returns the appearance position of a fictive concept $FC_i \in FC$ within the email e. SP is a sub-pattern $\Leftrightarrow l \leq m$ and $SP = [(FC_0, Pos(FC_0, e)), ..,(FC_{l-1}, Pos(FC_{l-1}, e))] \mid Pos(FC_i, e) < Pos(FC_j, e), \forall i, j \in [0, l-1]^2$.*

A sub-pattern appearing in an email e is then defined by a list of tuples (fictive concept, position of appearance in e). Theses tuples appear in an ascending order in respect to the fictive concepts positions.

Definition 7 *(Sub-Pattern Dispersion).* *Let $SP = [\{C_i, Pos(FC_0, e), i = 0..l-1\}]$ be a sub-pattern that belongs to a set of sub-patterns S and let $e \in E$ which is a set of emails. The Dispersion function is defined as follows: Dispersion : $S \times E \to IN^* \mid Dispersion(PC, e) = [\{|Pos(C_{i+1}) - Pos(C_i)|, i \in [0..l-2]\}]$.*

This function reflects the closeness of fictive concepts positions belonging to a sub-pattern SP within an email e. It returns a list of size l–1 referring to the distance between the appearance positions of each pair of successive fictive concepts composing the same sub-pattern.

Definition 8 *(Dispersion Constraints).* *Let $d = [d_1, .., d_{l-1}]$ be the dispersion of a sub-pattern SP within an email e and let thresh be a positive integer reflecting the minimum distance that separates successive concepts. The low dispersion constraints of coherent expressions are defined as follows:*

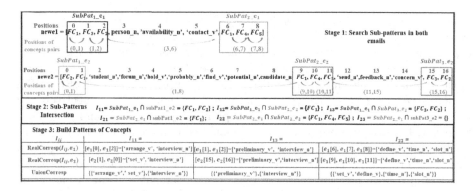

Fig. 3. Correlating e_1 and e_2

$$LowDispersionConstraints(PC, e) = [\{Boolean_i, i \in [1..l-1]\}] \mid Boolean_i = d_i < thresh \forall i \in [1..l-1].$$

In other words, the criterion of low dispersed patterns (Definition 8) will be verified if all the Boolean variables will be equal to True and without imposing additional constraints concerning the sequencing or the appearance order of the concepts (e.g; gap between concepts is allowed).

To find sub-patterns (line 7 & 8, Algorithm 2), we propose to iteratively construct them. We consider that it is mainly about finding the real positions of the concepts forming it in an email while verifying dispersion constraints. By scrolling through the corresponded list of concepts' positions, each element will be added to the same sub-pattern if and only if: (1) their concepts were not previously added, and (2) the distance that separates them from those preceding them does not exceed the defined threshold. Once having a concept position where dispersion constraints of the new sub-pattern will be not verified, this latter will mark the end of one potential sub-pattern and the beginning of another.

Getting back to our example of the two emails e_1 and e_2, with a value of threshold set to 3 (thresh = 3). After searching sub-patterns in both emails as illustrated in Fig. 3 (Stage1), we obtain from e_1, $subPat_1\_e_1$ and $subPat_2\_e_1$ and from e_1, we obtain $subPat_1\_e_2$, $subPat_2\_e_2$ and $subPat_3\_e_2$. Then, after calculating their intersections (Fig. 3, Stage2), we return I_{11}, I_{13} and I_{22} as well as their real positions as shared sub-patterns by the two emails (because they verify dispersion constraints and their size is greater than 2, while the sizes of the others intersections I_{12}, I_{21} and I_{23} are lower than 2 and can not form patterns).

Sub-Step3: Build Patterns of Concepts (line 9 & 10, Algorithm 2): The goal of this sub-step is to deduce the shared patterns of concepts (by the two emails) from the *intersection* generated from *Sub-Step2* (of line 8, Algorithm 2). To this end, for each sub-pattern, we aim to find the actual meanings that their fictive concepts express, depending on their local context of appearance in both emails. For this purpose, we retrieve first the real correspondences of the discovered sub-patterns (which means the actual words related to them in both emails, see

Stage 3 of Fig. 3, lines $RealCorresp(I_{ij}, e_1)$ and $RealCorresp(I_{ij}, e_2)$). Then, each concept composing a sub-pattern will be equal to the union of its actual correspondences in the two emails to find their unique meanings according to their local context (see Stage 3 of Fig. 3, line UnionCorresp).

Updating Dictionary of Patterns of Concepts (line 5, Algorithm 1): At each iteration, after correlating a new pair of emails, we use a dictionary ($Dic\_patternConcepts$) to store and update the obtained set of patterns. For each pattern of this set, if it is similar with another one obtained from another correlation (of another couple of emails), it will be updated in the case of having new synonyms that were not previously affected to it. Otherwise, this pattern will be affected to the dictionary with a new key. In the overall of updating step, the traceability of the appearance (emails IDs) of each detected pattern is preserved (using dic_ids) to estimate latter its frequency. Supposing at an iteration m, our dictionary contains a pattern of concepts $PC_1 = \{\{$'get_v', 'have_v'$\}$, $\{$'reservation_n'$\}\}$. Then, in another iteration p ($p > m$), we obtain $PC_2 = \{\{$ 'get_v', 'take_n'$\}$, $\{$'reservation_n'$\}\}$ among the shared patterns. As the elements of $\{$'get_v', 'take_v', 'have_v'$\}$ are reciprocally synonyms, PC_1 will be updated, in iteration p, by the new term 'take_v'; the dictionary $Dic\_patternConcepts$ will contain $PC_1 = \{\{$'get_v', 'have_v', 'take_v'$\}$, $\{$'reservation_n'$\}\}$.

4.3   Group Patterns into Activities

The goal here is to regroup patterns into activities after classifying them into action and non-action patterns. To this end, we proceed as follows:

Step1: Group action patterns by similar verb concepts: Considering two verb concepts $C_1 = [c_{11}, ...c_{1k}]$ and $C_2 = [c_{21}, .., c_{2m}]$ of two action patterns having t as syntactic function. C_1 and C_2 are similar $\Leftrightarrow C_1 \bigcap C_2 \neq \emptyset$ AND $\forall i, j$ $c_{1i} \in SynFc(c_{2j}, t)$ and $c_{2j} \in SynFc(c_{1i}, t)$ (For SynFc, see Definition 2).

Step2: Identify BD of action patterns: We identify here the highly correlated non-action patterns (that present the potential BD) with action patterns in terms of coexistance within the same email. Considering: (1) a non-action and an action pattern P_n and P_a appearing respectively in these sets of emails $E_n = [e_{n1}, .., e_{nm}]$ and $E_a = [e_{a1}, .., e_{ap}]$, and (2) $card(L)$ to denote the size of a list L, the coexistence fraction of two patterns is equal to $f(E_a, E_n)$ while:

$$f(L_1, L_2) = \frac{card(L_1 \bigcap L_2)}{min(card(L_1), card(L_2))} \tag{1}$$

Step3: Group action patterns by similar BD and deduce activity names: For each group of action patterns (returned by *Step1*), we group them by similar business context (that is characterized by their highly correlated BD). Finally, activity names will be deduced from the most recurrent one of each group.

Considering two action patterns P_1 and P_2 of business data $BD_1 = [bd_{11}, .., bd_{1n_1}]$ and $BD_2 = [bd_{21}, .., bd_{2n_2}]$, the similarity measure of P_1 and P_2 in terms of business context is equal to $f(BD_1, BD_2)$.

Step4: Inferring BD of activities: We keep here the non-action patterns that are highly correlated with all the action patterns forming each activity. Finally, we divide them into: (1) main BD patterns; they represent those that contain at least one named entity tag (e.g. numeric, personname, locname), and (2) optionally BD patterns; they do not contain named entity tags but give indications concerning business context activities' execution.

5   Evaluation

We evaluated the proposed algorithm on real emails belonging to Enron public dataset. We computed two metrics: (1) Recall which reflects its capacity of retrieving relevant activities from emails with unsupervised way; we have used here annotated emails of two employees and (2) Accuracy which reflects the precision degree of its discovered activities; we extended here the evaluation dataset to the emails of 5 employees as the accuracy metric does not require annotated emails. During such evaluations, we set the values of (1) the distance that defines dispersion constraints to 3, and (2) frequent patterns to 3. Different values of thresholds ($\in [5, 12]$) that defines the minimum number of activities' occurrence are studied. We present in this section only the results in relation to a threshold equal to 7. More information about the other values are in this link[2].

During this section, a pattern of concepts will be reflected by one string where: (1) concepts are separated by white spaces, and (2) concept's synonyms (without syntactic tags) are separated by '_' (e.g.,'buy_purchase power' reflects the pattern of concepts {'buy_v', 'purchase_v'}, {'power_n'}).

Evaluation Based on Annotated Emails

Methodology of annotation: We selected here all the emails sent by two employees (E_1 and E_2) having different business roles. The first one is a Managing Trader, he has 331 unique emails and his main task is to online trade energy (especially of electricity power). The second employee is an administrative assistant, he has 342 unique emails and he mainly arranges and coordinates interviews/meetings. For each email, we manually annotated the list of discussed activities and the related business data when they exist. In such process, we have not considered standard activities (like call, contact or meet persons) or those that can be discovered without the need of analysing textual content of emails (e.g; the activity 'attach file' without precising the type of the file can be inferred from the presence of an attached file in the email). Table 1 summarizes the list of frequent annotated activities that we obtained. For each employee, the table enumerates the list of annotated activity names as well as their associated Ids, their occurrence number across emails (Freq) and their related BD (e.g; in the context of trading activities ; hourEnding indicates the hour during which trades are conducted, counterpart indicates the seller/buyer that trades with Enron).

---

[2] http://www-inf.it-sudparis.eu/SIMBAD/tools/MailMining/.

Table 1. Annotated activities

Id	Activity Name	Freq	BD
	Managing Trader Activities		
1	sellPower	53	hourEnding
2	scheduleTrade	29	
3	buyPower	17	
4	longPower	16	zone,
5	enterDeal	13	quantity,
6	shortPower	12	price,
7	replaceDeal	10	dealNumber,
8	checkDeal	7	
9	runPlant	8	quantity
10	pay price	9	price,
11	receivePrice	7	counterpart
12	attendMeeting	8	
13	*attachResume*	9	
	Assistant Activities		
14	conductInterview	28	candidate
15	bringCandidate ForInterview	17	position
16	forwardResume	13	
17	*attachResume*	10	
18	scheduleInterview	15	time
19	makeReservation	11	date
20	scheduleMeeting	10	
21	reserveConfRoom	8	roomNumber
22	arrangeInterview	9	

Table 2. Discovered activities

	Discovered Name	N	id_a	R	BD
Managing Trader	sell_trade pricenumeric	45	1	0.83	hour0end numeric,
	schedule locname	25	2	0.86	
	buy_purchase pricenumeric	18	3	0.94	locname zone,
	long numeric0mw	18	4	1	numeric0mw,
	enter deal	14	5	1	orgname
	short numeric0mw	9	6	0.75	pricenumeric,
	replace deal_trade	10	7	1	deal
	check_see deal_trade	7	8	0.85	numeric,
	run plant	10	9	0.87	numeric0mw
	pay pricenumeric	12	10	0.91	pricenumeric
	receive price	13	11	0.85	orgname
	attend meeting	7	12	0.87	
	attach resume	7	13	0.77	
Assistant	conduct interview	19	14	0.67	
	bring interview	18	15	0.94	personname
	forward resume	12	16	0.92	
	arrange_set interview	9	22	1	
	make reservation	8	19	0.72	timenumeric
	schedule meeting	8	20	0.8	datenumeric
	schedule interview	11	18	0.73	
	hold_reserve eb0numeric	8	21	0.88	eb0numeric
	attach resume	11	17	0.8	
	Recall Average			0.86	

Experiments and results: We ran here our algorithm to discover frequent activities of both employees (using the same threshold as in the annotation phase). We used Python as implementation language and Spacy as a Natural Language Processing library to lemmatize and detect syntactic functions of words in emails. We have considered here the business role nature of the trader employee; we defined some syntactic function exceptions concerning some trading keywords 'short' and 'long' by setting their functions to verbs (to respect trading terminology because such terms refer to the creation of long/short trading positions). The obtained results are summarized in Table 2; we obtained 'attach resume' as a common activity with both employees and other non common activities. For each one, the table shows its automatic name generated by our algorithm, the number of emails where our algorithm identifies it, the Id of its related annotated activity (Id_a), the recall of its detection in emails (R) and finally, its related BD discovered by the algorithm. We define the recall of an activity as the ratio of the correct detection to the total number of emails that actually contain it. Finally, we calculated the recall average of the discovered activities within emails and we obtained a value of 0.86; this reflects that our algorithm is able to identify a good percentage of emails containing relevant activities. We calculated also the recall of the overall activities and their BD, which means the ratio of the correct discovered activities/BD to the total number of actual activities/BD that must be discovered from emails. We obtained a value of 1 for activities and a value of 0.91 for BD. This indicates that our algorithm was able to discover all the activities and a good percentage of their BD that must be discovered; We can notice that all the Ids in Table 1 appears in Table 2. Additionally, most the

BD in Table 1 were reflected in Table 2 by explicit patterns (e.g; 'dealNumber' was discovered in the form of 'deal numeric', the traded quantity of electricity power was discovered in the form of 'numeric0mw' while mw is the abbreviation of power unit (megawatt)). However, our algorithm does not recognize one BD ('position') that concerns interviewing candidates activities. Additionally, it miss the detection of 'personName' for 'conduct interview' activity.

Our algorithm can generate also highly detailed activities that were not considered in annotation phase. (Due to space constraints, we not show in Table 2 these activities , but we evaluate their accuracy in the second type of evaluation). This can be useful when aiming to discover BP with low granularity. For instance, in the context of setting interviews' schedules; we obtain 'give date0time' that reflects contacting candidates to ask them for their availability. As for BD, we obtained additional patterns that further specifies the context of activities execution (e.g. 'interview phone_telephone' refers to phone interviews).

Extended Evaluation: We aim here to evaluate the relevance of our discovered activities, which means the fraction of relevant activities among all the retrieved ones. Thus, we applied our algorithm on the emails of 5 employees of Enron. Then, we manually annotated the discovered activities for each employee (relevant or not). Table 3 summarizes the obtained results. For each employee, it shows the number of his/her unique sent emails, the number (Nbr) of the discovered activities and the obtained relevance ratio. Then, we calculated the average relevance ratio and we obtained a value of 0.87. This indicates that our approach is likely to select relevant activities and discard irrelevant information.

To visualize the obtained results, we have implemented a visualization tool that generates for each employee a figure resuming his main activities. Such figure hierarchically organizes activities' patterns in the form of tree; parent nodes present the different verb concepts appearing in the discovered activities ('replace' in Fig. 4) and child nodes present patterns containing them. Patterns that belong to the same activity (e.g., 'replace purchase', 'replace sale') are grouped through a rectangle where its generated name is indicated ('replace deal_trade'). BD are linked to each activity that it characterizes. Such example of the Fig. 4 shows the ability of our algorithm to group patterns expressing differently the same activity while sharing similar business context ('power',

Table 3. Accuracy

Employee	Size	N_{act}	Acc
E_1	331	16	1
E_2	342	12	1
E_3	375	19	0.83
E_4	589	24	0.79
E_5	1940	38	0.76
Average of accuracy			0.87

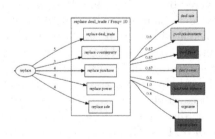

Fig. 4. A node of concept verb

'sale', 'purchase' and 'counterparty' are not synonyms but the patterns that they form share the same BD). We provide in this link (see footnote 2) the visualizations results that we have obtained with the other employees considered in our experiments.

These results show good performances in terms of unsupervised discovery of relevant activities. However, at evaluation perspective, we must study its error rate in terms of grouping patterns belonging to different activities, or inversely, separating patterns belonging to the same activity. Such error rate would be minimized if we integrate some rewording techniques, so that patterns referring to the same activity and are not composed of synonymous verb concepts would be recognized. Additionally, additional steps must be studied to contextualize some standards and non precise activities (e.g., contact persons, attach file) as their patterns can be actually discovered by our algorithm.

6   Conclusion

In this work, we have proposed an unsupervised approach for discovering frequent activities as well as their BD from emails without disposing prior knowledge about them. For this purpose, we have introduced a pattern discovery algorithm to discover frequent activities discussed through emails. The algorithm is based on several key features: (1) it analyzes per actor emails to reduce the variability of expressions. The goal is to capture their frequent patterns when expressing activities; (2) it takes into account the words meanings in the generation of the patterns; (3) it considers the business context, defined by the BD that co-occur with patterns in order to regroup them into a single activity. By characterizing activities in the form of pattern of concepts, our approach allows: (1) automatic generation of significant labels (names) close to the actual vocabulary of employees, and (2) detecting activities in emails even if they are expressed differently, and (3) discovering several activities per email; if multiple patterns belonging to different activities will be detected in one email (while verifying low dispersity criterion), they will be assigned to it.

We have tackled the discovery of data perspective (BD) of activities which has not been concretely handled to date (in the context of emails). We have carried also experiments on a public dataset and we have publicly shared our results (link (see footnote 2)), which is, in our knowledge, absent in related works (that's why comparison with them was not feasible when evaluating our proposals).

We are currently investigating how to regroup the discovered activities into BP and how to build their related event logs. Actually, emails can be used either as a support of carrying out, informing, planning or requesting the execution of an activity. As a result, timestamps of emails do not necessarily correspond to the actual timestamps of the corresponding activities' events. Other challenges are therefore added at this level: How to identify email's role for executing activities (e.g., request)? and how to estimate the actual timestamps of activity events?

References

1. Laga, N., Elleuch, M., Gaaloul, W., AlaouiIsmaili, O.: Emails analysis for business process discovery. In: ATAED Workshop, p. 54 (2019)
2. Kushmerick, N., Lau, T., Dredze, M., Khoussainov, R.: Activity-centric email: a machine learning approach. In: AAAI, vol. 21, pp. 1634–1637 (2006)
3. Ringger, E.K., Campbell, R., Corston-Oliver, S., Gamon, M.: Task-focused summarization of email (2004)
4. Cohen, W.W., Carvalho, V.R., Mitchell, T.M.: Learning to classify email into "speech acts". In: EMNLP, pp. 309–316 (2004)
5. Jeong, M., Lin, C.-Y., Lee, G.G.: Semi-supervised speech act recognition in emails and forums. In: EMNLP, vol. 3, pp. 1250–1259 (2009)
6. Qadir, A., Riloff, E.: Classifying sentences as speech acts in message board posts. In: EMNLP, pp. 748–758 (2011)
7. Soares, D.C., Santoro, F.M., Baião, F.A.: Discovering collaborative knowledge-intensive processes through e-mail mining. J. Netw. Comput. Appl. **36**(6), 1451–1465 (2013)
8. Jlailaty, D., Grigori, D., Belhajjame, K.: On the elicitation and annotation of business activities based on emails. In: Proceedings of the 34th ACM/SIGAPP Symposium on Applied Computing, pp. 101–103 (2019)
9. Nigam, A., Caswell, N.S.: Business artifacts: an approach to operational specification. IBM Syst. J. **42**(3), 428–445 (2003)
10. Maylawati, D., Aulawi, H., Ramdhani, M.: The concept of sequential pattern mining for text. In: IOP Conference Series: Materials Science and Engineering, vol. 434, p. 012042 (2018)
11. Ghose, A., Koliadis, G., Chueng, A.: Process discovery from model and text artefacts. In: IEEE Congress on Services (Services 2007), pp. 167–174. IEEE (2007)
12. Friedrich, F., Mendling, J., Puhlmann, F.: Process model generation from natural language text. In: Mouratidis, H., Rolland, Colette (eds.) CAiSE 2011. LNCS, vol. 6741, pp. 482–496. Springer, Heidelberg (2011). https://doi.org/10.1007/978-3-642-21640-4_36

Conformance Checking Using Activity and Trace Embeddings

Jari Peeperkorn[1]($\boxtimes$), Seppe vanden Broucke[1,2], and Jochen De Weerdt[1]

[1] Department of Decision Sciences and Information Management,
Faculty of Economics and Business, KU Leuven, Leuven, Belgium
{jari.peeperkorn,seppe.vandenbroucke,jochen.deweerdt}@kuleuven.be
[2] Department of Business Informatics and Operations Management,
Faculty of Economics and Business Administration,
Ghent University, Ghent, Belgium

Abstract. Conformance checking describes process mining techniques used to compare an event log and a corresponding process model. In this paper, we propose an entirely new approach to conformance checking based on neural network-based embeddings. These embeddings are vector representations of every activity/task present in the model and log, obtained via act2vec, a Word2vec based model. Our novel conformance checking approach applies the Word Mover's Distance to the activity embeddings of traces in order to measure fitness and precision. In addition, we investigate a more efficiently calculated lower bound of the former metric, i.e. the Iterative Constrained Transfers measure. An alternative method using trace2vec, a Doc2vec based model, to train and compare vector representations of the process instances themselves is also introduced. These methods are tested in different settings and compared to other conformance checking techniques, showing promising results.

Keywords: Process mining · Conformance checking · Representation learning · Word embedding

1 Introduction

Conformance checking is a set of process mining techniques capable of comparing event logs and corresponding process models. It can be used to compare the actual execution of a process (log) to the should-be execution (a normative model) or an automatically discovered model. Usually the degree of conformance is described over four quality dimensions: fitness, precision, simplicity and generalisation. Generally two different approaches for obtaining the fitness and precision are distinguished: log replay algorithms and trace alignment algorithms [1]. In this work, we propose an entirely new perspective on conformance checking, moving away from classical approaches relying on replay or alignments, but instead performing a fully data driven conformance analysis and subsequent global conformance measure development. Currently, our novel conformance checking technique relies on generating an event log of a model and

© Springer Nature Switzerland AG 2020
D. Fahland et al. (Eds.): BPM Forum 2020, LNBIP 392, pp. 105–121, 2020.
https://doi.org/10.1007/978-3-030-58638-6_7

comparing this with the actual event log using different representation learning techniques inspired by work in Natural Language Processing (NLP). As such, the presented fitness and precision metrics are computed as log-to-log metrics.

The remainder of this paper is structured as follows. First, we introduce the notions of activity and trace embeddings in Sect. 2. In Sect. 3, the embedding-based conformance checking technique is outlined, before we perform several experimental assessments in Sect. 4. The paper is concluded with a section discussing related work (Sect. 5) and our conclusions in Sect. 6.

2   Activity and Trace Embeddings

Representation or feature learning describes a set of techniques that are capable of extracting (useful) representations of objects for different types of input. This can help machine learning models perform better, using an input more suited or by reducing the dimensions of the input. The techniques proposed in this paper are based on Word2vec [20,21] and Doc2vec [17], two widely used and popular representation learning techniques from NLP. The Word2vec algorithm uses a two layer neural network and a large corpus of words in order to train a vector space in which each different word gets a specific vector value called a representation or embedding. The network tries to predict a certain word in a text by using the window of surrounding words (continuous bag-of-words or CBOW) or tries to predict the surrounding window of a certain word (skip-gram). The input and output are one-hot encoded vectors of the words (with the vocabulary size as dimension). The weight used for the sum in the hidden layer (and the weight used for the output layer) is a matrix with each word's embedding stored in its columns (rows). By updating the weights while training to predict the words better, the embeddings get more optimal/meaningful vector values in such a way that similar words (words used in a similar context) should get similar vector values. In Doc2vec each specific document (or sentence) containing the word also gets a vector representation that is used as input in the neural network. By training the network, these document embeddings are updated as well. During training the representation of the document (sentence) is determined by the information from its content. These representations can then be used to compare different documents (sentences) for e.g. classification purposes.

In process mining the use of representation learning applied to activities, process instances (traces), logs and models has been introduced by [9]. In this work the authors propose among others the algorithms *act2vec* and *trace2vec*. Act2vec works similarly to Word2vec using activity labels as words. During training, activities are predicted based on the activities occurring before and after it within a process instance. In this way (meaningful) vector representations for each type of activity are learned. Trace2vec works similarly to Doc2vec using activity labels as words and process instances as sentences. Therefore every trace ID gets an embedding, next to the activity embeddings. These trace embeddings can then be used for e.g. trace clustering.

3   Conformance Checking Techniques

The basic structure of the new techniques we introduce is illustrated in Fig. 1. The required input is, as usual in conformance checking, a (real) log and a model. The first step is playing out a so called "model log" from the model. The model can therefore adopt any usual structure or notation for which execution semantics can be defined (Petri net, Process trees, BPMN etc.). Using both of these logs (real and model), embeddings for the activities (and possibly for each unique trace) can be trained, as described above. Next a decision is to be made on a (dis)similarity function that measures the difference between two traces. Depending on whether one is using activity embeddings or trace embeddings different functions exist, which allows one to obtain a dissimilarity matrix. This matrix gives the function value of each trace from the real log (columns) with each trace from the model log (rows). We can now take the minimum of each column, meaning we find for each trace in the real log its best matching trace in the model log (allowing equal matches). If we take the average of these minima it gives us a measurement for the fitness of the model regarding the real log. The same can be done with finding the minimum in each row, i.e. finding for each trace in the model log its best matching trace in the real log. Again averaging these minima presents us with a measurement for the precision.

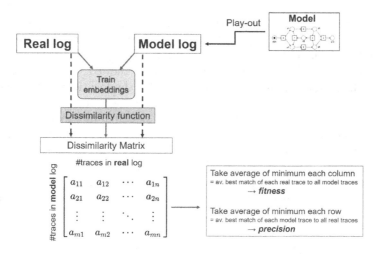

Fig. 1. The general structure of the proposed techniques.

The exact implementation of the algorithm can be decided upon by changing how embeddings are derived, changing the dissimilarity function or even by changing the way the model plays out the log. In this work it was opted to use the act2vec and trace2vec settings described above, in a Continuous Bag of Words (CBOW) and distributed bag of words (PV-DBOW) way respectively. For the time being, the window around the activity being trained is taken to be 3 and bi-directional, and the dimension of the embeddings is set to 16. Obviously, all of these settings could easily be changed if deemed appropriate. In this work, we develop conformance metrics by implementing two distance functions at the level of activity embeddings: the Words Mover's Distance and the Iterative Constrained Transfers method. A third algorithm, relying on trace2vec, applies the cosine distance between two trace embeddings as the dissimilarity function. In the current research we have built implementations on top of the Gensim-library for unsupervised semantic modelling from plain text [27].

3.1   Words Mover's Distance

A first dissimilarity function put forward is the Words Mover's Distance (or WMD) [16], which is a commonly used function in NLP, for instance for sentence similarity. It compares two sentences (of word embeddings) similarly to the Earth Mover's Distance (or EMD) [29]. The EMD is a distance measure between two distributions in a certain region. In plain words it describes how much work has to be done to go from one vector of distributions to another. Each can be looked at as a certain configuration of "piles of earth", with each element of the vector representing a certain amount of earth in a certain location. The effort to move earth from one location to another is determined by the amount of earth transported and the distance between the locations. The optimization to go from one configuration to another can be seen as a linear program minimizing the transportation cost while satisfying two constraints. The first (outflow) constraint is that from each pile (distribution) in the first configuration there cannot be more weight transported than present. The second (inflow) constraint requires not more weight being transported than allowed to each pile (distribution) in the second configuration. In the WMD the locations of the piles are the different words and the distribution/weight of each word is its normalized word count. The distance between the words is calculated by using the Euclidean distance between its embeddings. Applied to the act2vec environment, this function allows us to calculate the effort to go from one trace to another using the embeddings of the activities within. The method is described in Algorithm 1.

The main attraction of EMD-based approaches is their high accuracy in different applications like e.g. classification [4]. The downside is the inefficiency, as according to the authors the average time complexity of solving the WMD optimization problem scales $O(p^3 \log p)$, with p the size of the vocabulary (amount of unique words/activities). Therefore an alternative method, relaxing one of the constraints is also presented here. It has to be noted that the WMD does not take the order of the activities into account, but only the count of an activity within a trace. Order can have an influence during the training of the embeddings, but is not explicitly taken into account in the WMD (nor in its approximations).

3.2   Iterative Constrained Transfers

A faster alternative for WMD is Relaxed Word Mover's Distance (RWMD), which drops one of the two constraints of the WMD completely. This allows for

---

Algorithm 1: Method using WMD [16].

Result: Calculates the distance between two traces t_1 and t_2.
n = vocabulary size
d_i = normalized count of activity i within its trace
$c(i,j)$ = Euclidean distance between embeddings word i and word j.
Function $F_{WMD}(t_1, t_2, c)$

$\quad$ distance = $\min\limits_{\mathbf{T} \geq 0} \sum\limits_{i,j=1}^{n} T_{ij} c(i,j)$

$\quad$ subject to:

$\quad$ $\sum\limits_{j=1}^{n} T_{ij} = d_i \ \forall i = 1, \ldots, n$

$\quad$ $\sum\limits_{i=1}^{n} T_{ij} = d_j \ \forall j = 1, \ldots, n$

$\quad$ **return** $distance$

end

---

more transportation from or to certain words than actually correct (but with lower distance). The RWMD therefore gives a lower boundary to the WMD [16]. The Iterative Constrained Transfers (ICT) proposes the addition of an edge capacity constraint when relaxing one of the two constraints [4]. This additional constraint entails that when moving weight between two words along an edge, the total transportation always has to be smaller than the minimum of the distribution on both sides of the edge. In other words it cannot be bigger than the total distribution of the outgoing word (from the first sentence) and of the ingoing word (from the second sentence). The ICT replaces the second (inflow) constraint of the WMD by this new edge capacity constraint. The new problem can be solved optimally by considering the first sentence word by word. Each time sorting the edges leaving from it to the different words in the second sentence, in increasing order of transportation cost. Then iteratively transferring weights from the word in the first sentence to the words in the second sentence, under the edge constraints, until the outflow constraint is met. The ICT can be approximated by limiting the number of iterations than can be performed, called the Approximate Computation of ICT (or ACT). In this work it was opted to use this constraint with the maximum number of iterations being 3. Only if the outgoing trace has certain activities which occur significantly more than others, this number should be taken higher. The inflow constraint of the WMD is not necessarily met anymore, therefore the ICT (or ACT) provides a lower bound to the WMD, but due to the extra edge constraint more tight than the RWMD. The algorithm used to calculate the ICT between two traces (with activity embeddings) can be found in Algorithm 2.

3.3 Trace Embeddings

The third approach uses trace embeddings. In this work it was opted to train one embedding for each unique trace. This means that traces with the same activity

---

Algorithm 2: Method using ICT, referred to as ACT in [4].

Result: Calculates the distance between two traces p and q.
k = number of edges considered per activity (in this work 3)
$c(i,j)$ = Euclidean distance between embeddings word i and word j
h_p, h_q = amount of different activities in trace p and q
p_i = normalized weights of each activity in trace p with $i = 1, \ldots, h_p$
q_i = normalized weights of each activity in trace q with $i = 1, \ldots, h_q$
Function $F_{ICT}(p, q, c, k)$
   | Initialize transportation cost $t = 0$
   | **for** $i = 1 \ldots h_p$ **do**
   | | Find k smallest: $s = \arg\min_k(c(i, [1, \ldots, h_q]))$
   | | Initialize l = 1
   | | **while** $l < k$ **do**
   | | | Edge constraint: $r = \min(p_i, q_{s(l)})$
   | | | Transport weight: $p_i = p_i - r$
   | | | Update cost: $t = t + r \cdot c(i, j)$
   | | | $l = l + 1$
   | | **end**
   | | Solve for possible excess weight
   | | **if** $p_i \neq 0$ **then**
   | | | Move rest to $q_{s(k)}$:
   | | | $t = t + p_i \cdot c(i, s(k))$
   | | **end**
   | **end**
   | **return** t
end

---

sequence, have the same embedding. Once these embeddings are obtained one could use any distance metric in vector spaces deemed appropriate. In this work it was opted to use the cosine similarity. Using the cosine similarity means that the output of this algorithm will always be a number between 0 and 1, where 0 means a perfect value in both the fitness and the precision calculation. The algorithm to calculate the cosine distance between two trace embeddings is displayed in Algorithm 3.

Algorithm 3: Method using trace2vec.

Result: Calculates the distance between two traces embeddings p and q.
n = dimensions trace embeddings
Function $F_{t2v}(p, q)$
$$\cos = \frac{\sum\limits_{i=1}^{n} p_i q_i}{\sqrt{\sum\limits_{i=1}^{n} p_i^2}\sqrt{\sum\limits_{i=1}^{n} q_i^2}}$$
   return cos
end

---

4   Experimental Evaluation

The goal of the empirical evaluation in this work is twofold. On the one hand, we want to investigate the computational complexity of the proposed techniques. On the other hand, we want to provide evidence that the measures are indeed capable to reveal when models and logs become more discrepant.

4.1   Experimental Setup

In order to test the methods proposed, different experiments are performed[1]. A first experiment focuses solely on the scalability of our proposed methods, varying log size and dictionary size (number of activities). Next, in order to assess the conceptual appropriateness, another set of experiments was conducted. For these experiments different process trees, depicting different types of processes, are generated randomly using the implementation of the Python library PM4Py [6,15]. The different settings used for each tree can be found in Table 1. These trees vary in size (depicted by the minimum, mode and maximum number of visible activities) and the probabilities of adding sequence, parallel, choice and loop operators to the tree. Other parameters are left default.

Table 1. The different settings used to generate the process trees used in the experiments.

	Size (min - mode - max)	Sequence	Parallel	Choice	Loop
Tree 1	5-10-15	0.75	0.25	0	0
Tree 2	5-10-15	0.75	0	0.25	0
Tree 3	5-10-15	0.5	0.25	0.25	0
Tree 4	5-10-15	0.25	0.25	0.25	0.25
Tree 5	10-20-30	0.75	0.25	0	0
Tree 6	10-20-30	0.75	0	0.25	0
Tree 7	10-20-30	0.5	0.25	0.25	0
Tree 8	10-20-30	0.25	0.25	0.25	0.25
Tree 9	15-30-45	0.75	0.25	0	0
Tree 10	15-30-45	0.75	0	0.25	0
Tree 11	15-30-45	0.5	0.25	0.25	0
Tree 12	15-30-45	0.25	0.25	0.25	0.25

First, the ability of the methods to measure differences between two logs is assessed in a noise experiment depicted in Fig. 2. For each process tree, a ground

---

[1] The implementations of the algorithm, the tests and most of the synthetic data used can be found on https://github.com/jaripeeperkorn/Conformance-checking-using-activity-and-trace-embeddings.

truth log is played out. Then different levels of noise are introduced to this ground truth log to obtain a noisy log. The different methods are tested on these two logs, checking whether more noise actually equals less optimal values. We define three types of noise: replacing a random activity with another random activity label (form the vocabulary), swapping two random activities and removing a random activity. In a first noise experiment each of these noise types are added to only one activity (or two in the case of swapping). But the percentage of traces on which we apply each of these noise functions (one after the other) is varied from 10 to 50%. A second small noise experiment sets the percentage of traces on which we apply noise to 40%, but then varies the amount of activities within these traces we apply noise on. For this we are only using the noise function that randomly replaces activities, omitting the other two.

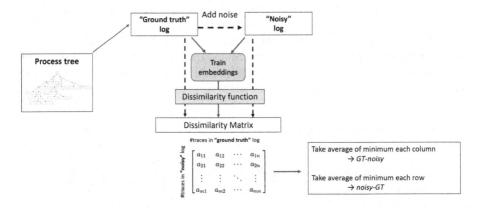

Fig. 2. The noise experiment.

Then, in a second experiment, the conceptual appropriateness is further investigated, again relying on synthetic logs. However, we now include different process discovery algorithms to obtain models with different fitness/precision and compare with well-established conformance metrics. The experiment is shown in Fig. 3. Instead of adding noise, we now discover a model from the ground truth logs using different discovery techniques. Once discovered models are obtained, the different proposed methods can be applied by first playing out the discovered model. Regarding the process discovery algorithm selection, the goal was to obtain a varied set of discovered models in terms of fitness and precision. Moreover, we wanted to restrict the number of discovery techniques in order to prevent blowing up the analysis. Therefore, we opted for the following algorithms: Alpha miner [31], Inductive Miner infrequent (IMi) with the noise parameter once set to 0 and once to 1 [19] and the ILP miner [33]. For the ILP miner settings the alpha parameter was used with the zero value and concurrency ratio left 0. We have used the implementation within the ProM framework [11] for each of the discovery techniques. The conformance checking techniques selected to compare are the behavioral negative event recall [14], alignment based fitness and precision [2] and the ETC precision [22]. All of these conformance checking algorithms were used as they were implemented in the CoBeFra framework [7].

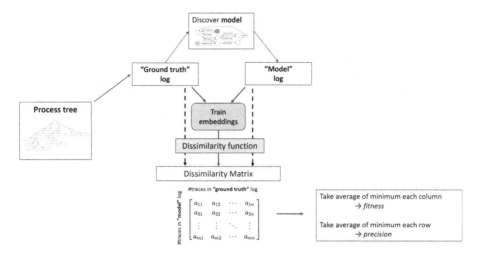

Fig. 3. The discovery experiment.

4.2   Results and Discussion

Scalability. The experiment each time compares two randomly generated logs, varying in size. The average trace length is left to 20, but the vocabulary size (amount of different activities) is also altered. The results of this experiment, as performed on an Intel(R) Core(TM) i7-9850h CPU @ 2.60ghz, can be found in Table 2 (performed three times, taking the average time). What is important to note is that the methods are, for now written in python. The WMD (EMD) implementation uses however pyemd [24,25], a wrapper that allows the use of numpy (C efficiency). The (own) ICT implementation used in this work is not C optimized, and could therefore still be enhanced significantly. From the results it can be seen that the run times depend on the log size significantly in all three methods. The dictionary size has a big influence on the WMD, smaller influence in the ICT and no influence on the run time of the trace2vec based method. If optimized the ICT could definitely be a computationally less demanding alternative to the WMD. The trace2vec based method is the fastest, as it only has to calculate the cosine distance of the trace embeddings. Training the embeddings does not take a long time.

Noise Experiment. In a first noise experiment the percentage of traces on which we apply each of the three noise functions is varied between 10 and 50%.

Table 2. Table showing the run times of the different variations of the algorithm.

Log size	Dictionary size	wmd	ict	t2v
100	10	1 s	1 s	1 s
	20	2 s	2 s	1 s
	30	3 s	2 s	1 s
500	10	20 s	26 s	12 s
	20	46 s	39 s	12 s
	30	1 m15 s	45 s	12 s
1000	10	1 m14 s	1 m43 s	42 s
	20	2 m57 s	2 m30 s	43 s
	30	4 m50 s	2 m51 s	43 s
5000	10	30 m38 s	37 m2 s	15 m20 s
	20	1 h10 m4 s	55 m23 s	15 m24 s
	30	1 h57 m12 s	1 h3 m44 s	15 m20 s

Again: we apply it in each of these traces only once. For each setting in Table 1, 5 different process trees are generated and the results over these 5 trees are averaged. The generated ground truth consists each time of 1000 traces. Because this number is usually very high as compared to the amount of different variants of traces, the noisy log often still contains at least once each (original) variant. This means that the ground truth - noisy value as shown in Fig. 2 is almost always 0. The values of the noisy - ground truth values of this noise experiment can be found in Table 3. We can see that for each setting and each method, adding noise adds to the distance. Note that you should not directly compare the activity embedding based methods and the trace2vec model, as they have a different scale. Figure 4 shows the average over all the different process tree generations of each noise level.

As mentioned earlier another small noise experiment was performed as well in order to show how the methods handle different levels of noise within the noisy traces. In this setting we are only using the noise of randomly replacing and activity. The process trees are generated using the 10th setting described in Table 1. The noise is each time applied on 40% of the traces in the log but on different amount of activities (1–15). The results shown in Table 4 and Fig. 5 show that for each of the three methods an increase in noise corresponds to a higher distance.

Discovery Experiment. For the time being the generated logs for both the original tree and the discovered models are generated randomly and are each time limited to 1000 traces. This also means that the discovery algorithms use a log of (only) 1000 traces. The embeddings are being trained each time again and thus are not used over the multiple discovered models (nor over the three different techniques). In the real life situation of comparing multiple models to

Table 3. Results of the noise experiment using three different types of noise, each on one (two for swapping) activity in different amounts of traces.

	Tree 1			Tree 2			Tree 3			Tree 4		
Noise	wmd	ict	t2v	wmd	ict	t2v	wmd	ict	t2v	wmd	ict	t2v
10%	0.135	0.126	0.044	0.101	0.095	0.023	0.150	0.142	0.030	0.155	0.148	0.038
20%	0.244	0.227	0.123	0.192	0.179	0.089	0.282	0.264	0.090	0.282	0.268	0.091
30%	0.349	0.327	0.194	0.276	0.260	0.143	0.387	0.365	0.142	0.397	0.378	0.147
40%	0.439	0.4121	0.249	0.349	0.328	0.190	0.492	0.465	0.194	0.499	0.472	0.199
50%	0.517	0.482	0.290	0.417	0.391	0.228	0.570	0.535	0.232	0.579	0.546	0.238
	Tree 5			Tree 6			Tree 7			Tree 8		
Noise	wmd	ict	t2v	wmd	ict	t2v	wmd	ict	t2v	wmd	ict	t2v
10%	0.151	0.128	0.027	0.083	0.074	0.010	0.119	0.110	0.014	0.121	0.106	0.016
20%	0.285	0.247	0.090	0.162	0.145	0.041	0.228	0.210	0.036	0.225	0.200	0.051
30%	0.407	0.344	0.149	0.237	0.212	0.067	0.321	0.295	0.067	0.314	0.282	0.088
40%	0.506	0.435	0.197	0.305	0.273	0.096	0.410	0.377	0.098	0.390	0.348	0.115
50%	0.594	0.513	0.224	0.371	0.335	0.117	0.484	0.445	0.118	0.476	0.421	0.144
	Tree 9			Tree 10			Tree 11			Tree 12		
Noise	wmd	ict	t2v	wmd	ict	t2v	wmd	ict	t2v	wmd	ict	t2v
10%	0.089	0.080	0.020	0.075	0.062	0.006	0.122	0.111	0.007	0.093	0.084	0.005
20%	0.173	0.157	0.074	0.147	0.121	0.021	0.231	0.213	0.018	0.182	0.164	0.013
30%	0.250	0.228	0.130	0.215	0.179	0.046	0.327	0.301	0.031	0.257	0.231	0.023
40%	0.322	0.293	0.174	0.281	0.232	0.067	0.420	0.386	0.048	0.331	0.300	0.038
50%	0.390	0.354	0.240	0.341	0.285	0.096	0.506	0.467	0.064	0.400	0.361	0.054

Table 4. Results of the noise experiment, varying only the amount of activities affected.

Amount of noise	1	3	5	7	9	11	13	15
wmd	0.134	0.326	0.451	0.542	0.611	0.659	0.692	0.702
ict	0.090	0.238	0.344	0.421	0.491	0.538	0.563	0.570
t2v	0.015	0.079	0.118	0.135	0.146	0.151	0.156	0.157

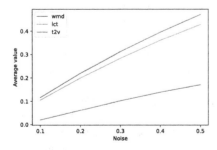

Fig. 4. The average of the first noise experiment.

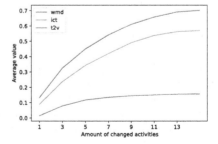

Fig. 5. The second noise experiment.

select the optimal one, it might be beneficial to train embeddings only once for all of the models together. The discovery experiment was performed for Trees 5–12 from Table 1. The results for each of the fitness and precision measurements can be found in Table 5. Beware that for the newly introduced WMD, ICT and t2v method a more optimal value corresponds to a value closer to 0 but for the other algorithms this corresponds to a value closer to 1. For the Tree setting 8 and 12 the alpha discovery algorithm produced a Petri Net not readable by the log replay algorithm used here due to its unconventional structure. Because the corresponding Petri Nets did not correspond to realistic (useful) models, they were omitted as a whole. The alignment based fitness and precision did also not produce any results (within reasonable time) when considering the net discovered by the infrequent inductive miner with noise parameter set to 0 in tree 12.

From the results in Table 5 it can be seen that when the alignment based fitness and ETC precision agree on perfect fitness or precision the WMD, ICT and t2v usually do as well. For the rest these results are harder to interpret as the methods from the literature do not always agree as well. It can be seen however that when alignment based fitness or ECT precision seem to pick one of the discovered models as having a significantly better (or worse) fitness or precision, the proposed measures seem to agree. Nevertheless, more rigorous testing should be put in place to get more conclusive insights into the algorithms proposed in this work.

5   Related Work

The first literature trying to quantify a relation between process models an event logs can be found in [8]. In later years multiple conformance checking techniques have been proposed in the field of Process Mining. Most of them consider log-model conformance techniques. For fitness some of the first noteworthy algorithms are proper completion and token based sequence replay [28]. Other approaches are e.g. behavioral recall [14], which uses a percentage of correctly classified positive events, or behavioral profile based metrics [32] which is based on different constraints a process model can impose on a log. Most of the recent research has been following the (average) alignment based trace fitness approach introduced in [2]. One of the first proposed precision measurements is advanced behavioral appropriateness [28]. Another approach, called behavioral specificity [14], replays the sequences and takes the percentage of correctly classified negative events. A similar approach using the amount of "false positives" (behavior allowed by model, but labeled a negative event based on the log) was defined by [10]. A different approach using log prefix automatons and the number of "escaping" edges, called the ETC precision, was introduced in [22]. Also alignment based precision methods have been proposed in [2], and later the one align precision and best align precision [3]. Further, in recent years fitness and precision models comparing Markovian abstractions of the models and event logs have been presented [5]. The Earth Mover's Distance has previously been used

Table 5. Results of the discovery experiment. For WMD, ICT and t2v a value closer to zero means a better fitness or precision. For the alignment based fitness and precision [2], behavioral fitness [14] and the ETC precision [22] a value closer to 1 means a better value.

		Fitness					Precision				
		wmd	ict	t2v	[2]	[14]	wmd	ict	t2v	[2]	[22]
Tree 5	alpha	0.000	0.000	0.000	1.000	1.000	0.000	0.000	0.000	1.000	1.000
	ind. 0	0.000	0.000	0.000	1.000	1.000	0.000	0.000	0.000	1.000	1.000
	ind. 1	0.000	0.000	0.000	1.000	1.000	0.000	0.000	0.000	1.000	1.000
	ilp	0.000	0.000	0.000	0.898	0.898	0.000	0.000	0.000	1.000	1.000
Tree 6	alpha	3.233	2.751	0.766	0.300	0.786	3.327	2.864	0.504	0.000	1.000
	ind. 0	0.000	0.000	0.015	0.933	0.071	0.000	0.000	0.011	0.410	0.274
	ind. 1	0.000	0.000	0.018	0.849	0.727	0.000	0.000	0.011	1.000	0.998
	ilp	0.000	0.000	0.290	0.766	0.804	0.000	0.000	0.000	0.969	1.000
Tree 7	alpha	0.459	0.459	0.178	0.882	0.866	1.718	1.670	0.316	0.923	0.989
	ind. 0	0.000	0.000	0.000	0.879	0.733	0.000	0.000	0.000	0.995	0.994
	ind. 1	0.000	0.000	0.000	0.879	0.733	0.000	0.000	0.000	0.995	0.994
	ilp	0.000	0.000	0.001	0.859	0.711	0.000	0.000	0.000	0.998	0.999
Tree 8	ind. 0	0.000	0.000	0.006	0.832	0.466	0.002	0.002	0.006	0.691	0.526
	ind. 1	0.183	0.172	0.048	0.809	0.542	0.000	0.000	0.007	0.954	0.909
	ilp	0.001	0.001	0.006	0.784	0.710	0.000	0.000	0.004	0.937	0.957
Tree 9	alpha	0.000	0.000	0.000	1.000	1.000	0.000	0.000	0.000	1.000	1.000
	ind. 0	0.000	0.000	0.000	1.000	1.000	0.000	0.000	0.000	1.000	1.000
	ind. 1	0.000	0.000	0.000	1.000	1.000	0.000	0.000	0.000	1.000	1.000
	ilp	0.000	0.000	0.000	1.000	0.920	0.000	0.000	0.000	1.000	0.917
Tree 10	alpha	0.000	0.000	0.000	1.000	1.000	0.000	0.000	0.000	1.000	1.000
	ind. 0	0.000	0.000	0.000	0.976	0.950	0.000	0.000	0.000	0.998	1.000
	ind. 1	0.000	0.000	0.000	0.976	0.950	0.000	0.000	0.000	0.998	1.000
	ilp	0.000	0.000	0.002	0.948	0.946	0.000	0.000	0.000	1.000	1.000
Tree 11	alpha	0.114	0.093	0.034	0.315	0.979	0.000	0.000	0.001	0.000	0.936
	ind. 0	0.000	0.000	0.002	0.802	0.721	0.273	0.237	0.041	0.586	0.352
	ind. 1	0.971	0.866	0.265	0.296	0.342	1.542	1.136	0.221	0.987	0.901
	ilp	0.210	0.177	0.061	0.811	0.713	0.000	0.000	0.001	0.901	0.952
Tree 12	ind. 0	0.243	0.149	0.136	/	0.238	0.400	0.232	0.192	/	0.001
	ind. 1	2.440	2.158	0.537	0.640	0.241	0.459	0.459	0.237	0.664	0.646
	ilp	1.441	1.281	0.564	0.640	0.428	0.000	0.000	0.001	0.944	1.000

in conformance checking in the recent work of [18], although their approach requires the Petri Nets to be stochastic (and does not use embeddings). For a more comprehensive overview of conformance checking techniques, interested readers are referred to [12].

6 Conclusion and Future Work

In this work a novel, fully data driven conformance analysis was introduced. The techniques are inspired by the work in Natural Language Processing (NLP) and rely on training meaningful embeddings for the different activities (and instances) of the process. In a first simple empirical assessment based on experiments with artificial logs and models, we obtain promising results. More specifically, we show that the newly introduced techniques are capable to correctly assess when logs start to differ from each other. In our discovery experiment, we do show that the metrics are intrinsically capable to detect fitness and/or precision problems respectively. Nonetheless, a more elaborate design for the discovery experiment is required in order to better understand and adjust the values obtained. More concretely the WMD and ICT based method could be standardized, in order to obtain a real metric with value between 0 and 1. In the future we are also planning to move away from artificial data only and test with real life logs. From a business point of view, it should be checked to what extend similar traces but one crucial difference (error) are detected as being not conforming. The information about conformance that could be extracted beyond fitness and precision will also be investigated. So far, we have demonstrated that embedding-based conformance checking can provide an interesting alternative to classical conformance checking. Significant opportunities exist to leverage the abstraction that neural networks can obtain, to gain insights into conformance problems and report to end users. While for now, we have solely focused on measurement development, it is important to realise that our methods have strong potential to be extended towards real-life application. Nonetheless, there is still plenty of future work that should be considered. First, an investigation on the application of wrappers on the code seems worthwhile, especially when calculating the ICT, as this should ameliorate its efficiency significantly. More research on the influence of different settings when training the embeddings could improve understanding of how this novel technique could potentially be used in real life. Optimizing the window size or possibly limiting the direction of the window could potentially improve the methods. Other possible improvements can be made in the inclusion of other n-grams in stead of only using 1-grams. Due to the low vocabulary size of business processes as compared to NLP, usual arguments to not use bag of n-grams based methods do not necessarily hold. In this way the order of activities can also be taken into account. Other word embeddings used in NLP like the deep contextualized Embeddings from Language Models (ELMo) [13] or Global Vectors for Word Representation (GloVe) [26] may be interesting to investigate. Also experimenting with the influence of how the model plays out the model log, could potentially prove some valuable insights. For the moment being, it was chosen to use random playing out but one could also use e.g. a fully covering approach. Another key improvement to the algorithm is related to a more explicit incorporation of the order of activities into the calculations. This could either be done by changing the trace embeddings, training them with e.g. a recurrent neural network [23], or, at the level of activity embeddings, e.g. by adding extra dimensions to the activity embeddings (after training) corresponding to the location in the trace. Evaluating the algorithm with the recently proposed Conformance Propositions [30], would also help to evaluate the quality of the technique. The evaluation of the technique regarding these axioms should be performed in the near future.

References

1. van der Aalst, W.: Process Mining: Data Science in Action, 2nd edn. (2016)
2. van der Aalst, W., Adriansyah, A., van Dongen, B.: Replaying history on process models for conformance checking and performance analysis. WIREs Data Min. Knowl. Discov. **2**(2), 182–192 (2012)
3. Adriansyah, A., Munoz-Gama, J., Carmona, J., van Dongen, B.F., van der Aalst, W.M.P.: Alignment based precision checking. In: La Rosa, M., Soffer, P. (eds.) BPM 2012. LNBIP, vol. 132, pp. 137–149. Springer, Heidelberg (2013). https://doi.org/10.1007/978-3-642-36285-9_15
4. Atasu, K., Mittelholzer, T.: Linear-complexity data-parallel earth mover's distance approximations. In: Proceedings of the 36th International Conference on Machine Learning. Proceedings of Machine Learning Research, vol. 97, pp. 364–373, 09–15 June 2019
5. Augusto, A., Armas-Cervantes, A., Conforti, R., Dumas, M., Rosa, M.L., Reissner, D.: Measuring fitness and precision of automatically discovered process models: a principled and scalable approach (2019)
6. Berti, A., van Zelst, S.J., van der Aalst, W.: Process mining for python (PM4Py): bridging the gap between process-and data science. In: Proceedings of the ICPM Demo Track 2019, co-located with 1st International Conference on Process Mining (ICPM 2019), Aachen, Germany, 24–26 June 2019, pp. 13–16 (2019)
7. vanden Broucke, S., Weerdt, J.D., Vanthienen, J., Baesens, B.: A comprehensive benchmarking framework (CoBeFra) for conformance analysis between procedural process models and event logs in prom. In: Proceedings of the 2013 IEEE Symposium on Computational Intelligence and Data Mining (CIDM 2013), pp. 254–261 (2013)
8. Cook, J.E., Wolf, A.L.: Software process validation: quantitatively measuring the correspondence of a process to a model. ACM Trans. Softw. Eng. Methodol. **8**(2), 147–176 (1999)
9. De Koninck, P., vanden Broucke, S., De Weerdt, J.: act2vec, trace2vec, log2vec, and model2vec: representation learning for business processes. In: Weske, M., Montali, M., Weber, I., vom Brocke, J. (eds.) BPM 2018. LNCS, vol. 11080, pp. 305–321. Springer, Cham (2018). https://doi.org/10.1007/978-3-319-98648-7_18
10. De Weerdt, J., De Backer, M., Vanthienen, J., Baesens, B.: A robust F-measure for evaluating discovered process models. In: IEEE Symposium Series on Computational Intelligence, pp. 148–155 (2011)
11. van Dongen, B.F., de Medeiros, A.K.A., Verbeek, H.M.W., Weijters, A.J.M.M., van der Aalst, W.M.P.: The ProM framework: a new era in process mining tool support. In: Ciardo, G., Darondeau, P. (eds.) ICATPN 2005. LNCS, vol. 3536, pp. 444–454. Springer, Heidelberg (2005). https://doi.org/10.1007/11494744_25
12. Dunzer, S., Stierle, M., Matzner, M., Baier, S.: Conformance checking: a state-of-the-art literature review. In: Proceedings of the 11th International Conference on Subject-Oriented Business Process Management, S-BPM ONE 2019 (2019)

13. Gardner, M., et al.: AllenNLP: a deep semantic natural language processing platform (2017)
14. Goedertier, S., Martens, D., Vanthienen, J., Baesens, B.: Robust process discovery with artificial negative events. J. Mach. Learn. Res. **10**, 1305–1340 (2009)
15. Jouck, T., Depaire, B.: PTandLogGenerator: a generator for artificial event data. In: Proceedings of the BPM Demo Track 2016 (BPMD 2016), vol. 1789, pp. 23–27. CEUR Workshop Proceedings, Rio de Janeiro (2016)
16. Kusner, M.J., Sun, Y., Kolkin, N.I., Weinberger, K.Q.: From word embeddings to document distances. In: Proceedings of the 32nd International Conference on International Conference on Machine Learning, ICML 2015, vol. 37, pp. 957–966 (2015)
17. Le, Q., Mikolov, T.: Distributed representations of sentences and documents. In: Proceedings of the 31st International Conference on Machine Learning. Proceedings of Machine Learning Research, vol. 32, pp. 1188–1196, 22–24 June 2014
18. Leemans, S.J.J., Syring, A.F., van der Aalst, W.M.P.: Earth movers' stochastic conformance checking. In: Hildebrandt, T., van Dongen, B.F., Röglinger, M., Mendling, J. (eds.) BPM 2019. LNBIP, vol. 360, pp. 127–143. Springer, Cham (2019). https://doi.org/10.1007/978-3-030-26643-1_8
19. Leemans, S.J.J., Fahland, D., van der Aalst, W.M.P.: Discovering block-structured process models from event logs containing infrequent behaviour. In: Lohmann, N., Song, M., Wohed, P. (eds.) BPM 2013. LNBIP, vol. 171, pp. 66–78. Springer, Cham (2014). https://doi.org/10.1007/978-3-319-06257-0_6
20. Mikolov, T., Chen, K., Corrado, G., Dean, J.: Efficient estimation of word representations in vector space. In: 1st International Conference on Learning Representations, ICLR 2013 (2013)
21. Mikolov, T., Sutskever, I., Chen, K., Corrado, G.S., Dean, J.: Distributed representations of words and phrases and their compositionality. In: Advances in Neural Information Processing Systems 26, pp. 3111–3119 (2013)
22. Muñoz-Gama, J., Carmona, J.: A fresh look at precision in process conformance. In: Hull, R., Mendling, J., Tai, S. (eds.) BPM 2010. LNCS, vol. 6336, pp. 211–226. Springer, Heidelberg (2010). https://doi.org/10.1007/978-3-642-15618-2_16
23. Palangi, H., et al.: Deep sentence embedding using the long short term memory network: analysis and application to information retrieval. CoRR abs/1502.06922 (2015)
24. Pele, O., Werman, M.: A linear time histogram metric for improved SIFT matching. In: Forsyth, D., Torr, P., Zisserman, A. (eds.) ECCV 2008. LNCS, vol. 5304, pp. 495–508. Springer, Heidelberg (2008). https://doi.org/10.1007/978-3-540-88690-7_37
25. Pele, O., Werman, M.: Fast and robust earth mover's distances. In: 2009 IEEE 12th International Conference on Computer Vision, pp. 460–467, September 2009
26. Pennington, J., Socher, R., Manning, C.D.: Glove: global vectors for word representation. In: Empirical Methods in Natural Language Processing (EMNLP), pp. 1532–1543 (2014)
27. Řehůřek, R., Sojka, P.: Software framework for topic modelling with large corpora. In: Proceedings of the LREC 2010 Workshop on New Challenges for NLP Frameworks, pp. 45–50, May 2010
28. Rozinat, A., van der Aalst, W.: Conformance checking of processes based on monitoring real behavior. Inf. Syst. **33**(1), 64–95 (2008)
29. Rubner, Y., Tomasi, C., Guibas, L.: The earth mover's distance as a metric for image retrieval. Int. J. Comput. Vis. **40**, 99–121 (2000)

30. Syring, A.F., Tax, N., van der Aalst, W.M.P.: Evaluating conformance measures in process mining using conformance propositions. In: Koutny, M., Pomello, L., Kristensen, L.M. (eds.) Transactions on Petri Nets and Other Models of Concurrency XIV. LNCS, vol. 11790, pp. 192–221. Springer, Heidelberg (2019). https://doi.org/10.1007/978-3-662-60651-3_8

31. van der Aalst, W., Weijters, T., Maruster, L.: Workflow mining: discovering process models from event logs. IEEE Trans. Knowl. Data Eng. **16**(9), 1128–1142 (2004)

32. Weidlich, M., Polyvyanyy, A., Desai, N., Mendling, J., Weske, M.: Process compliance analysis based on behavioural profiles. Inf. Syst. **36**, 1009–1025 (2011)

33. van der Werf, J.M.E.M., van Dongen, B.F., Hurkens, C.A.J., Serebrenik, A.: Process discovery using integer linear programming. In: van Hee, K.M., Valk, R. (eds.) PETRI NETS 2008. LNCS, vol. 5062, pp. 368–387. Springer, Heidelberg (2008). https://doi.org/10.1007/978-3-540-68746-7_24

Privacy-Preserving Data Publishing in Process Mining

Majid Rafiei$^{(\boxtimes)}$ and Wil M. P. van der Aalst

Chair of Process and Data Science, RWTH Aachen University, Aachen, Germany
majid.rafiei@pads.rwth-aachen.de

Abstract. Process mining aims to provide insights into the actual processes based on event data. These data are often recorded by information systems and are widely available. However, they often contain sensitive private information that should be analyzed responsibly. Therefore, privacy issues in process mining are recently receiving more attention. Privacy preservation techniques obviously need to modify the original data, yet, at the same time, they are supposed to preserve the data utility. Privacy-preserving transformations of the data may lead to incorrect or misleading analysis results. Hence, new infrastructures need to be designed for publishing the privacy-aware event data whose aim is to provide metadata regarding the privacy-related transformations on event data without revealing details of privacy preservation techniques or the protected information. In this paper, we provide formal definitions for the main anonymization operations, used by privacy models in process mining. These are used to create an infrastructure for recording the privacy metadata. We advocate the proposed privacy metadata in practice by designing a privacy extension for the XES standard and a general data structure for event data which are not in the form of standard event logs.

Keywords: Responsible process mining · Privacy preservation · Privacy metadata · Process mining · Event logs

1 Introduction

No one doubts that data are extremely important for people and organizations and their importance is growing. Hence, the interest in *data science* is rapidly growing. Of particular interest are the so-called event data used by the *process mining* techniques to analyze end-to-end processes. Process mining bridges the gap between traditional model-based process analysis, e.g., simulation, and data-enteric analysis, e.g., data mining. It provides fact-based insights into the actual processes using event logs [1]. The three basic types of process mining are: *process discovery*, *conformance checking*, and *enhancement* [1].

An event log is a collection of events, and each event is described by its attributes. The main attributes required for process mining are *case id*, *activity*, *timestamp*, and *resource*. Table 1 shows an event log recorded by an information

© Springer Nature Switzerland AG 2020
D. Fahland et al. (Eds.): BPM Forum 2020, LNBIP 392, pp. 122–138, 2020.
https://doi.org/10.1007/978-3-030-58638-6_8

system in a hospital, where $\perp$ indicates that the corresponding attribute was not recorded. Some of the event attributes may refer to individuals. For example, in the health-care context, the *case id* may refer to the patients whose data are recorded, and the *resource* may refer to the employees performing activities for the patients, e.g., nurses or surgeons. When the individuals' data are explicitly or implicitly included, *privacy issues* arise. According to regulations such as the European General Data Protection Regulation (GDPR) [25], organizations are compelled to consider the privacy of individuals while analyzing their data.

The *privacy* and *confidentiality* issues in process mining are recently receiving more attention [9,14,20,22]. The proposed methods cover a range of solutions from privacy/confidentiality frameworks to privacy guarantees. Privacy preservation techniques often apply some anonymization operations to modify the data in order to fulfill desired *privacy requirements*, yet, at the same time, they are supposed to preserve the *data utility*. The transformed event log may only be suitable for specific analyses. For example, in [20], the privacy requirement is to *discover social networks of the individuals (resources) involved in a process without revealing their activities*, and the resulting event log is only suitable for the *social network discovery*. Moreover, the original event log may be transformed to another form of event data which does not have the structure of a standard event log. For example, in [22], the privacy requirement is to *discover processes without revealing the sequence of activities performed for the individuals (cases)*, where the transformed event data are not in the form of event logs and contain only *directly follows relations between activities*, which is merely suitable for the *process discovery*. Therefore, the modifications made by privacy preservation techniques need to be reflected in the transformed event data to inform the data analysts.

In this paper, for the first time, we formalize the main anonymization operations on the event logs and exploit them as the basis of an infrastructure for proposing privacy metadata, we also design a privacy extension for the XES standard and a general data structure to cope with the event data generated by some privacy preservation techniques which are not in the form of an event log. The proposed metadata, along with the provided tools support, supply privacy-aware event data publishing while avoiding inappropriate or incorrect analyses.

The remainder of the paper is organized as follows. In Sect. 2, we explain the motivation of this research. Section 3 outlines related work. In Sect. 4, formal models for event logs are presented. We explain the privacy-preserving data publishing in process mining in Sect. 5, where we formalize the main anonymization operations, privacy metadata are proposed, and the tools support is presented. Section 6 concludes the paper.

2   Motivation

Compare Table 2 with Table 1, they both look like an original event log containing all the main attributes to apply process mining techniques. However, Table 2

Table 1. An event log (each row represents an event).

Table 2. An anonymized event log (each row represents an event).

Case Id	Act.	Timestamp	Res.	Age	Disease
1	a	01.01.2019–08:30:10	E1	22	Flu
1	b	01.01.2019–08:45:00	D1	22	Flu
2	a	01.01.2019–08:46:15	E1	30	Infection
3	a	01.01.2019–08:50:01	E1	32	Infection
4	a	01.01.2019–08:55:00	⊥	29	Poisoning
1	e	01.01.2019–08:58:15	E2	22	Flu
4	b	01.01.2019–09:10:00	D2	29	Poisoning
4	r	01.01.2019–09:30:00	B1	29	Poisoning
2	d	01.01.2019–09:46:00	E3	30	Infection
3	d	01.01.2019–10:00:25	E3	32	Infection
2	f	01.01.2019–10:00:05	N1	30	Infection
3	f	01.01.2019–10:15:22	N1	32	Infection
4	e	01.01.2019–10:30:35	E2	29	Poisoning
2	f	01.02.2019–08:00:45	N1	30	Infection
2	b	01.02.2019–10:00:00	D2	30	Infection
3	b	01.02.2019–10:15:30	D1	32	Infection
2	e	01.02.2019–14:00:00	E2	30	Infection
3	e	01.02.2019–14:15:00	E2	32	Infection

Case Id	Act.	Timestamp	Res.	Age	Disease
1	a	01.01.2019–08:30:00	E1	22	Flu
1	b	01.01.2019–08:45:00	D1	22	Flu
2	a	01.01.2019–08:46:00	E1	30	Infection
3	a	01.01.2019–08:50:00	E1	32	Infection
4	a	01.01.2019–08:55:00	⊥	29	Poisoning
1	e	01.01.2019–08:58:00	E2	22	Flu
4	b	01.01.2019–09:10:00	D2	29	Poisoning
4	r	01.01.2019–09:30:00	⊥	29	Poisoning
2	d	01.01.2019–09:46:00	E3	30	Infection
3	d	01.01.2019–10:00:00	E3	32	Infection
2	g	01.01.2019–10:00:00	N1	30	Infection
3	g	01.01.2019–10:15:00	N1	32	Infection
4	e	01.01.2019–10:30:00	E2	29	Poisoning
2	k	01.02.2019–08:00:00	N1	30	Infection
2	b	01.02.2019–10:00:00	D2	30	Infection
3	b	01.02.2019–10:15:00	D1	32	Infection
2	e	01.02.2019–14:00:00	E2	30	Infection
3	e	01.02.2019–14:15:00	E2	32	Infection

is derived from Table 1 by randomly substituting some activities (f was substituted with g and k), generalizing the timestamps (the timestamps got generalized to the *minutes* level), and suppressing some resources (B1 was suppressed). Hence, a performance analysis based on Table 2 may not be as accurate as the original event log, the process model discovered from Table 2 contains some fake activities, and the social network of resources is incomplete. The main motivation of this research is to provide concrete privacy metadata for process mining without exposing details of privacy/confidentiality techniques or the protected sensitive information so that the analysts are aware of the changes and avoid inappropriate analyses.

Process mining benefits from a well-developed theoretical and practical foundation letting us perform this research. In theory, event logs, as the input data types, have a concrete structure by the definition. In practice, the IEEE Standard for eXtensible Event Stream (XES)[1] is defined as a grammar for a tag-based language whose aim is to provide a unified and extensible methodology for capturing systems behaviors by means of event logs, e.g., Fig. 1 shows the first case of the event log Table 1 in the XES format. In this paper, the XES standard will be used to show the concrete relation between the theory and practice, but the concepts are general.

3 Related Work

In process mining, the research field of confidentiality and privacy received rather little attention, although the *Process Mining Manifesto* [4] already pointed out

---

[1] http://www.xes-standard.org/.

```
<?xml version="1.0" encoding="UTF-8" ?>
<log xes.version="1.0" xes.features="nested-attributes" openxes.version="1.0RC7"
    xmlns="http://www.xes-standard.org/">
    <extension name="Organizational" prefix="org" uri="http://www.xes-standard.org/org.xesext"/>
    <extension name="Time" prefix="time" uri="http://www.xes-standard.org/time.xesext"/>
    <extension name="Concept" prefix="concept" uri="http://www.xes-standard.org/concept.xesext"/>
    <global scope="trace">
        <string key="concept:name" value="UNKNOWN"/>
    </global>
    <global scope="event">
        <string key="concept:name" value="UNKNOWN"/>
    </global>
    <classifier name="Event Name" keys="concept:name"/>
    <string key="concept:name" value="sample event log"/>
    <trace>
        <string key="concept:name" value="1"/>
        <string key="age" value="22"/>
        <string key="disease" value="Flu"/>
        <event>
            <string key="concept:name" value="a"/>
            <date key="time:timestamp" value="2019-01-01T08:30:10.000+01:00"/>
            <string key="org:resource" value="E1"/>
        </event>
        <event>
            <string key="concept:name" value="b"/>
            <date key="time:timestamp" value="2019-01-01T08:45:00.000+01:00"/>
            <string key="org:resource" value="D1"/>
        </event>
        <event>
            <string key="concept:name" value="e"/>
            <date key="time:timestamp" value="2019-01-01T08:58:15.000+01:00"/>
            <string key="org:resource" value="E2"/>
        </event>
    </trace>
    <trace>
    <trace>
    <trace>
</log>
```

Fig. 1. The XES format for the event log Table 1, showing only the first case (trace). In XES, the log contains traces and each trace contains events. The log, traces, and events have attributes, and *extensions* may define new attributes. The log declares the extensions used in it. The *global attributes* are the ones that are declared to be mandatory with a default value. The *classifiers* assign identity to each event, which makes it comparable to the other events.

the importance of privacy. In [2], *Responsible Process Mining* (RPM) is introduced as the sub-discipline focusing on possible negative side-effects of applying process mining. In [15], the aim is to provide an overview of privacy challenges in process mining in human-centered industrial environments. In [7], a possible approach toward a solution, allowing the outsourcing of process mining while ensuring the confidentiality of dataset and processes, has been presented. In [16], the authors propose a privacy-preserving system design for process mining, where a user-centered view is considered to track personal data. In [22,23], a framework is introduced, which provides a generic scheme for confidentiality in process mining. In [20], the aim is to provide a privacy-preserving method for discovering roles from event data, which can also be exploited for generalizing *resources* as individuals in event logs. In [13], the authors consider a cross-organizational process discovery context, where public process model fragments are shared as safe intermediates. In [9], the authors apply k-anonymity and t-closeness [12] on event data to preserve the privacy of *resources*. In [14], the authors employ the notion of differential privacy [8] to preserve the privacy of *cases* in event logs. In [21], an efficient algorithm is introduced applying k-anonymity and *confidence bounding* to preserve the privacy of *cases* in event logs.

Most related to our work are [19] and [18] which are focused on healthcare event data. In [19], the authors analyze data privacy and utility requirements for healthcare event data and the suitability of privacy preservation techniques is assessed. In [18], the authors extend their previous research [19] to demonstrate the effect of applying some anonymization operations on various process mining results, privacy metadata are advocated, and a privacy extension for the XES standard is proposed. However, formal models, possible risks, the tools support, and the event data which are not in the form of event logs are not discussed. In this paper, we provide a comprehensive infrastructure for recording privacy metadata which considers possible risks for data leakage, and the data utility. Our infrastructure is enriched by the formal models in theory and the tools support in practice.

4 Preliminaries

In this section, we provide formal definitions for event logs used in the remainder. Let A be a set. A^* is the set of all finite sequences over A, and $\mathcal{B}(A)$ is the set of all multisets over the set A. A finite sequence over A of length n is a mapping $\sigma \in \{1, ..., n\} \rightarrow A$, represented as $\sigma = \langle a_1, a_2, ..., a_n \rangle$ where $\sigma_i = a_i = \sigma(i)$ for any $1 \leq i \leq n$, and $|\sigma| = n$. $a \in \sigma \Leftrightarrow a = a_i$ for $1 \leq i \leq n$. For $\sigma \in A^*$, $\{a \in \sigma\}$ is the set of elements in σ, and $[a \in \sigma]$ is the multiset of elements in σ, e.g., $[a \in \langle x, y, z, x, y \rangle] = [x^2, y^2, z]$. $\sigma \downarrow_X$ is the projection of σ onto some subset $X \subseteq A$, e.g., $\langle a, b, c \rangle \downarrow_{\{a,c\}} = \langle a, c \rangle$. $\sigma \cdot \sigma'$ appends sequence σ' to σ resulting a sequence of length $|\sigma| + |\sigma'|$.

Definition 1 (Event, Attribute). *Let $\mathcal{E}$ be the event universe, i.e., the set of all possible event identifiers. Events can be characterized by various attributes, e.g., an event may have a timestamp and corresponds to an activity. Let $\mathcal{N}_{event}$ be the set of all possible event attribute names. For any $e \in \mathcal{E}$ and name $n \in \mathcal{N}_{event}$, $\#_n(e)$ is the value of attribute n for event e. For an event e, if e does not have an attribute named n, then $\#_n(e) = \bot$ (null).*

We assume three standard explicit attributes for events: *activity*, *time*, and *resource*. We denote $\mathcal{U}_{act}$, $\mathcal{U}_{time}$, and $\mathcal{U}_{res}$ as the universe of activities, timestamps, and resources, respectively. $\#_{act}(e) \in \mathcal{U}_{act}$ is the *activity* associated to event e, $\#_{time}(e) \in \mathcal{U}_{time}$ is the *timestamp* of event e, and $\#_{res}(e) \in \mathcal{U}_{res}$ is the *resource* associated to event e.

Definition 2 (Case, Trace). *Let C be the case universe, i.e., the set of all possible case identifiers, and $\mathcal{N}_{case}$ be the set of case attribute names. For any case $c \in C$ and name $n \in \mathcal{N}_{case}$: $\#_n(c)$ is the value of attribute n for case c. For a case c, if c does not have an attribute named n, then $\#_n(c) = \bot$ (null). Each case has a mandatory attribute "trace" such that $\#_{trace}(c) \in \mathcal{E}^*$. A trace σ is a finite sequence of events such that each event appears at most once, i.e., for $1 \leq i < j \leq |\sigma| : \sigma(i) \neq \sigma(j)$.*

Definition 3 (Event Log). *Let $\mathcal{U}$ be a universe of values including a designated null value ($\bot$), and $\mathcal{N} = \mathcal{N}_{event} \cup \mathcal{N}_{case}$ such that $\mathcal{N}_{event}$ and $\mathcal{N}_{case}$ are disjoint. An event log is a tuple $L = (C, E, N, \#)$, in which; $C \subseteq \mathcal{C}$ is a set of case identifiers, $E \subseteq \mathcal{E}$ is a set of event identifiers such that $E = \{e \in \mathcal{E} \mid \exists_{c \in C} \; e \in \#_{trace}(c)\}$, $N \subseteq \mathcal{N}$ is a set of attribute names such that $N \cap \mathcal{N}_{event}$ are the event attributes, and $N \cap \mathcal{N}_{case}$ are the case attributes. For $n \in N$, $\#_n \in (C \cup E) \to \mathcal{U}$ is a function which retrieves the value of attribute n assigned to a case or an event ($\bot$ is used if an attribute is undefined for the given case or event). In an event log, each event appears at most once in the entire event log, i.e., for any $c_1, c_2 \in C$ such that $c_1 \neq c_2 : \{e \in \#_{trace}(c_1)\} \cap \{e \in \#_{trace}(c_2)\} = \emptyset$. If an event log contains timestamps, then the ordering in a trace should respect the timestamps, i.e., for any $c \in C$, i and j such that $1 \leq i < j \leq |\#_{trace}(c)| : \#_{time}(\#_{trace}(c)_i) \leq \#_{time}(\#_{trace}(c)_j)$. We denote $\mathcal{U}_L$ as the universe of event logs.*

5 Privacy-Preserving Data Publishing

Privacy-preserving data publishing is the process of changing the original data before publishing such that the published data remain practically useful while individual privacy is protected [10]. Note that the assumption is that the data analysts (process miners) are not trusted and may attempt to identify sensitive personal information. Various privacy models could be applied to the original data to provide the desired privacy requirements before data publishing. However, the transformations applied should be known in order to interpret the data. Therefore, we propose the privacy metadata in process mining based on the main anonymization operations. We consider *suppression (sup)*, *addition (add)*, *substitution (sub)*, *condensation (con)*, *swapping (swa)*, *generalization (gen)*, and *cryptography (cry)* as the main anonymization operation types in process mining. In this section, we define these operations and demonstrate how the original event log is modified by them.

5.1 Anonymization Operations

Anonymization operations are the main functions which modify the original event log to provide the desired privacy requirements and may have any number of parameters. Moreover, the operations can be applied at the case or event level, and the target of the operation could be a case, an event, or attributes of such an object. We define the *anonymizer* as a function which applies an anonymization operation to an event log.

Definition 4 (Anonymizer). *An anonymizer anon $\in \mathcal{U}_L \to \mathcal{U}_L$ is a function mapping an event log into another one by applying an anonymization operation.*

Definition 5 (Anonymizer Signature). *Let $OT = \{sup, add, sub, con, swa, gen, cry\}$ be the set of operation types. A signature $sign = (ot, level, target) \in OT \times \{case, event\} \times (\{case, event\} \cup \mathcal{N})$ characterizes an anonymizer by its type,*

*the level (case or event), and the target (case, event, or case/event attribute).
We denote $\mathcal{U}_{sign}$ as the universe of signatures.*

Note that an anonymizer signature is not supposed to uniquely distinguish the anonymization operations applied to an event log, i.e., the same type of operation can be applied at the same level and to the same target multiple times, but it is designated to reflect the *type* and the *direct purpose* of the corresponding operation. This is considered as the minimum information required to make the analysts aware of the modifications w.r.t. the data utility and risks for data leakage. We say the *direct purpose* due to the interdependency of cases and events through traces, i.e., modifying cases may affect events and vice versa. This is demonstrated by some of the examples in the following. We assume that the only correlation between cases and events is the trace attribute of cases, and all the other attributes are independent. Certainly, the operations can be more accurately characterized by adding more information to the corresponding signatures, but this may lead to the data leakage in the sense of revealing the protected information or details of the anonymization operation which is orthogonal to the ultimate goal of privacy preservation. In the following, we demonstrate the anonymization operations using some examples.

Suppression replaces some values, specified by the target of the operation and probably some conditions, with a special value, e.g., null value. The reverse operation of suppression is called *disclosure*. The group-based privacy models such as k-anonymity and its extensions [12] often utilize suppression. In the following, we provide three examples to demonstrate the effect of suppression on an event log. In [9], a group-based privacy model is proposed for discovering processes while preserving privacy.

Example 1 (Event-event suppression based on the activity attribute)
*Let $L = (C, E, N, \#)$ be an event log. We want to suppress the events
$e \in E$, if their activity attribute value is $a \in \mathcal{A}$. $anon_1^a(L) = L'$ such
that: $L' = (C', E', N', \#')$, $E' = \{e \in E \mid \#_{act}(e) \neq a\}$, $C' = C$, and
$N' = N$. $\forall_{e \in E'} \forall_{n \in N'} \#'_n(e) = \#_n(e)$, $\forall_{c \in C'} \forall_{n \in N' \setminus \{trace\}} \#'_n(c) = \#_n(c)$, and
$\forall_{c \in C'} \#'_{trace}(c) = \#_{trace}(c) \downarrow_{E'}$. The anonymizer signature of this operation is
$sign = (sup, event, event)$.*

Example 2 (Case-case suppression based on the trace length)
*Let $L = (C, E, N, \#)$ be an event log. We want to suppress the cases $c \in C$,
if the trace length of the case is $k \in \mathbb{N}_{\geq 1}$. $anon_2^k(L) = L'$ such that: $L' =
(C', E', N', \#')$, $C' = \{c \in C \mid |\#_{trace}(c)| \neq k\}$, $E' = \{e \in \mathcal{E} \mid \exists_{c \in C'} e \in
\#_{trace}(c)\}$, and $N' = N$. $\forall_{e \in E'} \forall_{n \in N'} \#'_n(e) = \#_n(e)$ and $\forall_{c \in C'} \forall_{n \in N'} \#'_n(c) =
\#_n(c)$. The corresponding signature for this operation is $sign = (sup, case, case)$.*

Example 3 (Event-resource suppression based on the activity attribute)
*Let $L = (C, E, N, \#)$ be an event log. We want to suppress the resource
attribute value of the events $e \in E$, if their activity attribute value is $a \in \mathcal{A}$.*

$anon_3^a(L) = L'$ such that: $L' = (C', E', N', \#')$, $C' = C$, $E' = E$, and $N' = N$. $\forall_{e \in E'} \forall_{n \in N' \setminus \{res\}} \#_n'(e) = \#_n(e)$. For all $e \in E'$: $\#_{res}'(e) = \bot$ if $\#_{act}(e) = a$, otherwise $\#_{res}'(e) = \#_{res}(e)$. $\forall_{c \in C'} \forall_{n \in N'} \#_n'(c) = \#_n(c)$. The corresponding anonymizer signature is $sign = (sup, event, resource)$.

As above-mentioned, modifying cases may affect events and vice versa. In Example 1, event suppression modifies the cases through the trace attribute, which is an indirect effect not shown by the signature. Similarly, in Example 2, suppressing cases results in event suppression, i.e., all the events in the trace of the suppressed cases are removed. This is also an indirect effect which is not reflected in the corresponding signature.

Addition is often used by the *noise addition* techniques where the noise which is drawn from some distribution is randomly added to the original data to protect the sensitive values. In process mining, the noise could be added to cases, events, or the attribute values. In [14], the notion of *differential privacy* is used as a noise addition technique to perform privacy-aware process discovery.

Example 4 (Add an event to the end of traces based on the activity attribute of the last event in the trace)
Let $L = (C, E, N, \#)$ be an event log, and $N = \{trace, act, res, time\}$ such that "trace" is the case attribute and "act", "res", and "time" are the event attributes. We want to add an event $e' \in \mathcal{E} \setminus E$ with the activity attribute value $a_2 \in \mathcal{A}$ and the resource attribute value $r \in \mathcal{R}$ at the end of the trace of a case $c \in C$, if the activity of the last event in the trace is $a_1 \in \mathcal{A}$. $anon_4^{a_1, a_2, r}(L) = L'$ such that: $L' = (C', E', N', \#')$, $C' = C$, $N' = N$, $C_{cond} = \{c \in C \mid \#_{act}(\#_{trace}(c)_{|\#_{trace}(c)|}) = a_1\}$, and $f \in C_{cond} \to \mathcal{E} \setminus E$ is a total injective function which randomly assigns unique event identifiers to the cases having the desired condition. We denote $E_{add} = range(f)$ as the set of added events. Hence, $E' = E \cup E_{add}$. $\forall_{e \in E} \forall_{n \in N' \setminus \{trace\}} \#_n'(e) = \#_n(e)$. For all $c \in C_{cond}$: $\#_{act}'(f(c)) = a_2$, $\#_{time}'(f(c)) = \#_{time}(\#_{trace}(c)_{|\#_{trace}(c)|}) + 1$, and $\#_{res}'(f(c)) = r$. $\forall_{c \in C' \setminus C_{cond}} \#_{trace}'(c) = \#_{trace}(c)$, and $\forall_{c \in C_{cond}} \#_{trace}'(c) = \#_{trace}(c) \cdot \langle f(c) \rangle$. The corresponding signature for this operation is $sign = (add, case, trace)$.

Substitution replaces some values with some substitutions specified by a set, i.e., a set of substitutions. The substitution could be done randomly or could follow some rules, e.g., a *round robin* manner. In [20], a substitution technique is used in order to mine roles from event logs while preserving privacy.

Example 5 (Event-activity substitution based on a set of sensitive activities and a set of activity substitutions)
Let $L = (C, E, N, \#)$ be an event log, $A_L = \{a \in \mathcal{A} \mid \exists_{e \in E} \#_{act}(e) = a\}$ be the set of activities in L, $A_x \subseteq \mathcal{A} \setminus A_L$ be the set of activities used as the substitutions, $rand(X) \in X$ be a function which randomly returns an element from the set X, and $A_s \subset A_L$ be the set of sensitive activities. We want to randomly substitute the activity attribute value of the events $e \in E$, if the activity is sensitive. $anon_5^{A_s, A_x}(L) = L'$ such that: $L' = (C', E', N', \#')$, $C' = C$, $E' = E$,

and $N' = N$. For all $e \in E'$: $\#'_{act}(e) = rand(A_x)$ if $\#_{act}(e) \in A_s$, otherwise $\#'_{act}(e) = \#_{act}(e)$. $\forall_{e \in E'} \forall_{n \in N' \setminus \{act\}} \#'_n(e) = \#_n(e)$, $\forall_{c \in C'} \forall_{n \in N'} \#'_n(c) = \#_n(c)$. $sign = (sub, event, activity)$ is the signature corresponds to this operation.

Condensation first condenses the cases into similar clusters based on the sensitive attribute values, then in each cluster the sensitive attribute values are replaced with a collective statistical value, e.g., mean, mode, median, etc, of the cluster. In [5], the authors introduce condensation-based methods for privacy-preserving data mining. The following example shows how the condensation operates on the event logs assuming that there exists a sensitive case attribute.

Example 6 (Case-attribute condensation based on a set of case clusters, a cluster finder function, and using mode as the collective value)
Let $L = (C, E, N, \#)$ be an event log, $CL = \{cl_1, cl_2, ..., cl_n\}$ be the set of case clusters, whose sensitive attribute value is similar, such that for $1 \le i \le n$: $cl_i \subseteq C$ and for $1 \le i, j \le n$, $i \ne j$: $cl_i \cap cl_j = \emptyset$. Also, let $f \in C \to CL$ be a function which retrieves the cluster of a case, and $n' \in N$ be the sensitive case attribute, e.g., disease. For a set X, $mode(X) \in X$ retrieves the mode of the set. We want to replace the value of n' for each case $c \in C$ with the mode of n' in the cluster of the case. $anon_6^{CL, f, n'}(L) = L'$ such that: $L' = (C', E', N', \#')$, $C' = C$, $N' = N$, $E' = E$, $\forall_{c \in C'} \#'_{n'}(c) = mode(\{\#_{n'}(c') \mid c' \in f(c)\})$, $\forall_{c \in C'} \forall_{n \in N' \setminus \{n'\}} \#'_n(c) = \#_n(c)$, and $\forall_{e \in E'} \forall_{n \in N'} \#'_n(e) = \#_n(e)$. $sign = (con, case, n')$ is the signature corresponds to this operation.

Swapping aims to anonymize data by exchanging values of a sensitive attribute between individual cases. The individual cases which are chosen to exchange the sensitive attribute values are supposed to have similar sensitive attribute values. Therefore, cases need to be clustered into the clusters with the similar sensitive attribute values. The cases for swapping in the same cluster could be chosen either randomly or by some methods, e.g., the *rank swapping* method [17]. The following example shows how the swapping operates on the event logs assuming that there exists a sensitive case attribute.

Example 7 (Case-attribute swapping based on a set of case clusters and a cluster finder function)
Let $L = (C, E, N, \#)$ be an event log, $CL = \{cl_1, cl_2, ..., cl_n\}$ be the set of case clusters, whose sensitive attribute value is similar, such that for $1 \le i \le n$: $cl_i \subseteq C$ and for $1 \le i, j \le n$, $i \ne j$: $cl_i \cap cl_j = \emptyset$. Also, let $f \in C \to CL$ be the function which retrieves the cluster of a case, and $n' \in N$ be the sensitive case attribute, e.g., disease. For a set X, $rand(X) \in X$ is a function which randomly retrieves an element from the set. We want to randomly swap the value of n' for each case $c \in C$ with the n' value of another case in the same cluster. $anon_7^{CL, f, n'}(L) = L'$ such that: $L' = (C', E', N', \#')$, $C' = C$, $N' = N$, $E' = E$, $\forall_{c \in C'} \#'_{n'}(c) = rand(\{\#_{n'}(c') \mid c' \in f(c) \setminus \{c\}\})$, $\forall_{c \in C'} \forall_{n \in N' \setminus \{n'\}} \#'_n(c) = \#_n(c)$, and $\forall_{e \in E'} \forall_{n \in N'} \#'_n(e) = \#_n(e)$. The anonymizer signature of this operation is $sign = (swa, case, n')$.

Cryptography includes a wide range of techniques such as *encryption, hashing, encoding,* etc. In [22], the *connector method* is introduced as an encryption-based technique which breaks the relation between the events in the traces, while discovering the directly follows graph (DFG) [11] from an event log. In the following example, we demonstrate the effect of applying an encryption technique on the activity attribute of the events in an event log.

Example 8 (Event-activity encryption based on an encryption method)
Let $L = (C, E, N, \#)$ be an event log, ENC be the universe of encryption method names, and KEY be the universe of keys. We want to encrypt the activity attribute of all the events $e \in E$ using a method $m \in ENC$ and a key $k \in KEY$. Let $\mathcal{U}$ be a universe of values, and $\mathcal{U}_{enc}$ be a universe of encrypted values. $enc \in \mathcal{U} \times ENC \times KEY \nrightarrow \mathcal{U}_{enc}$ is a partial function which encryptes a value $u \in \mathcal{U}$ given the name of method and a key. Given $k \in KEY$, and $m \in ENC$, $anon_8^{k,m}(L) = L'$ such that: $L' = (C', E', N', \#')$, $C' = C$, $E' = E$, and $N' = N$. $\forall_{e \in E'} \#'_{act}(e) = enc(\#_{act}(e), m, k)$, $\forall_{e \in E'} \forall_{n \in N' \setminus \{act\}} \#'_n(e) = \#_n(e)$, and $\forall_{c \in C'} \forall_{n \in N'} \#'_n(c) = \#_n(c)$. The signature of this operation is $sign = (cry, event, activity)$.

Generalization replaces some values, indicated by the target of operation and probably some conditions, with a parent value in the taxonomy tree of an attribute. The reverse operation of generalization is called *specialization*. The four main generalization schemes are *full-domain, subtree, sibling, cell* [10]. In the *full-domain* scheme, all values in an attribute are generalized to the same level of the taxonomy tree. In the *subtree* scheme, at a non-leaf node, either all child values or none are generalized. The *sibling* generalization is similar to the *subtree*, but some siblings may remain ungeneralized. In the *cell* generalization, some instances of a value may remain ungeneralized while in all the other schemes if a value is generalized, all its instances are generalized.

The generalization techniques are often used by group-based anonymization techniques. In the following, we demonstrate the effect of applying the generalization operation on the event logs by a simple example that uses the *full-domain* scheme to generalize the time attribute of the events.

Example 9 (Event-time generalization based on a time generalization method)
Let $L = (C, E, N, \#)$ be an event log and $TL = \{seconds, minutes, hours, days, months, years\}$ be the level of time generalization. $g_{tl} \in \mathcal{T} \to \mathcal{T}$ is a function that generalizes timestamps to the given level of time. We want to generalize the time attribute of all the events $e \in E$ to the level $tl \in TL$. $anon_9^{tl}(L) = L'$ such that: $L' = (C', E', N', \#')$, $C' = C$, $E' = E$, and $N' = N$. $\forall_{e \in E'} \#'_{time}(e) = g_{tl}(\#_{time}(e))$, $\forall_{e \in E'} \forall_{n \in N' \setminus \{time\}} \#'_n(e) = \#_n(e)$, $\forall_{c \in C'} \forall_{n \in N'} \#'_n(c) = \#_n(c)$. $sign = (gen, event, time)$ is the anonymizer signature of this operation.

Privacy-preserving data publishing in process mining, is done by applying a sequence of the anonymization operations to fulfill a desired privacy requirement. Note that the operations are often applied by the privacy models, where the data utility preservation is also taken into account.

Definition 6 (Privacy Preserving Data Publishing in Process Mining - ppdp). *Let pr be the desired privacy requirement and $\mathcal{U}_L$ be the universe of event logs, $ppdp^{pr} : \mathcal{U}_L \rightarrow \mathcal{U}_L$ is a privacy method that gets an event log and applies $i \in \mathbb{N}_{\geq 1}$ anonymization operations to the event log in order to provide an anonymized event log which satisfies the given privacy requirement pr. If we assume that pr is satisfied at the step (layer) n of the anonymization process, then for $1 \leq i \leq n$, $L_i = anon_i(L_{i-1})$ such that $L_0 = L$ and L_n is the anonymized event log which satisfies pr.*

5.2 Data Utility and Data Leakage

We define *potential original event logs* to show how the anonymizer signature can be exploited to narrow down the set of possible original event logs.

Definition 7 (Potential Original Event Log). *Let $\mathcal{U}_L$ be the universe of event logs and $\mathcal{U}_{sign}$ be the universe of signatures. $po \in \mathcal{U}_L \times \mathcal{U}_L \times \mathcal{U}_{sign} \rightarrow \mathbb{B}$ is a function that for a given event log, an anonymized event log, and the signature of the anonymized event log checks whether the given event log could be an original event log. $ol \in \mathcal{U}_L \times \mathcal{U}_{sign} \rightarrow 2^{\mathcal{U}_L}$ retrieves a set of potential original event logs, s.t., for $L' \in \mathcal{U}_L$ and $sign \in \mathcal{U}_{sign}$: $ol(L', sign) = \{L \in \mathcal{U}_L \mid po(L, L', sign)\}$.*

To demonstrate the effect of the anonymizer signature, as a privacy metadata candidate, on *data utility* and *data leakage*, we analyze the set of potential original event logs. Here, the analysis is performed for Example 1. However, the concept is general and can be applied to all the operations. Let $L' = (C', E', N', \#')$ be an anonymized event log and $sign = (sup, event, resource)$. For an event log $L = (C, E, N, \#)$: $po(L, L', sign) = true \iff (C = C' \wedge E = E' \wedge N = N' \wedge \forall_{e \in E'} \forall_{n \in N' \setminus \{res\}} \#_n(e) = \#'_n(e) \wedge \forall_{\{e \in E' | \#'_{res}(e) = \perp\}} \#_{res}(e) \in \mathcal{U}_{res} \wedge \forall_{\{e \in E' | \#'_{res}(e) \neq \perp\}} \#_{res}(e) = \#'_{res}(e) \wedge_{c \in C'} \forall_{n \in N'} \#_n(c) = \#'_n(c))$. $ol(L', sign) = \{L \in \mathcal{U}_L \mid po(L, L', sign)\}$ is the set of potential original event logs. Intuitively, $|ol(L', sign)| = |\{e \in E' | \#'_{res}(e) = \perp\}| \times |\mathcal{U}_{res}|$, where $\mathcal{U}_{res}$ can be narrowed down to a few resources based on some background knowledge.

If we ignore the target information in the signature, i.e., $sign = (sup, event)$, the uncertainty regarding the original event log and the results of analyses would be much higher, since the suppressed data could be some events, or any event attribute n, s.t., $\exists_{e \in E} \#'_n(e) = \perp$. That is, the uncertainty regarding the results of analyses is expanded from the resource perspective to all the perspectives. In contrast, if we add more information to the signature, reconstructing the original event log could be easier for an adversary. For

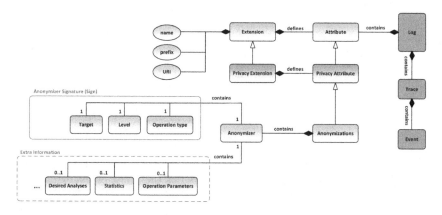

Fig. 2. The meta model of the proposed privacy metadata which is presented as an extension in XES. The privacy metadata attributes are added as the log attributes.

example, if we add the condition of resource suppression to the signature, i.e., $sign = (sup, event, resource, (activity = a))$, then $|ol(L', sign)| = |\{e \in E'|(\#'_{res}(e) = \bot \wedge \#'_{act}(e) = a)\}| \times |\mathcal{U}_{res}|$. That is, the only information that an adversary needs to reconstruct the original event log, with high confidence, is to realize the set of resources who perform the activity a. These analyses demonstrate that privacy metadata should contain the minimum information that preserves a balance between data utility and data leakage.

5.3 Privacy Metadata

Privacy metadata in process mining should correspond to the privacy-preserving data publishing and are supposed to capture and reflect the modifications, made by the anonymization operations. For event data in the form of an event log, privacy metadata are established by the means of XES. Figure 2 shows the meta model of the proposed privacy metadata as an extension in the XES. The privacy metadata attributes are defined by an extension at the *log level*. Note that the level (log, trace, or event) that is chosen to include the privacy metadata is one of the important risk factors. Although the anonymization operations are applied at the case and event levels, adding the corresponding metadata to the same level is of high risk and may lead to the protected data leakage. Assume that we add the privacy metadata regarding the resource suppression operation applied to Table 1 at the event level. By doing so, we expose the exact event that has been affected, and consequently, all the other information about the event are exposed, e.g, the activity performed by the removed resource is "r", and background knowledge could connect this information to the protected sensitive data. A similar risk scenario can be explored for the activity substitution operation applied to Table 1. If the corresponding metadata is added at the event level, the substitution set of the activities could be partially or entirely exposed.

```
<list key="privacy:anonymizations">
    <container key="privacy:anonymizer">
        <string key="privacy:operation type" value="substitution" />
        <string key="privacy:level" value="event" />
        <string key="privacy:target" value="concept:name" />
    </container>
    <container key="privacy:anonymizer">
        <string key="privacy:operation type" value="generalization" />
        <string key="privacy:level" value="event" />
        <string key="privacy:target" value="time:timestamp" />
    </container>
    <container key="privacy:anonymizer">
        <string key="privacy:operation type" value="suppression" />
        <string key="privacy:level" value="event" />
        <string key="privacy:target" value="org:resource" />
    </container>
</list>
```

Fig. 3. Privacy metadata recorded after transforming Table 1 to Table 2 in order to satisfy a privacy requirement. *privacy* is considered as the *prefix* of the extension.

In Fig. 2, *anonymizations* is the main privacy metadata element which contains at least one *anonymizer* element. Each *anonymizer* element corresponds to one anonymizer which is applied at a layer of the anonymization process in Definition 6. The *anonymizer* element contains two types of attributes: *mandatory* and *optional*. The mandatory attributes, residing in the solid box in Fig. 2, correspond to the *anonymizer signature* (Definition 5) and should be reflected in the XES. However, the optional attributes, residing in the dash box in Fig. 2, are the attributes which could be included in the XES. *Operation parameters* could contain the parameters which are passed to an anonymization operation, e.g., in Example 8, the name of encryption method is an operation parameter. *Statistics* could contain some statistics regarding the modification, e.g., the number of modified events, and *desired analyses* could contain the appropriate process mining activities which are applicable after modifying the event log. Note that the more unnecessary information included in the privacy metadata, the more risk is accepted. For example, if the *statistics* is included in an XES file where it is indicated that only 10 out of 1000 events have been modified, then an adversary makes sure that the transformed event log is almost the same as the original one (99% similarity).

Figure 3 shows the privacy metadata recorded after transforming Table 1 to Table 2. Note that *privacy* is considered as the *prefix* of the extension. At the first layer of the anonymization, a *substitution* operation has been applied which corresponds to Example 5, where $A_s = \{f\}$ and $A_x = \{g, k\}$. The metadata at this layer notices the analysts that some activity substitutions have been done at the event level, but it reveals neither the set of sensitive activities nor the substitution set. At the second layer of the anonymization, a *generalization* operation has been done which generalizes the timestamps to the minutes level and corresponds to Example 9, where $tl = minutes$. At the last layer of the anonymization ($n = 3$ in Definition 6), a *suppression* operation has been done which suppresses some resources and corresponds to Example 3, where the activity attribute value is r. This lets the analysts know that some resource suppression have been done without revealing the corresponding event. Note that the concept of *layer* is implicitly established by the means of list, which imposes an order on the keys.

```xml
<?xml version="1.0" encoding="UTF-8" ?>
<ELA>
    <header>
        <origin>BPI Challenge 2012</origin>
        <method>Connector Method</method>
        <desired_analyses>
            <field name="1">directly follows graph</field>
            <field name="2">process discovery</field>
        </desired_analyses>
    </header>
    <data>
        <item name="0">
            <field name="activity">W_Completeren aanvraag</field>
            <field name="prev_activity">W_Completeren aanvraag</field>
            <field name="relation_depth">1</field>
            <field name="trace_id">18</field>
            <field name="trace_length">40</field>
            <field name="connector">2be62f95e84d9bc75bc87e33aed2e64a</field>
        </item>
        <item name="1">
            <field name="activity">W_Nabellen incomplete dossiers</field>
            <field name="prev_activity">W_Nabellen incomplete dossiers</field>
            <field name="relation_depth">1</field>
            <field name="trace_id">80</field>
            <field name="trace_length">115</field>
            <field name="connector">f4489d7904f61899b9260e581be3b261</field>
        </item>
        <item name="2">
            <field name="activity">A_PREACCEPTED</field>
            <field name="prev_activity">A_PARTLYSUBMITTED</field>
            <field name="relation_depth">1</field>
            <field name="trace_id">75</field>
            <field name="trace_length">28</field>
            <field name="connector">764b7740c6ca99327ac62dd433dc1cb7</field>
        </item>
        <item name="3">
        <item name="4">
```

Fig. 4. The event log abstraction derived from the event log "BPI Challenge 2012" after applying the *connector* method. Only the first 3 items are shown.

So far, we only focused on the privacy preservation techniques applying the main anonymization operations on the original event log and transform it to another event log. However, there could be the techniques that do not preserve the structure of a standard event log. For example, in [22], the original event log is transformed to another form of event data where *directly-follows relations* are extracted from the original traces for discovering *directly-follows graph*. Such intermediate results derived from event logs and intended to relate logs and models are called *abstractions* [3]. We introduce *Event Log Abstraction* (ELA) to deal with the intermediate results created by some privacy preservation techniques which are not in the form of standard event logs. ELA is an XML tag-based language composed of two main parts: *header* and *data*. The *header* part contains the privacy/confidentiality metadata, and the *data* part contains the data derived from the original event log. The privacy metadata in ELA includes: *origin*, *method*, and *desired analyses*. The *origin* tag shows name of the event log the abstraction derived from, the *method* tag contains name of the method, and the *desired analyses* contains list of the appropriate analyses w.r.t. the abstraction. The *data* part represents the data derived from the log in a tabular manner. Figure 4 shows the ELA derived from the event log "BPI Challenge 2012" [24] by applying the *connector* method [22].

5.4 Tool Support

Since most of the privacy preservation techniques in process mining have been developed in Python, in order to support the tools, we have developed a Python library which is published as a Python package.[2,3] The library is based on *PM4Py* [6] and includes two classes correspond to two types of event data generated by the privacy preservation techniques in process mining: *privacyExtension* and *ELA*. The general advantage of the privacy metadata library is that the privacy experts do not need to deal with the tag-based files in order to read (write) the privacy metadata. Another crucial requirement provided by the *privacyExtension* class is that it keeps the consistency of the privacy metadata by managing the order of the *anonymizers* in the *anonymizations* list. This class provides four main methods: *set_anonymizer*, *set_optional_anonymizer*, *get_anonymizations*, and *get_anonymizer*. The *set_anonymizer* method gets the mandatory attributes and appends them to the *anonymizations* list as an *anonymizer* if there already exists a privacy extension, otherwise it first creates a privacy extension and an *anonymizations* list, then adds the attributes to the list. The *set_optional_anonymizer* method is responsible to add the optional attributes and should be called after setting the mandatory attributes. This method gets the layer, which is an index in the *anonymizations* list, and optional attributes and adds the attributes to the given layer. The *get_anonymizations* method returns the whole *anonymizations* tag in the XES file as a Python *dictionary*. The *get_anonymizer* method gets a layer and returns the metadata of the layer.

6 Conclusions

Due to the fact that event logs could contain highly sensitive personal information, and regarding the rules imposed by the regulations, e.g., GDPR, privacy preservation in process mining is recently receiving more attention. Event logs are modified by privacy preservation techniques, and the modifications may result in the event data which are not appropriate for all the process mining algorithms. In this paper, we discussed types of event data generated by the privacy preservation techniques. We provided formal definitions for the main anonymization operations in process mining. Privacy metadata were proposed for event logs which are supported by formal definitions in order to demonstrate the completeness of the proposed infrastructure. The ELA (Event Log Abstraction) was introduced to cope with event data which are not in the form of standard event logs. We employed the IEEE XES standard in order to consistently develop our proposal for privacy metadata of event logs in practice, where a privacy extension was introduced. We also provided a Python library to support the privacy preservation tools for process mining which have been often developed in Python.

---

[2] pip install p-privacy-metadata.

[3] https://github.com/m4jidRafiei/privacy_metadata.

Acknowledgment. Funded under the Excellence Strategy of the Federal Government and the Länder. We also thank the Alexander von Humboldt (AvH) Stiftung for supporting our research.

References

1. van der Aalst, W.M.P.: Process Mining - Data Science in Action, Second edn. Springer, Heidelberg (2016). https://doi.org/10.1007/978-3-662-49851-4
2. van der Aalst, W.M.P.: Responsible data science: using event data in a "people friendly" manner. In: Hammoudi, S., Maciaszek, L.A., Missikoff, M.M., Camp, O., Cordeiro, J. (eds.) ICEIS 2016. LNBIP, vol. 291, pp. 3–28. Springer, Cham (2017). https://doi.org/10.1007/978-3-319-62386-3_1
3. van der Aalst, W.M.P.: Process discovery from event data: relating models and logs through abstractions. Wiley Interdisc. Rev.: Data Mining Knowl. Discov. **8**(3), e1244 (2018)
4. van der Aalst, W.M.P., et al.: Process mining manifesto. In: Daniel, F., Barkaoui, K., Dustdar, S. (eds.) BPM 2011. LNBIP, vol. 99, pp. 169–194. Springer, Heidelberg (2012). https://doi.org/10.1007/978-3-642-28108-2_19
5. Aggarwal, C.C., Yu, P.S.: On static and dynamic methods for condensation-based privacy-preserving data mining. ACM Trans. Database Syst. **33**(1) (2008)
6. Berti, A., van Zelst, S.J., van der Aalst, W.M.P.: Process mining for python (PM4Py): bridging the gap between process-and data science. arXiv preprint arXiv:1905.06169 (2019)
7. Burattin, A., Conti, M., Turato, D.: Toward an anonymous process mining. In: 2015 3rd International Conference on Future Internet of Things and Cloud (FiCloud), pp. 58–63. IEEE (2015)
8. Dwork, C.: Differential privacy: a survey of results. In: Agrawal, M., Du, D., Duan, Z., Li, A. (eds.) TAMC 2008. LNCS, vol. 4978, pp. 1–19. Springer, Heidelberg (2008). https://doi.org/10.1007/978-3-540-79228-4_1
9. Fahrenkrog-Petersen, S.A., van der Aa, H., Weidlich, M.: PRETSA: event log sanitization for privacy-aware process discovery. In: International Conference on Process Mining, ICPM 2019, Aachen, Germany, 24–26 June 2019, pp. 1–8 (2019)
10. Fung, B.C., Wang, K., Fu, A.W.C., Philip, S.Y.: Introduction to Privacy-Preserving Data Publishing: Concepts and Techniques. Chapman and Hall/CRC, Boca Raton (2010)
11. Leemans, S.J.J., Fahland, D., van der Aalst, W.M.P.: Scalable process discovery and conformance checking. Softw. Syst. Modeling **17**(2), 599–631 (2016). https://doi.org/10.1007/s10270-016-0545-x
12. Li, N., Li, T., Venkatasubramanian, S.: t-closeness: privacy beyond k-anonymity and l-diversity. In: Proceedings of the 23rd International Conference on Data Engineering, ICDE 2007, The Marmara Hotel, Istanbul, Turkey, 15–20 April (2007)
13. Liu, C., Duan, H., Qingtian, Z., Zhou, M., Lu, F., Cheng, J.: Towards comprehensive support for privacy preservation cross-organization business process mining. IEEE Trans. Serv. Comput. **1**, 1–1 (2016)
14. Mannhardt, F., Koschmider, A., Baracaldo, N., Weidlich, M., Michael, J.: Privacy-preserving process mining - differential privacy for event logs. Bus. Inf. Syst. Eng. **61**(5), 595–614 (2019)
15. Mannhardt, F., Petersen, S.A., Oliveira, M.F.: Privacy challenges for process mining in human-centered industrial environments. In: 2018 14th International Conference on Intelligent Environments (IE), pp. 64–71. IEEE (2018)

16. Michael, J., Koschmider, A., Mannhardt, F., Baracaldo, N., Rumpe, B.: User-centered and privacy-driven process mining system design for IoT. In: Cappiello, C., Ruiz, M. (eds.) CAiSE 2019. LNBIP, pp. 194–206. Springer, Heidelberg (2019). https://doi.org/10.1007/978-3-030-21297-1_17

17. Nin, J., Herranz, J., Torra, V.: Rethinking rank swapping to decrease disclosure risk. Data Knowl. Eng. **64**(1), 346–364 (2008)

18. Pika, A., Wynn, M.T., Budiono, S., ter Hofstede, A.H., van der Aalst, W.M.P., Reijers, H.A.: Privacy-preserving process mining in healthcare. Int. J. Environ. Res. Public Health **17**(5), 1612 (2020)

19. Pika, A., Wynn, M.T., Budiono, S., ter Hofstede, A.H.M., van der Aalst, W.M.P., Reijers, H.A.: Towards privacy-preserving process mining in healthcare. In: Di Francescomarino, C., Dijkman, R., Zdun, U. (eds.) BPM 2019. LNBIP, vol. 362, pp. 483–495. Springer, Cham (2019). https://doi.org/10.1007/978-3-030-37453-2_39

20. Rafiei, M., van der Aalst, W.M.P.: Mining roles from event logs while preserving privacy. In: Di Francescomarino, C., Dijkman, R., Zdun, U. (eds.) BPM 2019. LNBIP, vol. 362, pp. 676–689. Springer, Cham (2019). https://doi.org/10.1007/978-3-030-37453-2_54

21. Rafiei, M., Wagner, M., van der Aalst, W.M.P.: *TLKC*-privacy model for process mining. In: Dalpiaz, F., Zdravkovic, J., Loucopoulos, P. (eds.) RCIS 2020. LNBIP, vol. 385, pp. 398–416. Springer, Cham (2020). https://doi.org/10.1007/978-3-030-50316-1_24

22. Rafiei, M., von Waldthausen, L., van der Aalst, W.M.P.: Ensuring confidentiality in process mining. In: Proceedings of the 8th International Symposium on Data-Driven Process Discovery and Analysis (SIMPDA 2018), Seville, Spain (2018)

23. Rafiei, M., von Waldthausen, L., van der Aalst, W.M.P.: Supporting confidentiality in process mining using abstraction and encryption. In: Ceravolo, P., van Keulen, M., Gómez-López, M.T. (eds.) SIMPDA 2018-2019. LNBIP, vol. 379, pp. 101–123. Springer, Cham (2020). https://doi.org/10.1007/978-3-030-46633-6_6

24. Van Dongen, B.F.: BPIC 2012. Eindhoven University of Technology (2012)

25. Voss, W.G.: European union data privacy law reform: general data protection regulation, privacy shield, and the right to delisting. Bus. Lawyer **72**(1) (2016)

Predictions and Recommendations

Explainability in Predictive Process Monitoring: When Understanding Helps Improving

Williams Rizzi[1,2] , Chiara Di Francescomarino[2(✉)] , and Fabrizio Maria Maggi[1]

[1] Free University of Bozen-Bolzano, Bolzano, Italy
maggi@inf.unibz.it
[2] Fondazione Bruno Kessler, Trento, Italy
{wrizzi,dfmchiara}@fbk.eu

Abstract. Predictive business process monitoring techniques aim at making predictions about the future state of the executions of a business process, as for instance the remaining execution time, the next activity that will be executed, or the final outcome with respect to a set of possible outcomes. However, in general, the accuracy of a predictive model is not optimal so that, in some cases, the predictions provided by the model are wrong. In addition, state-of-the-art techniques for predictive process monitoring do not give an explanation about what features induced the predictive model to provide wrong predictions, so that it is difficult to understand why the predictive model was mistaken. In this paper, we propose a novel approach to explain why a predictive model for outcome-oriented predictions provides wrong predictions, and eventually improve its accuracy. The approach leverages post-hoc explainers and different encodings for identifying the most common features that induce a predictor to make mistakes. By reducing the impact of those features, the accuracy of the predictive model is increased. The approach has been validated on both synthetic and real-life logs.

Keywords: Predictive process monitoring · Process mining · Machine learning · Explainable artificial intelligence

1 Introduction

Predictive (business) process monitoring is a family of techniques that use event logs to make predictions about the future state of the executions of a business process [8,13]. For example, a predictive monitoring technique may seek to predict the remaining execution time of each ongoing case of a process [22], the next activity that will be executed in each case [9], or the final outcome of a case with respect to a set of possible business outcomes [20]. For example, in an order-to-cash process, the possible outcomes of a case may be that the purchase order is closed satisfactorily (i.e., the customer accepted the products and paid)

© Springer Nature Switzerland AG 2020
D. Fahland et al. (Eds.): BPM Forum 2020, LNBIP 392, pp. 141–158, 2020.
https://doi.org/10.1007/978-3-030-58638-6_9

or unsatisfactorily (e.g., the order was canceled or withdrawn). In this paper, we focus on the latter category of methods.

In many applications of predictive process monitoring techniques, users are asked to trust a model helping them making decisions. However, a doctor will certainly not operate on a patient simply because "the model said so". And, even in lower-stakes situations, such as when choosing a movie to watch from Netflix, a certain measure of trust is required before wasting hours of our time watching a movie we do not like. Despite the fact that many predictive models are black-boxes, understanding the rationale behind the predictions would certainly help users decide when to trust or not to trust them.

In this paper, we propose a novel approach to explain why a predictive model is wrong and eventually use these explanations to improve its accuracy. The first step of our approach consists in training a classifier using a training set of historical execution traces of a process recorded in an event log. Then, if we have available a validation set and a gold standard providing the correct labels for each trace in the validation set, we can make predictions over trace prefixes derived from the validation set and check if those predictions are correct based on the gold standard. Since we want to understand when the predictive model is mistaken, we focus on the false positives provided by the predictor (i.e., negative outcomes predicted as positive) and false negatives (i.e., positive outcomes predicted as negative). Using post-hoc explainers such as LIME [16] or SHAP [12], we find the most common features that induce the predictor to provide wrong predictions.

Then, to increase the accuracy of the classifier, once the reasons why the mistakes were made have been provided, we retrain the classifier by randomizing the values of all the features that are very important when the predictor provides a false positive and/or a false negative, but have low or no importance when it provides the correct predictions. In our experimentation, we show that the accuracy of the retrained classifier is higher (in many cases significantly higher) than the one of the original classifier.

The rest of the paper is structured as follows. Section 2 presents some background notions; Sect. 3 introduces the problem we want to address and Sect. 4 the proposed solution; Sect. 5 discusses the evaluation of the approach using both synthetic and real-life logs; Sect. 6 summarizes the related work; and Sect. 7 concludes the paper and spells out directions for future work.

2 Background

In this section, we introduce some background notions about how explainable machine learning techniques work and about event logs.

2.1 Explainable Machine Learning Techniques

Basic machine learning algorithms like decision trees can be easily explained, e.g., by following the tree path which led to a prediction. However, predictions based on more complex algorithms are often incomprehensible by human intuition and, without understanding the rationale behind the predictions, users may simply not trust them. Explainability is motivated by lacking transparency of the black-box approaches, which do not foster trust and acceptance of machine learning algorithms. In the literature, there are two main sets of techniques used to develop explainable systems, a.k.a. explainers: post-hoc and ante-hoc techniques. Post-hoc techniques allow models to be trained as usual, with explainability only being incorporated at testing time. Ante-hoc techniques entail integrating explainability into a model from the beginning.

An example of post-hoc explainer is Local Interpretable Model-Agnostic Explanations (LIME) [16], which explains the prediction of any classifier in an interpretable manner. LIME learns an interpretable model locally around the prediction and explains the predictive models presenting individual explanations of individual predictions. Explanations are generated by approximating the underlying model by an interpretable one, learned on perturbations of the original instance. In particular, each feature is assigned with an *importance value* that represents the influence of that feature on a particular prediction.

Another post-hoc explainer is SHapley Additive exPlanations (SHAP) [12]. SHAP is a game-theoretic approach to explain the output of any machine learning model. It connects optimal credit allocation with local explanations using the classic Shapley values from game theory and their related extensions. SHAP provides local explanations based on the outcome of other explainers and representing the only possible consistent and locally accurate additive feature attribution method based on expectations.

An example of ante-hoc technique is Bayesian deep learning (BDL) [5]. BDL gives a measure of how uncertain a neural network is about its predictions. BDL models typically form uncertainty estimates by either placing distributions over model weights, or by learning a direct mapping to probabilistic outputs. By knowing the weight distributions of various predictions and classes, BDL can understand what features led to what decisions and the relative importance.

2.2   Event Logs

Event logs record the execution of business processes, i.e., *traces*, and traces consist of events. Each event in a log refers to an *activity* (i.e., a well-defined step in a business process) and is related to a particular trace. Events that belong to a trace are *ordered* and constitute a single "run" of the process. For example, in trace $\sigma_i = \langle event_1, event_2, \ldots event_n \rangle$, the first activity to be executed is the activity associated to $event_1$.

Events may be characterized by multiple *attributes*, e.g., an event refers to an *activity*, may have a *timestamp*, may be executed or initiated by a given *resource*, and may have associated other *data attributes*, i.e., data produced by the activities of the process. We indicate the value of an attribute a for an event e with $\pi_a(e)$. Standard attributes are, for instance, $\pi_{\mathsf{activity}}(e)$ representing the activity associated to event e; $\pi_{\mathsf{time}}(e)$ representing the timestamp associated to e; $\pi_{\mathsf{resource}}(e)$ representing the resource associated to e. Traces can also have attributes. Trace attributes, differently from event attributes, are not associated to a single event but to the whole trace. Data associated to events and traces in event logs are also called *data payloads*.

3   Problem

To illustrate the problem we want to address in this paper, we use the process model depicted in Fig. 1 using the BPMN language and introduced in [14]. The process considers two different checks of claims, a basic one and a complex one, depending on the value of the claim. After the decision is taken about the claim, the claimant is notified using different communication means depending on some specific conditions. In parallel, a questionnaire is sent to the claimant which can be received back before a certain deadline.

We can consider a labeling that classifies process executions in *accepted claims* (positive) and *rejected claims* (negative) based on whether the trace contains a claim acceptance (Accept Claim) or a claim rejection (Reject Claim). It is well known that the predictions provided by the classifier about the outcome of an ongoing process execution are not always correct and the accuracy of the classifier is not always optimal. This is due to the fact that, to make a prediction, the wrong information could be taken into consideration by the classifier. This is because the training set can contain noise or because the learning algorithm could not discover the "perfect" classifier (for example if the data provided in the training set is not sufficient to identify the correct correlation function between feature vectors and labels).

For example, we can assume that the dataset used to train the predictive model to make predictions at runtime about the acceptance or rejection of a claim is noisy and that the classifier learns that a process execution is positive also in those cases in which a claim is rejected and the questionnaire response is received. If this correlation is learned by the predictive model, when, at runtime, the claim is rejected and the questionnaire response is received the predictive model will return a false positive.

Using an explainer such as LIME [16] – which provides a ranking of the contributions of each feature towards a given prediction – the user can understand the features that have influenced the most the classifier when giving a wrong prediction. Therefore, the first problem we try to solve in this paper is to show to the user *why* a classifier makes a mistake. Then, by taking into account the features that have the highest influence on the classifier when it provides the wrong predictions according to the explainer, we try to improve the accuracy of the classifier by reducing the impact of these features. To this aim, we retrain the classifier after randomizing the values of all the features that are important when the predictor fails.

In our example, LIME would return that a false positive occurs when a claim is rejected and the questionnaire response is received. Therefore, we randomize the features related to the occurrence of Reject Claim and Receive Questionnaire Response when these activities occur together. Then, we retrain the classifier.

Note that the claim rejection is very likely an important feature in the case a trace is correctly predicted as negative. Therefore, we keep unaltered the feature related to the occurrence of Reject Claim when this activity occurs alone.

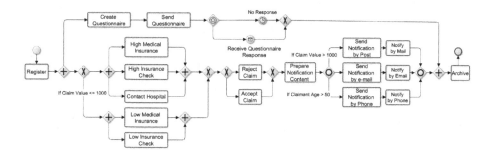

Fig. 1. Running example.

4   Approach

In this section, the proposed approach is described. In particular, first, the approach to extract a confusion matrix from an event log is illustrated. Then, the core part of the proposed approach is introduced, i.e., starting from false positives and negatives in the confusion matrix, how to identify features that induce the classifier to make wrong predictions. Then, a mechanism to retrain the classifier, based on the explanations indicating why the classifier provided the wrong predictions, is proposed.

4.1   Building the Confusion Matrix

Figure 2 shows the approach we use to extract a confusion matrix from an event log split into a training and a validation set. In the approach, from the training set, all the prefixes of a given length n are extracted and, in turn, encoded using different feature encodings (see Sect. *Encodings*). In addition, these sequences are labeled using a binary or categorical value according to their outcome. These "historical feature vectors" are used to train a classifier. Prefixes of length n are extracted also from the validation set and used to query the classifier that returns the label that is the most probable outcome for the current prefix according to the information derived from the historical prefixes. The predicted labels are compared with the labels available in the validation set (the *gold standard*) and a *confusion matrix* is built. The confusion matrix classifies predictions in four categories, i.e., (i) true-positive (T_P: positive outcomes correctly predicted); (ii) false-positive (F_P: negative outcomes predicted as positive); (iii) true-negative (T_N: negative outcomes correctly predicted); (iv) false-negative (F_N: positive outcomes predicted as negative).

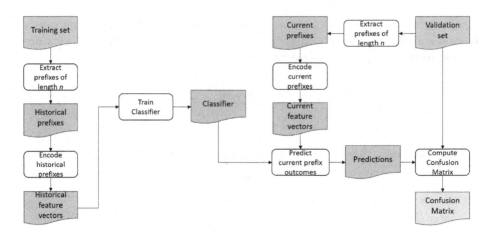

Fig. 2. How to extract the confusion matrix from a log.

Encodings. Each trace of a log corresponds to a sequence σ_i of events describing its control flow. Each event is also associated with data in the form of attribute-value pairs. Moreover, each completed trace is associated with an outcome - a *label*, which can assume binary or categorical values. We represent a trace in the following form:

$$(\text{event}_1\{\text{associated data}\},...,\text{event}_n\{\text{associated data}\}): \text{label}$$

The log in Fig. 3 pertains to the process described in Sect. 3. Each trace relates to a different claim and the corresponding sequence of events indicates the activities executed for processing that claim.

In the example, Register is the first event of sequence σ_1. Its data payload "$\{33, \text{financial}\}$" corresponds to the data associated to attributes age and department. Note that the value of age is static: it is the same for all the events in a trace (trace attribute), while the value of department is different for every event (event attribute). In case for some event the value for a specific attribute is not available, the value *unknown* is specified for it.

σ_1 (Register$\{33$, financial$\}$,...,Accept Claim$\{33$, assessment$\}$,...):true

...

σ_k (Register$\{56$, financial$\}$,..., Send Questionnaire$\{56$, secretary$\}$,...):false

---

Fig. 3. Example event log.

The goal of predictive business process monitoring is to build a classifier that learns from a set of historical traces L how to discriminate classes of traces and predict as early as possible the outcome of a new, unlabeled trace. More specifically, we are interested in automatically deriving a function f that, given an ongoing sequence σ_x provides a label for it, i.e., $f : (L, \sigma_x) \rightarrow \{label_x\}$. To achieve this goal, a classifier is trained on all sequence prefixes of the same length of σ_x derived from historical traces in L. In order to train the classifier, each (prefix) sequence σ_i, $i = 1...k$ has to be represented through a feature vector $g_i = (g_{i1}, g_{i2}, ...g_{ih})$. In some encodings, sequences are treated as simple symbolic sequences, while additional information related to data and data flow is neglected. The most straightforward of these encodings is based on information about how many times an activity occurs in the sequence. Some encodings, instead, combine and exploit both the control and the data flow dimension by considering the sequences as complex symbolic sequences [11]. In the following sections, we introduce control-flow and data-aware encodings.

Table 1. Frequency and simple-index encoding.

	Register	Accept Claim	...	Archive	label
σ_1	1	1	...	1	true
...					
σ_k	1	0	...	1	false

(a) Frequency-based encoding.

	event_1	...	event_i	...	event_m	label
σ_1	Register		Accept Claim		Archive	true
...						
σ_k	Register		Send Questionnaire		Archive	false

(b) Simple-index encoding.

Control-flow-based encodings. The *frequency-based* encoding describes sequences of events as feature vectors, where each feature corresponds to an activity from the log. Then the control flow in a trace is encoded using the frequency of each activity in the trace. Table 1a shows the frequency-based encoding for the example in Fig. 3. A second way of encoding a sequence using control-flow information is the *simple-index* encoding that takes into account information about the order in which events occur in the sequence. Here, each feature corresponds to a position in the sequence and the possible values for each feature are the activities. By using this type of encoding the example in Fig. 3 would be encoded as reported in Table 1b.

Table 2. Complex-index encoding.

	age	event_1	...	event_i	...	event_m	...	department_1	...	department_i	...	department_m	label
σ_1	33	Register		Accept Claim		Archive		financial		assessment dept		Archive	true
...													
σ_j	56	Register		Send Questionnaire		Archive		financial		secretary		Archive	false

Complex-index encoding. In the *complex-index* encoding, the data associated with events in a sequence is divided into static and dynamic information. Static information is the same for all the events in the sequence (e.g., the information contained in trace attributes), while dynamic information changes for different events (e.g., the information contained in event attributes). The resulting feature vector g_i, for a sequence σ_i, is:

$$g_i = (s_i^1, .., s_i^u, event_{i1}, event_{i2}, ..event_{im}, h_{i1}^1, h_{i2}^1...h_{im}^1, ..., h_{i1}^r, h_{i2}^r, ...h_{im}^r),$$

where each s_i is a static feature, each $event_{ij}$ is the activity at position j and each h_{ij} is a dynamic feature associated to an event. The example in Fig. 3 is transformed into the encoding shown in Table 2.

4.2 Using Explanations to Improve the Accuracy of a Classifier

For each trace in the validation set, we use an explainer like LIME or SHAP to extract the most important feature-value pairs impacting the prediction. The explainer takes as input the trained predictive model and the trace whose outcome we are predicting, encoded as described in Sect. 4.1, and returns as output a vector of *explanations*. Each explanation contains a feature-value pair and a score value ranging between -1 and 1, which represents the impact of the feature-value pair on the prediction. A negative value means that the feature-value pair influences the classifier towards the *false* prediction, while a positive value means that the feature-value pair influences the classifier towards the *true* prediction. The absolute value of the score denotes the strength of the impact of the feature-value pair towards the *true* or the *false* prediction.

In each vector returned by the explainer (for a given trace), explanations are filtered out based on a user-defined threshold, so as to keep only the feature-value pairs influencing the most the predictions. However, since explainers provide a vector of explanations for each trace separately, we need an abstraction mechanism to provide general explanations valid for all traces belonging to a certain prediction category (e.g., for false negatives). We use association rule mining [3] to identify which explanations are the most frequently provided by the explainer for a certain prediction category. We call these frequent itemsets *frequent explanation itemsets*.

If a frequent explanation itemset characterizes the false positives and/or the false negatives, but it does not characterize the correct predictions, this itemset identifies a set of features that lead the classifier to make mistakes. Once we have identified the "bad" frequent explanation itemsets, we can make the classifier learn from those explanations. To this aim, we randomize the values of the features in the frequent explanation itemset influencing the wrong predictions, when the frequent explanation itemset occurs in a trace of the training set. We use then the randomized training set to re-train the predictive model. In this way, we aim at neutralizing the effect of the combinations of feature-value pairs that were determining the wrong predictions. Note that if a frequent explanation itemset characterizing the wrong predictions is a subset of a frequent explanation itemset characterizing the correct predictions, when the superset occurs the subset must not be randomized.

5   Evaluation

In this section, we evaluate the proposed approach that has been implemented in the Nirdizati tool [17].

5.1   Datasets

For the evaluation, we used both a synthetic and a real-life log. We used a synthetic log to validate the proposed approach and a real-life log to test it in a real scenario. By labeling each log with different labelings, we derived 4 datasets from each log.

The synthetic log - CLAIM MANAGEMENT - is related to the example reported in Sect. 3. It consists of 4800 traces with an average trace length of 11 events. The log contains 52 935 events and 16 different activities. The 4 different conditions used for labeling the log are reported in column 'Condition' of Table 3. To be able to evaluate the approach, we have systematically introduced some noise in the data used to train the classifier so as to induce the classifier to make mistakes in a controlled way. In particular, we slightly changed the labeling of the training set using the conditions shown in column 'Noise condition' of Table 3.

The real-life log is the one provided for the BPI Challenge 2011 [1] - BPIC11. This log pertains to a healthcare process related to the treatment of patients diagnosed with cancer in a large Dutch academic hospital. The whole event log contains 1143 traces and 150 291 events distributed across 623 activities, with an average trace length of 121 events. Each trace refers to the treatment of a different patient. The event log contains domain specific attributes that are both trace attributes and event attributes. For example, Age, Diagnosis, and Treatment code are trace attributes and Activity code, Number of executions, Specialism code, Producer code, and Group are event attributes. We use the satisfaction of 4 different temporal properties for labeling the log (reported in Table 3).

Table 3. Dataset labelings (a trace is positive if the condition is satisfied).

Log	Label.	Condition	Noise condition
CLAIM MGMT	L11	$\#(\text{Accept Claim}) > 0$	$\#(\text{Accept Claim}) > 0$ and $\#(\text{Send Questionnaire}) = 0$
	L12	$\#(\text{Accept Claim}) > 0$	$\pi_{\text{activity}}(\text{event}_5) = \text{Accept Claim}$
	L13	$(\text{Age} < 60)$ and $(\text{CType} = \text{'Gold' or CType} = \text{'Silver'})$	$(\text{Age} < 60)$ and $(\text{CType} = \text{'Gold'})$
	L14	$\pi_{\text{activity}}(\text{event}_5) = \text{Accept Claim}$ or $(\text{CType} = \text{'Gold' or CType} = \text{'VIP' or CType} = \text{'Silver'})$	$\pi_{\text{activity}}(\text{event}_5) = \text{Accept Claim}$ and $(\text{CType} = \text{'Gold' or CType} = \text{'VIP'})$
BPIC11	L21	tumor marker CA-19-9 or ca-125 using meia occur	
	L22	CEA − tumor marker using meia followed by squamous cell carcinoma using eia	
	L23	histological examination - biopsies nno preceded by squamous cell carcinoma using eia	
	L24	histological examination - big resectiep occurs	

5.2   Experimental Setting

In order to evaluate the proposed approach, we apply the following procedure on each dataset:

- we split the dataset in a training set (60%), a validation set (20%) and a testing set (20%);
- we change the labeling of the training set using the noise condition (only for the synthetic datasets);
- we use the training set for building a Random Forest classifier, by using $\sim 2\%$ of the set to optimize the hyperparameters [6]; in particular, the hyperparameter procedure is set to run 1000 trials optimizing the AUC metric;
- we use the validation set for extracting false positives and false negatives together with the frequent explanation itemsets that characterize them (and do not characterize the correct predictions), based on the explanations provided by LIME.
- we retrain the classifier using the same training set, now randomizing the values of the frequent explanation itemsets characterizing false positives and false negatives;
- we use the testing set for comparing the accuracy of the original classifier and the one of the re-trained classifier.

For trace encoding, we used four different types of encodings: frequency, simple-index and complex-index encoding (described in Sect. 4). Since each condition used to inject noise in the labelings of the synthetic datasets is tailored to affect a specific trace encoding, to test each synthetic dataset, we used the encoding that is affected by the corresponding noise condition. For example, for L11, in which only the (non-)occurrence of a certain activity has to be captured, we used the frequency encoding. For L12, in which both the occurrence and the position of a certain activity characterize the wrong predictions, we used the simple-index encoding. For L13 and L14, in which the noise condition affects the data attributes, we used the complex-index encoding.

5.3   Results

In this section, we first report the results related to the characterization of the wrong predictions showing the frequent explanation itemsets for false positives and false negatives. We then report the results obtained on the testing sets when retraining the classifier.

Characterizing the Wrong Predictions. Table 4 reports, for each synthetic dataset, the characterizations of the false positives and/or the false negatives. In particular, the table reports (i) the encoding used for each dataset; (ii) the frequent explanation itemsets identified; (iii) for each frequent explanation itemset, whether it characterizes false positive (F_P) or false negative (F_N) predictions; and (iv) for each frequent explanation itemset, the relative support (i.e., the

Table 4. Explanation: synthetic results.

Label.	Encoding	Characterisation	Set	Rel. support
L11	Frequency	#(Send Questionnaire) = 1	F_N	1
		#(Send Questionnaire) = 1 and #(Receive Questionnaire Response) = 0	F_N	0.544
		#(Send Questionnaire)=1 and #(Reject Claim) = 0	F_N	0.398
L12	Simple	$\pi_{activity}$(event$_5$)=Create Questionnaire	F_N	0.545
		$\pi_{activity}$(event$_4$) = Accept Claim	F_N	0.372
		$\pi_{activity}$(event$_5$)=Create Questionnaire and $\pi_{activity}$(event$_4$)=Accept Claim	F_N	0.256
L13	Complex	CType = 'Silver'	F_N	1
		CType = 'Silver' and PClaims = 'No'	F_N	0.941
		CType = 'Silver' and Age=[50–59]	F_N	0.457
L14	Complex	**CType = 'Silver'**	F_N	1
		CType = 'Silver' and PClaims = 'No'	F_N	0

relative frequency) of the itemset over the traces in the F_P/F_N set. We report in the table the two itemsets with the highest relative support score and, in bold, the itemset that best explains the F_P/F_N predictions according to the noise condition injected in the labeling of the training set.

For example, in the case of L11, the most important itemset characterizing the F_N predictions is the occurrence of Send Questionnaire (at least once) with a relative support of 1. This means that Send Questionnaire occurs at least once in all traces in the set of the false negatives and that, in all of them, it contributes towards a (wrong) negative prediction. The itemset with the second highest relative support (0.544) concerns the occurrence of Send Questionnaire and the non-occurrence of Receive Questionnaire Response. Among the frequent explanation itemsets characterizing the F_N predictions for L11, we also find the noise condition injected in the labeling of the training set (with a relative support of 0.398) indicating that the F_N predictions are the ones in which Send Questionnaire occurs and Reject Claim does not. Being aware that Accept Claim and Reject Claim are mutually exclusive, this condition is perfectly in line with the injected condition. Indeed, the classifier has been trained to classify traces in which Accept Claim and Send Questionnaire both occur as negative, while, according to the gold standard, these traces should be positive, thus resulting in a F_N.

For L12, we used the simple-index encoding, which is the most suitable encoding to characterize the F_N predictions according to the noise condition injected in the labeling of the training set for this dataset. Also in this case, together with the two itemsets with the highest relative support score, we also report the itemset that best fits the expected characterization of the F_N predictions, i.e., $\pi_{\text{activity}}(\text{event}_5) = \text{Create Questionnaire}$ and $\pi_{\text{activity}}(\text{event}_4) = \text{Accept Claim}$. Indeed, based on the noise condition injected in the labeling of the training set, we expect that the classifier learns to classify as negative the traces in which Accept Claim occurs in a position different from position 5. However, according to the gold standard, these traces are labeled as positive. The F_N predictions are hence related to those traces in which Accept Claim occurs ($\pi_{\text{activity}}(\text{event}_4) = $ Accept Claim), but in a position different from position 5 ($\pi_{\text{activity}}(\text{event}_5) = $ Create Questionnaire). Although this characterization is a subset of the original one, it can be considered as a good starting point to understand the rationale behind the wrong predictions.

Concerning L13, the condition we expect to capture is CType = 'Silver' and Age< 60. Indeed, for these traces, the classifier learns that it has to predict a negative outcome, while, in the gold standard, they are labeled as positive. The characterization that we are able to capture (CType = 'Silver' and Age = [50–59]) is hence in line with the expectations. Again, although this is not exactly the original noise condition, it offers a clue on the problematic cases.

For L14, we are able to capture only a part of the condition that we expect to find. Indeed, while we would expect that the F_N predictions are those traces in which Accept Claim occurs in position 5 and Ctype has value 'Silver', we are not able to capture the fact that Accept Claim has to occur at position 5. This is mainly due to the fact that, according to LIME, the occurrence of Accept Claim in different positions of the trace is a feature-value pair that contributes towards a positive prediction rather than towards a negative one.

The reason why we are not always able to identify the exact condition we used to inject noise in the labeling of the training set is related to the abstraction mechanism we use to aggregate the explanations. In particular, since the explainer returns a set of explanations for every single trace, we use association rule mining as an abstraction mechanism to "discover" explanations that frequently occur together. This type of abstraction is however not always sufficient to reconstruct the original condition used to inject noise in the labeling. In addition, any abstraction mechanism would not be able to rediscover the noise condition if each part of the condition is not covered by a sufficient amount of observations (occurrences of feature-value pairs) supporting it.

Table 5 reports the frequent explanation itemsets characterizing the F_P/F_N predictions related to the BPIC2011 with the four different labelings and with different encodings. In the table, we omit the encodings for which we did not discover any itemset (i.e., frequency for L22 and L23). Overall, we can observe that the relative support is higher for frequency and complex-index encodings. This can be due to the fact that with the simple-index encoding data is more sparse, as the position in which an activity occurs in a trace can largely vary.

Table 5. Explanation: BPIC 2011.

Label.	Encoding	Characterisation	Set	Rel. support
L21	Frequency	#(administrative fee $-$ the first pol)=0 and #(outpatient follow $-$ up consultation) $= 0$ and #(assumption laboratory) $= 3$	F_P	0.897
	Simple	$\pi_{\text{activity}}(\text{event}_4) = $ assumption laboratory	F_N	0.333
		$\pi_{\text{activity}}(\text{event}_4) = $ assumption laboratory and $\pi_{\text{activity}}(\text{event}_6) = $ unconjugated bilirubin	F_N	0.286
	Complex	Specialism code $= $ 'SC86' and Section $= $ 'Section 4' and group $= $ 'General Lab Clinical Chemistry'	F_N	0.863
		$\pi_{\text{lifecycle:transition}}(\text{event}_2) = $ 'complete' and $\pi_{\text{Producer code}}(\text{event}_7) = $ 'CRLA' and $\pi_{\text{Number of executions}}(\text{event}_4) = $ 'CRLA'	F_P	0.75
L22	Simple	$\pi_{\text{activity}}(\text{event}_5) = $ bilirubin $-$ total	F_P	0.275
		$\pi_{\text{activity}}(\text{event}_5) = $ bilirubin $-$ total and $\pi_{\text{activity}}(\text{event}_2) = $ assumption laboratory and $\pi_{\text{activity}}(\text{event}_4) = $ unconjugated bilirubin	F_P	0.261
	Complex	$\pi_{\text{Activity code}}(\text{event}_2) = $ 'AC370000'	F_P	0.688
		$\pi_{\text{Activity code}}(\text{event}_2) = $ 'AC370000' and Section $= $ 'Section 4'	F_P	0.672
L23	Simple	$\pi_{\text{activity}}(\text{event}_4) = $ assumption laboratory	F_P	0.413
	Complex	$\pi_{\text{Specialism code}}(\text{event}_6) = $ 'SC86' and $\pi_{\text{Specialism code}}(\text{event}_5) = $ 'SC86'	F_P	0.933
		$\pi_{\text{Specialism code}}(\text{event}_6) = $ 'SC86' and $\pi_{\text{Specialism code}}(\text{event}_5) = $ 'SC86' and Section $= $ 'Section 4'	F_P	0.867
L24	Frequency	#(verlosk. $-$ gynaec. short $-$ out cardcost) $= 0$ and #(order rate) $= 0$ and #(thorax)=0	F_N	0.961
	Simple	$\pi_{\text{activity}}(\text{event}_1) = $ ultrasound $-$ internal genitals	F_N	0.143
		$\pi_{\text{activity}}(\text{event}_6) = $ unconjugated bilirubin	F_N	0.143
	Complex	$\pi_{\text{lifecycle:transition}}(\text{event}_3) = $ 'complete'	F_P	0.833
		$\pi_{\text{Producer code}}(\text{event}_3) = $ 'CRLA' and Section $= $ 'Section 4'	F_P	0.833

Indeed, if we look at the itemsets with the highest relative support identified using the complex-index encoding, we can see that they never include conditions related to activities occurring in a certain position, but they mainly refer to trace or event attributes characterized by a low variability throughout the process execution.

Table 6. Re-training: synthetic results

Label.	Encoding	Baseline			Re-training$_1$			Re-trainining$_2$		
		fm	acc	auc	fm	acc	auc	fm	acc	auc
L11	Frequency	0.872	0.875	0.886	**0.984**	**0.984**	**0.99**	0.886	0.886	0.896
L12	Simple	0.686	0.698	0.728	**0.692**	**0.704**	**0.731**	**0.692**	**0.704**	**0.731**
L13	Complex	0.758	0.799	0.746	**0.8**	**0.829**	0.789	0.792	0.823	**0.816**
L14	Complex	0.867	0.93	0.817	**0.9**	**0.945**	**0.869**	0.881	0.936	0.853

Retraining. Table 6 and Table 7 report the results related to the retraining (and evaluation on the testing set) carried out by reducing the impact of the frequent explanation itemsets characterizing the wrong predictions on the synthetic and on the real-life dataset, respectively. In particular, for each labeling and encoding, we selected the two frequent explanation itemsets with highest relative support (itemset$_1$ and itemset$_2$) and we retrained the classifier with a dataset in which we replace (i) itemset$_1$ (Retraining$_1$) as well as (ii) when itemset$_2$ is available, itemset$_1$ and itemset$_2$ (Retraining$_2$) with random values, so as to reduce their impact on the predictions. By looking at the results related to the synthetic dataset (Table 6), we can observe that the retraining always improves the results, although the best improvements are usually obtained when randomizing itemset$_1$ alone.

Table 7. Re-training: BPIC11 results

Label.	Encoding	Baseline			Retraining$_1$			Retraining$_2$		
		fm	acc	auc	fm	acc	auc	fm	acc	auc
L21	Frequency	0.603	0.618	0.644	**0.608**	**0.623**	**0.649**			
	Simple	0.565	0.592	0.554	**0.764**	**0.776**	**0.858**	0.565	0.592	0.554
	Complex	0.548	0.588	0.621	**0.585**	**0.597**	**0.638**	0.549	0.557	0.563
L22	Simple	0.412	**0.667**	**0.604**	0.412	**0.667**	**0.604**	**0.446**	0.633	0.603
	Complex	0.433	0.662	0.299	**0.754**	**0.807**	**0.761**	0.425	0.645	0.289
L23	Simple	0.548	0.548	0.649	**0.569**	**0.583**	**0.657**			
	Complex	0.584	0.605	0.578	**0.634**	**0.645**	**0.661**	0.614	0.623	0.632
L24	Frequency	0.379	**0.61**	0.635	**0.464**	**0.61**	**0.664**			
	Simple	0.387	0.605	0.578	**0.408**	**0.61**	**0.656**	0.408	0.61	0.648
	Complex	**0.796**	**0.811**	0.799	**0.796**	**0.811**	**0.806**	0.792	0.807	0.797

The results obtained with the BPIC11 dataset reported in Table 7 suggest that, also in this case, the randomization of itemset$_1$ improves (or does not decrease) the accuracy of the classifier. However, in some cases, when the randomization is applied not only to itemset$_1$, but also to itemset$_2$, the results are worse than the baseline. This could be due to the lower relative support of itemset$_2$ that, in most of the cases, deteriorates the effect of randomizing itemset$_1$ only.

6   Related Work

The body of previous work related to this paper is the one concerning predictive process monitoring. A first group of approaches deals with time predictions. In [2], the authors present a family of approaches in which annotated transition systems, containing time information extracted from event logs, are used to: (i) check time conformance; (ii) predict the remaining processing time; (iii) recommend appropriate activities to end users. In [10], an ad-hoc clustering approach for predicting process performance measures is presented, in which context-related execution scenarios are discovered and modeled through state-aware performance predictors. In [18], stochastic Petri nets are used to predict the remaining execution time of a process.

Other works focus on approaches that generate predictions and recommendations to reduce risks. For example, in [4], the authors present a technique to support process participants in making risk-informed decisions with the aim of reducing process failures. In [15], the authors make predictions about time-related process risks by identifying and exploiting statistical indicators that highlight the possibility of transgressing deadlines. In [19], an approach for Root Cause Analysis through classification algorithms is presented.

A third group of approaches in the predictive process monitoring field predicts the outcome (e.g., the satisfaction of a business objective) of a process [20]. In [13] an approach is introduced, which is able to predict the fulfilment (or the violation) of a boolean predicate in a running case, by looking at: (i) the sequence of activities already performed in the case; (ii) the data payload of the last activity of the running case. The approach, which provides accurate results at the expense of a high runtime overhead, has been enhanced in [7] by introducing a clustering preprocessing step that reduces the prediction time. In [11], the authors compare different feature encoding approaches where cases are treated as complex symbolic sequences, i.e., sequences of activities each carrying a data payload consisting of attribute-value pairs. In [21], the approach in [11] has been extended by clustering the historical cases before classification. In this paper, we focus on this latter group of predictions and we aim at explaining why predictors are wrong and at leveraging these explanations for eventually improving the performance of the predictive model.

7   Concluding Remarks

In this paper, we propose an approach to explain why a predictive model for outcome-oriented predictions sometimes provides wrong predictions and we use this information to eventually improve the model accuracy. The approach leverages post-hoc explainers for identifying the most common features that induce a predictor to make mistakes. By reducing the impact of these features, the accuracy of the predictive model can be increased.

Despite the obtained results are encouraging, the proposed approach presents some limitations. Indeed, the mechanism currently used to abstract explanations related to single traces is not able to handle complex conditions, by aggregating them in a simple and compact representation. For instance, the current approach is unable to deal with inequalities between categorical attributes (e.g., CType! = 'Silver') or comparisons of numerical attributes (e.g., age < 60). In the future, we plan to implement more sophisticated abstraction mechanisms able to discover such complex conditions, while limiting the redundancy and the complexity of the resulting abstraction representation. Furthermore, we plan to extend the experimentation to other post-hoc explainers and classifiers, so as to compare the results and identify whether a given explainer or classifier is more suitable than others to characterize wrong predictions and to improve accuracy results.

Acknowledgments. We thank Marco Maisenbacher and Matthias Weidlich for providing us with the synthetic log used in the paper.

References

1. 3TU Data Center: BPI Challenge 2011 Event Log (2011). https://doi.org/10.4121/uuid:d9769f3d-0ab0-4fb8-803b-0d1120ffcf54
2. van der Aalst, W.M.P., Schonenberg, M.H., Song, M.: Time prediction based on process mining. Inf. Syst. **36**(2), 450–475 (2011). https://doi.org/10.1016/j.is.2010.09.001
3. Agrawal, R., Imielinski, T., Swami, A.N.: Mining association rules between sets of items in large databases. In: Proceedings of the 1993 ACM SIGMOD International Conference on Management of Data, Washington, DC, USA, 26–28 May 1993, pp. 207–216. ACM Press (1993). https://doi.org/10.1145/170035.170072
4. Conforti, R., de Leoni, M., La Rosa, M., van der Aalst, W.M.P.: Supporting risk-informed decisions during business process execution. In: Salinesi, C., Norrie, M.C., Pastor, Ó. (eds.) CAiSE 2013. LNCS, vol. 7908, pp. 116–132. Springer, Heidelberg (2013). https://doi.org/10.1007/978-3-642-38709-8_8
5. Denker, J.S., et al.: Large automatic learning, rule extraction, and generalization. Complex Syst. **1**(5) (1987). http://www.complex-systems.com/abstracts/v01_i05_a02.html
6. Di Francescomarino, C., Dumas, M., Federici, M., Ghidini, C., Maggi, F.M., Rizzi, W.: Predictive business process monitoring framework with hyperparameter optimization. In: Nurcan, S., Soffer, P., Bajec, M., Eder, J. (eds.) CAiSE 2016. LNCS, vol. 9694, pp. 361–376. Springer, Cham (2016). https://doi.org/10.1007/978-3-319-39696-5_22
7. Di Francescomarino, C., Dumas, M., Maggi, F.M., Teinemaa, I.: Clustering-based predictive process monitoring. IEEE Trans. Serv. Comput. **12**(6), 896–909 (2019). https://doi.org/10.1109/TSC.2016.2645153
8. Di Francescomarino, C., Ghidini, C., Maggi, F.M., Milani, F.: Predictive process monitoring methods: which one suits me best? In: Weske, M., Montali, M., Weber, I., vom Brocke, J. (eds.) BPM 2018. LNCS, vol. 11080, pp. 462–479. Springer, Cham (2018). https://doi.org/10.1007/978-3-319-98648-7_27
9. Evermann, J., Rehse, J.-R., Fettke, P.: A deep learning approach for predicting process behaviour at runtime. In: Dumas, M., Fantinato, M. (eds.) BPM 2016. LNBIP, vol. 281, pp. 327–338. Springer, Cham (2017). https://doi.org/10.1007/978-3-319-58457-7_24

10. Folino, F., Guarascio, M., Pontieri, L.: Discovering context-aware models for predicting business process performances. In: Meersman, R., Panetto, H., Dillon, T., Rinderle-Ma, S., Dadam, P., Zhou, X., Pearson, S., Ferscha, A., Bergamaschi, S., Cruz, I.F. (eds.) OTM 2012. LNCS, vol. 7565, pp. 287–304. Springer, Heidelberg (2012). https://doi.org/10.1007/978-3-642-33606-5_18

11. Leontjeva, A., Conforti, R., Di Francescomarino, C., Dumas, M., Maggi, F.M.: Complex symbolic sequence encodings for predictive monitoring of business processes. In: Motahari-Nezhad, H.R., Recker, J., Weidlich, M. (eds.) BPM 2015. LNCS, vol. 9253, pp. 297–313. Springer, Cham (2015). https://doi.org/10.1007/978-3-319-23063-4_21

12. Lundberg, S.M., Lee, S.: A unified approach to interpreting model predictions. In: Advances in Neural Information Processing Systems 30: Annual Conference on Neural Information Processing Systems 2017, 4–9 December 2017, Long Beach, CA, USA, pp. 4765–4774 (2017). http://papers.nips.cc/paper/7062-a-unified-approach-to-interpreting-model-predictions

13. Maggi, F.M., Di Francescomarino, C., Dumas, M., Ghidini, C.: Predictive monitoring of business processes. In: Jarke, M., Mylopoulos, J., Quix, C., Rolland, C., Manolopoulos, Y., Mouratidis, H., Horkoff, J. (eds.) CAiSE 2014. LNCS, vol. 8484, pp. 457–472. Springer, Cham (2014). https://doi.org/10.1007/978-3-319-07881-6_31

14. Maisenbacher, M., Weidlich, M.: Handling concept drift in predictive process monitoring. In: 2017 IEEE International Conference on Services Computing, SCC 2017, Honolulu, HI, USA, 25–30 June 2017, pp. 1–8 (2017). https://doi.org/10.1109/SCC.2017.10

15. Pika, A., van der Aalst, W.M.P., Fidge, C.J., ter Hofstede, A.H.M., Wynn, M.T.: Predicting deadline transgressions using event logs. In: La Rosa, M., Soffer, P. (eds.) BPM 2012. LNBIP, vol. 132, pp. 211–216. Springer, Heidelberg (2013). https://doi.org/10.1007/978-3-642-36285-9_22

16. Ribeiro, M.T., Singh, S., Guestrin, C.: "why should I trust you?": explaining the predictions of any classifier. In: Krishnapuram, B., Shah, M., Smola, A.J., Aggarwal, C.C., Shen, D., Rastogi, R. (eds.) Proceedings of the 22nd ACM SIGKDD International Conference on Knowledge Discovery and Data Mining, San Francisco, CA, USA, 13–17 August 2016, pp. 1135–1144. ACM (2016). https://doi.org/10.1145/2939672.2939778

17. Rizzi, W., Simonetto, L., Di Francescomarino, C., Ghidini, C., Kasekamp, T., Maggi, F.M.: Nirdizati 2.0: new features and redesigned backend. In: Proceedings of the Dissertation Award, Doctoral Consortium, and Demonstration Track at BPM 2019 co-located with 17th International Conference on Business Process Management, BPM 2019, Vienna, Austria, 1–6 September 2019, pp. 154–158 (2019). http://ceur-ws.org/Vol-2420/paperDT8.pdf

18. Rogge-Solti, A., Weske, M.: Prediction of remaining service execution time using stochastic petri nets with arbitrary firing delays. In: Basu, S., Pautasso, C., Zhang, L., Fu, X. (eds.) ICSOC 2013. LNCS, vol. 8274, pp. 389–403. Springer, Heidelberg (2013). https://doi.org/10.1007/978-3-642-45005-1_27

19. Suriadi, S., Ouyang, C., van der Aalst, W.M.P., ter Hofstede, A.H.M.: Root cause analysis with enriched process logs. In: La Rosa, M., Soffer, P. (eds.) BPM 2012. LNBIP, vol. 132, pp. 174–186. Springer, Heidelberg (2013). https://doi.org/10.1007/978-3-642-36285-9_18

20. Teinemaa, I., Dumas, M., La Rosa, M., Maggi, F.M.: Outcome-oriented predictive process monitoring: review and benchmark. ACM Trans. Knowl. Discov. Data **13**(2), 17:1–17:57 (2019). https://doi.org/10.1145/3301300
21. Verenich, I., Dumas, M., La Rosa, M., Maggi, F.M., Di Francescomarino, C.: Complex symbolic sequence clustering and multiple classifiers for predictive process monitoring. In: Reichert, M., Reijers, H.A. (eds.) BPM 2015. LNBIP, vol. 256, pp. 218–229. Springer, Cham (2016). https://doi.org/10.1007/978-3-319-42887-1_18
22. Verenich, I., Dumas, M., La Rosa, M., Maggi, F.M., Teinemaa, I.: Survey and cross-benchmark comparison of remaining time prediction methods in business process monitoring. ACM Trans. Intell. Syst. Technol. **10**(4), 34:1–34:34 (2019). https://doi.org/10.1145/3331449

Bayesian Network Based Predictions
of Business Processes

Stephen Pauwels$^{(\boxtimes)}$ ⬤ and Toon Calders ⬤

University of Antwerp, Antwerp, Belgium
{stephen.pauwels,toon.calders}@uantwerpen.be

Abstract. Predicting the next event(s) in Business Processes is becoming more important as more and more systems are getting automated. Predicting deviating behaviour early on in a process can ensure that possible errors are identified and corrected or that unwanted delays are avoided. We propose to use Bayesian Networks to capture dependencies between the attributes in a log to obtain a fine-grained prediction of the next activity. Elaborate comparisons show that our model performs at par with the state-of-the-art methods. Our model, however, has the additional benefit of explainability; due to its underlying Bayesian Network, it is capable of providing a comprehensible explanation of why a prediction is made. Furthermore, the runtimes of our learning algorithm are orders of magnitude lower than those state-of-the-art methods that are based on deep neural networks.

Keywords: Business Process · Event prediction · Dynamic Bayesian Network

1 Introduction

Process Mining [1] is the field that studies the creation of understandable models given data from a Business Process (BP). With more and more systems that automate process execution it becomes more important to be able to predict as soon as possible when deviations occur in a log or when we can already predict that the deadline will be exceeded with high probability. *Di Francescomarino et al.* [6] show the growing importance and interest in predicting events within Business Processes. Different methods have been proposed already, using probabilistic models, machine learning, and Deep Learning. All these methods have in common that they use historical data for creating a reference model that subsequently can be used in new cases to determine the next activity before it actually occurs.

The commonly used process models such as workflow nets or BPMN diagrams are not suitable for next-event prediction as they fail to capture transition probabilities and as a result have inaccurate predictions. Therefore, most of the state-of-the-art next-event prediction models rely on Deep Artificial Neural Networks. Despite that these Deep Learning methods are able to provide accurate

© Springer Nature Switzerland AG 2020
D. Fahland et al. (Eds.): BPM Forum 2020, LNBIP 392, pp. 159–175, 2020.
https://doi.org/10.1007/978-3-030-58638-6_10

predictions, they also have some issues. A first issue is that training the models requires extensive resources such as time and computation power, necessitating the use of specialized hardware. Typically these methods also require excessive amounts of training data for learning an accurate model. Secondly, neural networks are so-called black-box models, in the sense that the predictions they make cannot be explained easily, making them unsuitable for many applications.

Therefore, in this paper we propose an alternative method based on *Dynamic Bayesian Networks* for predicting the next event in BP logs. In comparison with the state-of-the-art deep learning methods, our method has the advantage that it produces predictions that can be explained and takes orders of magnitude less time and resources to train, while performing with comparable predictive accuracy. An elaborated use case showing the explainability can be found in [15].

In this paper we start from our previously defined eDBNs that were developed for detecting anomalies in BP logs [16]. Using Bayesian Networks and their Conditional Dependencies we can create a probabilistic model that is able to predict next events. The contributions of our paper are:

- We propose a fast and comprehensible algorithm that is able to accurately predict the next events in a BP log. The learned model can easily be modified by a domain expert.
- We performed an elaborated survey and comparison of the state-of-the-art techniques, which was not yet done in literature. We compare the methods proposed by *Tax et al.* [19], *Camargo et al.* [4], *Lin et al.* [14], and *Di Mauro et al.* [7]. We both compare accuracy and runtimes for these algorithms. All code used to run the experiment is provided in a publicly available Github repository.

The remainder of the paper is structured as follows: in the next section we explain existing methods and indicate some of their strengths and weaknesses. Section 3 introduces the concepts we need in order to explain in Sect. 4 how we use Dynamic Bayesian Networks for next activity prediction. An extensive evaluation and comparison is performed in Sect. 5, where we take a look at both next activity prediction problem and suffix prediction. Next to comparing the accuracy of the different methods, we also compare runtime of both training and testing.

2   Related Work

We can distinguish three different types of predictions that are researched within the BP field. The first type only considers predicting the next activity, the second predicts the entire suffix for a given sequence of events, and the third type predicts the (remaining) duration of a case until completion. Another difference we can see between different methods is the type of input data they consider. Some methods only use the sequence of activities while others may use some extra attributes (e.g. resource). Table 1 contains an overview of which of the

Table 1. Overview of related work.

Paper	Model type	Input data	Predict Next Activity	Predict Suffix	Predict Duration
Our paper	DBN	multivariate	✓	✓	
Becker et al.	HMM	activity	✓		
Breuker et al.	PFA	activity	✓		
Lakshmanan et al.	PPM	activity	✓		
Everman et al.	LSTM	activity, resource	✓	✓	
Tax et al.	LSTM	activity, duration	✓	✓	✓
Lin et al.	LSTM	multivariate	✓	✓	
Camargo et al.	LSTM	activity, resource and duration	✓	✓	✓
Di Mauro et al.	CNN	activity, duration	✓		✓
Hinkka et al.	Clustering RNN	multivariate	✓		

described methods solves which type of prediction problem and what type of input they require.

There already exists ample research on applying probabilistic models for predictions in Business Processes. *Becker et al.* [2] proposed a framework for real-time prediction based on a probabilistic model learned from historical data. They use Hidden Markov Models (HMM), which are state machines where the next state depends on the current state and current event, and Probabilistic Finite Automatons (PFAs), where every transition outputs the probability for going to a specific state. An advantage of these models is their explainability and possibility of visualization, but on the other hand, their model does not take multiple attributes into account and is not able to find long-term dependencies. The same techniques were used by *Breuker et al.* [3], they improve this model by incorporating more event data.

Lakshmanan et al. [13] propose an instance-specific Probabilistic Process Model (PPM) that calculates the likelihood for all possible events to occur next. They show that under some non-restrictive assumptions, their model can be transformed into a Markov Chain, allowing them to rely on standard Markov techniques. To create their model, they first mine a Petri Net process model based on the given traces. This model is then extended by adding a probabilistic model for all OR and AND splits that allows to predict the next event.

An important factor in these probabilistic methods is that they are comprehensible and are able to explain why certain predictions were made, but only use limited information of the events. As a result, they encounter the same challenges as most process models; i.e.: these models fail at detecting and correctly predicting long-term dependencies.

With the growing popularity of Deep Learning, achieving great results in the field of Natural Language Processing (NLP) and image processing, researchers also investigated the usage of Neural Networks in Business Process Mining. Most of the state-of-the-art methods for next event prediction use so-called Long Short

Term Memory (LSTM) cells in their neural networks [4,11,12,14,19]. LSTM cells are cells that are able to selectively remember parts of the sequence and use this memory for determining the output. Most of these methods are very similar and differ mainly with respect to the architecture of the network; e.g., extra LSTM cells can be added or they are organised in different ways. Where the probabilistic methods only predict the next activity, the neural network solutions usually also include prediction of the duration of an event. This duration prediction, however, is outside the scope of this paper.

LSTMs have the downside of being hard to train, due to their sequential nature. To counter this disadvantage, *Di Mauro et al.* [7] propose to use Convolutional Neural Networks (CNNs) instead. The resulting networks also take the sequence of events into account but can be trained more efficiently.

Most deep learning methods use a one-hot encoding feature vector as input for the network. Nevertheless some approaches try to take multiple attributes of the events into account. For instance *Camargo et al.* [4] first creates equal-size embeddings for both activity and resource. Next, they use the dot product to combine these two features into a single feature vector which has lower dimensionality than the original feature vectors. *Lin et al.* [14] on the other hand, have adapted their network to take multiple attributes into account. In order to do so they introduced a new type of layer: the Modulator. This modulator combines the values of all attributes and uses a weighted sum to determine how much an attribute contributes to predicting another attribute. Having a large amount of event attributes and event values is one of the biggest challenges for Recurrent Neural nets. To tackle this problem, *Hinkka et al.* [12] propose a Recurrent Neural Net method were they first cluster events according to their attribute values to lower the overall complexity.

As the used Neural Networks need sequences of the same length, two solutions were proposed. The first solution is to pad all sequences with 0 until they all match the length of the longest sequence [7,19]. The other solution is to divide the sequences in different windows with a fixed window-size [4]. The advantage of this second solution is that the complexity of the neural network does not depend on the length of the longest sequence, as we now the length of the input sequences (windows) beforehand.

3   Background

Before we explain how we extend and use Bayesian Networks for event prediction, we first introduce the basic problem setting and some concepts we need to create the model. First, we formally define an event and a logfile:

Definition 1. *Let $e = (eID, a_{act}, desc)$ be an event with a unique identifier eID, an associated activity a_{act} and an event-descriptor desc. The event-descriptor is a tuple $(a_0, \ldots, a_n)$ denoting the values of (possible) extra attributes $\mathcal{A} = (A_0, \ldots, A_n)$, an ordered set of attributes.*

A case $c = (cID, [e_0, e_1, \ldots, e_{m_c}])$ has a unique identifier cID and consists of an ordered list of events. The order in the list determines the order in which the events were executed. m_c denotes the number of events of a case c.

A logfile is a finite set of cases originating from the same process or institution.

We express the problem of finding the next activity as follows:

Definition 2. Given a partially completed case $C = (cID, [e_0, \ldots, e_i])$, with $0 \leq i < m_c$, we want to predict the activity of the immediately next event e_{i+1}.

Predicting the suffix of a case is defined as follows:

Definition 3. Given a partially completed case $C = (cID, [e_0, \ldots, e_i])$, with $0 < i < m_c$, we want to predict the list $[a_{act}^{i+1}, \ldots, a_{act}^{m_c}]$ of the activities of all following events.

3.1 Dynamic Bayesian Network

We use Bayesian Networks (BNs) [17] to model the dependencies between attributes present in a logfile. A BN is a directed acyclic graph where every node represents an attribute and a directed edge (A_i, A_j) exists when attribute A_j depends on A_i. We call variable A_i the *parent* of variable A_j. We denote the set of all parents of a node A as $Pa(A)$.

The BN with nodes $X_0, \ldots, X_n$ represents a *joint probability* distribution that can be decomposed as follows:

$$P(X_0, \ldots, X_n) = \prod_{i=0}^{n} P(X_i | Pa(X_i)) \tag{1}$$

To learn a BN we first need to learn the structure of the model (the dependencies between attributes). At the same time we learn the Conditional Probability Tables (CPTs) that capture $P(X_i | Pa(X_i))$ for the attributes X_i. These tables contain the probabilities for all possible values given the occurrence of values of the parents. In the literature there already exist algorithms that can learn the structure of such a BN given a reference dataset [18].

Dynamic Bayesian Networks (DBNs) [18] incorporate the sequential aspect by adding extra variables to the network. For the previous k time steps, variables are added representing the values of the attributes in the different time steps we call them the *past-variables*. Because we are interested in learning how variables in the current time step depend on variables from the previous time steps, we only allow arrows going from the past-variables to the variables in the current time step. An example of a DBN with $k = 1$ can be found in Fig. 1. In this diagram one can easily see the different time steps. Dependencies may exist within one time step (*Activity* → *Type*) or between time steps (*Activity_previous* → *Activity_current*).

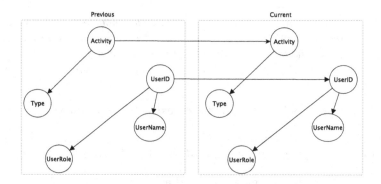

Fig. 1. Example Dynamic Bayesian Network

In [16] we use our model to detect anomalies in (partially) finished cases. Hence, all values of the attributes are available when checking for consistency with the model. In this paper, however, we want to predict the values of the current time step using only the values of the past-variables. Therefore, we do not allow arrows between variables in the current time step.

4   Predicting

We use historical input data to train our probabilistic model, details about how to learn the model can be found in [16]. In the next sections we explain how we use this model for event prediction. Note that, although we only consider event prediction using categorical attributes, DBNs can be extended to also be able to handle numerical attributes and predict numerical values as is shown in [16].

4.1   Next Activity Prediction

To predict the next activity, we calculate $P(a_{act})$ for all possible (known) values of the activity attribute. Following Eq. 1 we estimate the probability for the activity as $P(A_{act}|Pa(A_{act}))$. Calculating the most likely activity, given the values of the parent attributes, comes down to looking at the CPT associated with the activity attribute. This table contains the known combinations of values of parent attributes. Every combination is linked with possible values for the activity with their respective probability. We then select the activity which has the highest probability as the predicted next activity.

Example 1. Suppose our activity attribute depends on activity and resource from the previous time step. The probability $P(activity|activity_1, resource_1)$ gets described by the CPT shown in Table 2.

Table 2. Example CPT; assume activity depends on activity and resource from the previous time step. This CPT gives the empirical conditional probabilities $P(activity \mid activity_1, resource_1)$, where $activity_1$ and $resource_1$) denote respectively the activity and resource of the previous timestamp.

$activity_1$	$resource_1$	activity	P(activity)
start	resource1	a	1.0
start	resource2	a	1.0
a	resource1	b	0.7
a	resource1	c	0.3
a	resource2	d	1.0
b	resource3	c	0.4
b	resource3	end	0.6
c	resource1	end	1.0
c	resource3	end	1.0

Given the values *b* and *resource3* for activity and resource respectively from the previous time step, we can see in the table that two possible values for the next activity are possible: *c* (with a probability of 0.4) and *end* (with a probability of 0.6). We return the value with the highest probability, which is *end*, as the prediction for the next event.

Because we use a conditional probability table with all combinations which have occurred in the training data, it is possible that the combination of parent values we encounter in the test data does not occur in the CPT. In that case we have to make an estimation based on the data and probabilities present in the CPT, to avoid having to use a probability of 0. We use marginalization techniques in order to come up with the most likely next activity. We make a distinction between two cases: the first one where there is at least one parent value that we did not encounter in the training data and the other where all values have been observed in the training data, but not together.

For the first case, when there are new values for some attributes, that have not been seen in the training set, we calculate the marginalized probability where we iterate over all known values for these attributes. We let the new attributes vary over all possible values, which occur in the CPT, while we keep the known attributes the same.

That is: Let A_n be the set of parent attributes of A_{act} for which we observe an unseen value and A_k the other parents with known attribute values. For determining the probability of the activities we need to calculate $P(A_{act}|A_n, A_k)$. We calculate alternatively $P(A_{act}|A_k)$ as follows:

$$\sum_{a_n \in dom(A_n)} P(a_n|A_k) * P(A_{act}|a_n, A_k) \tag{2}$$

Notice that we used $a_n \in dom(A_n)$ to denote all previously observed combinations of the attributes A_n.

Example 2. Consider the CPT from Table 2, suppose that we need to predict the next activity, based on the values e and *resource3* for attribute and resource. The value for resource has already been seen, but the value for the activity is new. We thus marginalize over the activity attribute. The seen attribute *resource* only occurs together with the values b and c. We want to calculate the following formula for all possible values of activity, note that we only take into account combinations of parent values that do occur in the CPT:

$$P(A_{act}|resource3) = P(b|resource3) * P(A_{act}|b, resource3)$$
$$+ P(c|resource3) * P(A_{act}|c, resource3)$$

Assume $P(b|resource3) = 0.4$ and $P(c|resource3) = 0.6$. We can now calculate the probabilities for all activities as follows:

$$P(start|e, resource3) = 0.4 * 0 + 0.6 * 0 = 0$$
$$P(a|e, resource3) = 0.4 * 0 + 0.6 * 0 = 0$$
$$P(b|e, resource3) = 0.4 * 0 + 0.6 * 0 = 0$$
$$P(c|e, resource3) = 0.4 * 0.4 + 0.6 * 0 = 0.16$$
$$P(d|e, resource3) = 0.4 * 0 + 0.6 * 0 = 0$$
$$P(end|e, resource3) = 0.4 * 0.6 + 0.6 * 1 = 0.84$$

From these results we can see that the next activity with the highest probability is *end*.

For the second case, when all values occur in the training data but the combination of values does not occur in the CPT, we take the average probability of marginalizing over every attribute. We use the following formula to estimate the probabilities for the activities:

$$\frac{\sum_{A \in Pa(A_{act})} \sum_{a \in dom(A)} P(a|Pa(A_{act}) \setminus A) * P(A_{act}|a, Pa(A_{act}) \setminus A)}{\#Pa(A_{act})} \tag{3}$$

Again, we calculate the probability for every possible activity to determine the most likely next activity.

Example 3. Consider again the CPT from Table 2, suppose this time we have the values a and *resource3* for activity and resource. Both values already occur in the CPT but never together. We thus calculate the following formula for every possible activity:

$$P(X_{act}|a, resource3) = (P(b|resource3) * P(X_{act}|b, resource3)$$
$$+ P(c|resource3) * P(X_{act}|c, resource3)$$
$$+ P(resource1|a) * P(X_{act}|a, resource1)$$
$$+ P(resource2|a) * P(X_{act}|a, resource2))/2$$

Assume $P(b|resource3) = 0.7$, $P(c|resource3) = 0.3$, $P(resource1|a) = 0.8$ and, $P(resource2|a) = 0.2$. We then get the following probabilities for the possible activities:

$$P(start|a, resource3) = (0.7 * 0 + 0.3 * 0 + 0.8 * 0 + 0.2 * 0)/2 = 0$$
$$P(a|a, resource3) = (0.7 * 0 + 0.3 * 0 + 0.8 * 0 + 0.2 * 0)/2 = 0$$
$$P(b|a, resource3) = (0.7 * 0 + 0.3 * 0 + 0.8 * 0.7 + 0.2 * 0)/2 = 0.28$$
$$P(c|a, resource3) = (0.7 * 0.4 + 0.3 * 0 + 0.8 * 0.3 + 0.2 * 0)/2 = 0.26$$
$$P(d|a, resource3) = (0.7 * 0 + 0.3 * 0 + 0.8 * 0 + 0.2 * 1)/2 = 0.10$$
$$P(end|a, resource3) = 0.7 * 0.6 + 0.3 * 1 + 0.8 * 0 + 0.2 * 0)/2 = 0.36$$

From these results we can see that in this case the next activity with the highest probability is *end*.

4.2   Suffix Prediction

To predict the suffix of activities we also have to predict other attributes, as we need these attributes to be able to keep predicting the activities in the suffix. First, we recursively determine all attributes the activity depends on. Apart from the parents of the activity attribute we also need to take the parents of the parents etc. into account.

Example 4. Consider the example from the previous section. In this example, *activity* depends on *activity* and *resource* in the previous time step. Assume now that *resource* itself also depends on *activity* and *resource* from the previous time step. To be able to keep predicting the next activity, we also need to predict values for the resource by using the activity and resource from the previous time step.

To predict the value for an attribute we use the same formulas as were introduced in Sect. 4.1. As these formulas are attribute independent we can also use them for predicting other attributes. We can use the same marginalization techniques when new combinations of parent values occur.

Example 5. To predict the suffix for our example we have to predict values for activity and resource. We thus need to calculate the following probabilities:

$$P(activity|activity_1, resource_1)$$
$$P(resource|activity_1, resource_1)$$

The disadvantage of always selecting the activity with highest probability is that we can get stuck in an infinite self-loop when the most likely next activity is the same as the current one. To solve this issue, we introduce a restriction on the size of these self-loops. During the learning phase of the algorithm we count the length of all self-loops for the activity. We use the average length of these self-loops as the maximum size for self-loops in the prediction phase. When we

Table 3. CPT for activity used in example 6

activity₁	activity	P(activity)
start	a	1.0
a	a	0.6
a	b	0.4
b	end	0.8
b	c	0.2

predict the same activity as the current activity, we increase a self-loop counter. When this self-loop counter reaches the maximum size for self-loops we select the second best activity as our prediction. When we predict a different activity we reset the self-loop counter to 0.

Example 6. Consider a situation where activity only depends on the previous activity. The CPT for activity is shown in Table 3. Assume we have $[start]$ as the current prefix, when we would predict the suffix without limitations on the self-loops we would get the following sequence: $[start, a, a, a, a, a, \ldots, a]$ until we reach the maximum allowed suffix length. When we set the maximum size for self-loops to 3, we get the following sequence: $[start, a, a, a, b, end]$.

5   Evaluation

In this section, we describe the different experiments we performed to assess the quality of our method. First, we describe the different datasets used for performing the experiments. Next, we give a description of how we performed the experiments. The results are split in an evaluation of our method, where we test the influence of the number of time steps used, and a comparison with state-of-the-art methods. We perform both runtime and accuracy experiments to fully compare the different methods.

5.1   Datasets

For our evaluation we selected different datasets that are most used in related work.

– **BPIC12** [9]: a log from a Dutch financial institution, containing applications for personal loans. The logfile contains three intertwined processes, one of which was used to create an additional logfile (**BPIC12W**).
– **BPIC15** [10]: a log containing building permit applications from five different Dutch municipalities, where the data of every municipality is saved in a single log file. We denote these 5 datasets respectively **BPIC15_1** to **BPIC15_5**.
– **Helpdesk** [20]: a log containing ticket requests of the helpdesk from an Italian Software Company.

Table 4. Characteristics of the datasets after preprocessing

Dataset	# Events	# Cases	# Activities	Avg. activities per case	Max. length of a case
BPIC12	171,175	7,785	23	22.0	106
BPIC12W	61,449	4,848	6	12.6	74
BPIC15_1	52,217	1,199	398	43.5	101
BPIC15_2	44,347	829	410	53.5	132
BPIC15_3	59,533	1,377	383	43.2	124
BPIC15_4	47,271	1,046	356	45.2	116
BPIC15_5	59,068	1,153	389	51.2	154
Helpdesk	20,722	4,371	14	4.7	15
BPIC18	2,514,266	43,809	41	57.4	2,973

- **BPIC18** [8]: a log containing applications for EU agricultural grants. This is the most challenging dataset and not all state-of-the-art methods could be executed for this logfile.

Following the same preprocessing steps as in [4], we first remove all cases with less than 5 events from the logs. Only for the Helpdesk dataset we set this threshold to 3, because of the smaller average length of the cases in this dataset.

Table 4 gives the characteristics of the used datasets after preprocessing. The table shows that the datasets used in our experiments exhibit a wide variety of characteristics leading to an evaluation as complete as possible.

5.2 Method

We ran our experiments on compute nodes with 2 Xeon E5 processors, which have 14 cores each, and 128 GB RAM. We also used GPU based nodes having 2 NVIDIA P100 GPUs with 16 GB of memory for some of the experiments. Most of the experiments where conducted using only CPU nodes, as our experiments showed that training LSTMs on a GPU is less efficient than using a CPU. We include some of these results in our general runtime results. We explicitly mention it when the experiment was run on a GPU node.

We split all datasets into a training and testing part. We have chosen to split the datasets as described by Camargo et al. [4]. The first 30% of cases are used for testing and the remaining 70% for training.

In this paper we are only interested in predicting activities, therefore we have chosen to only use architectures that don't use time if available for a method.

All Neural Network learning algorithms use an early stopping method with a patience of 42, meaning that when no improvement occurred for 42 epochs the learning algorithm stopped. Increasing or decreasing this parameter has a direct influence on runtimes, but has also an influence on the accuracy. The value of 42 was chosen because it was used in the original code of the other implementations.

To measure the performance of the next activity prediction we use accuracy, which gives us the fraction of activities that were predicted correctly. Events that did not get a prediction, because of unknown values are considered to be wrong. For the performance of the suffix prediction we use the Damerau-Levenstein distance [5] between the predicted trace s_p and the correct trace s_e as basis for our measure. As we want a measure that gives 1 when two traces are completely identical and 0 when they are completely different we use the following formula for the accuracy of the suffix prediction:

$$S(s_p, s_e) := 1 - \frac{DL\_distance(s_p, s_e)}{max(len(s_p), len(s_e))} \qquad (4)$$

An implementation of our method and code to execute the experiments can be found in our GitHub repository[1].

5.3   Evaluation

In this section, we examine the influence of k, which determines how many previous time steps the model takes into account, on the accuracy. We performed this test both for next event prediction and suffix prediction.

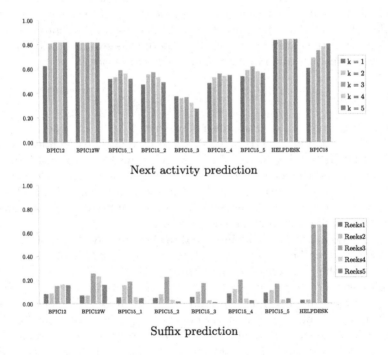

Fig. 2. Results for varying k-values

---

[1] https://github.com/StephenPauwels/edbn.

Table 5. Comparison of the accuracies for next event prediction

Dataset	Our method	Camargo et al.	Lin et al.	Tax et al.	Di Mauro et al.
BPIC12	0.82	0.80	**0.83**	0.81	0.82
BPIC12W	**0.82**	0.80	0.81	0.80	0.74
BPIC15_1	0.53	0.50	0.60	**0.64**	0.60
BPIC15_2	0.55	0.45	0.45	**0.56**	**0.56**
BPIC15_3	0.36	0.35	0.38	**0.46**	0.40
BPIC15_4	0.56	0.48	0.52	**0.59**	0.52
BPIC15_5	0.62	0.47	0.61	**0.64**	0.58
Helpdesk	**0.84**	0.80	0.82	0.80	0.83
BPIC18	**0.80**	0.79	0.79	-	-

Figure 2a and 2b show the performance of our method for different k values. Given that the optimal value of k for both prediction tasks widely differs between datasets, we have opted to include finding the optimal value of k as part of the training process. That is we iterate over different k values and use a separate validation set, extracted from the training data, to determine the best k value.

5.4 Comparison

We compare our method with four state-of-the-art methods, all of which use neural networks. We used the implementations provided by Camargo et al. [4], Tax et al. [19] and Di Mauro et al. [7], which were slightly adapted to match our input format. We implemented the method described by Lin et al. [14] as correctly as possible according to the paper, as no implementation was available. All used implementations have been added to the GitHub repository. We used the activity and the resource of an event as input data for all datasets.

Table 6. Comparison of runtimes to train the model (in seconds)

Dataset	Our method	Camargo et al.	Lin et al.	Tax et al.	Di Mauro et al.
BPIC12	66	16,681–13,251	5,736	14,123	1,418
BPIC12W	30	4,957–4,722	1,734	6,331	489
BPIC15_1	16	4,309–4,369	1,660	5,742	1,524
BPIC15_2	13	3,198–2,698	2,089	3,724	712
BPIC15_3	17	5,292–4,298	2,360	6,204	1,233
BPIC15_4	15	4,300–4,552	3,339	4,118	1,103
BPIC15_5	18	4,412–3,868	3,134	9,280	5,107
Helpdesk	11	1,951–3,516	730	1,273	45
BPIC18	1,704	107,574–117,707	32,249	-	-
BPIC12 (GPU)	-	> 21,600	> 21,600	13,969	328

Next Activity Prediction. When looking at the comparison in Table 5 we see that our method performs at par with current state-of-the-art methods for predicting the next activity. Important to note is that BPIC12, BPIC12W and Helpdesk are the most used datasets for evaluating prediction algorithms. When comparing the results for these datasets we see that all recent methods perform equally well. Only for the BPIC15 datasets, which are more complex in nature, we can see a larger difference in performance.

When comparing the runtimes in Table 6 we can clearly see that our method needs orders of magnitude less time to train than other methods. The training of our model consists of learning the structure and populating the different CPTs. The table confirms that a CNN can be trained more efficiently than the LSTM-based networks. The CNN has the added advantage that it can be efficiently learned using a GPU, whereas training LSTMs on a GPU is less efficient.

We also tested to train the models using GPU compute nodes. The runtimes for the BPIC12 dataset can also be found in Table 6. As our compute time on the nodes was limited we did not get results for some of the methods. The results show that there is, except for the CNN, no benefit in using GPUs over CPUs.

Table 7. Comparison of runtimes to predict the next events (in seconds)

Dataset	Our method	Camargo et al.	Lin et al.	Tax et al.	Di Mauro et al.
BPIC12	39	447–456	204	902	12
BPIC12W	8	166–174	82	243	8
BPIC15_1	9	133–145	71	258	9
BPIC15_2	8	144–149	72	386	10
BPIC15_3	10	163–166	79	389	9
BPIC15_4	9	116–116	63	269	9
BPIC15_5	12	165–165	78	510	10
Helpdesk	5	56–61	37	27	6
BPIC18	490	4,994–5,396	2,410	-	-

Suffix Prediction. Table 8 shows that our method is not suitable for predicting the entire suffix of a running case. The runtimes shown in Table 9 are orders of magnitude higher than predicting the next activity. We need to remark that much of the code provided by the different methods has significant room for efficiency improvements. Currently most suffix prediction methods predict every next event separately, although the libraries used are optimised for batch prediction. An improvement would be to alter the code in order to facilitate batch prediction for all cases which have not yet been ended. This includes adding some

Table 8. Comparison of the accuracies for suffix prediction

Dataset	Our method	Camargo et al.	Lin et al.	Tax et al.	Di Mauro et al.
BPIC12	0.15	**0.58**	0.18	0.15	-
BPIC12W	0.25	**0.47**	0.12	0.09	-
BPIC15_1	0.18	**0.57**	0.55	0.50	-
BPIC15_2	0.22	**0.57**	0.51	0.39	-
BPIC15_3	0.17	**0.55**	0.54	0.42	-
BPIC15_4	0.20	0.53	**0.58**	0.37	-
BPIC15_5	0.16	0.51	**0.58**	0.44	-
Helpdesk	0.66	**0.90**	0.62	0.87	-

extra housekeeping, but the cost for this housekeeping is insignificant in comparison with the gained efficiency of not having to call the prediction function for every event separately.

Table 9. Comparison of runtimes to predict the suffixes (in seconds)

Dataset	Our method	Camargo et al.	Lin et al.	Tax et al.	Di Mauro et al.
BPIC12	911	6,666–6,913	10,731	58,386	-
BPIC12W	273	2,369–2,586	3,833	66,752	-
BPIC15_1	34	1,113	1,014	778	-
BPIC15_2	33	1,090–1326	1,102	9,501	-
BPIC15_3	35	1,295–1,451	1,076	9,080	-
BPIC15_4	35	952–1,057	758	9,760	-
BPIC15_5	49	1,387–1,278	1,204	19,763	-
Helpdesk	26	50–54	67	76	-

6 Conclusion

In this paper we used Dynamic Bayesian Networks in the context of modeling known behavior in Business Processes. This method has the advantage of being comprehensible and returning explainable results. This is due to the fact that the DBN explicitly models dependencies between attributes. We used the DBN both for predicting the next event and for predicting the suffix of a case.

We performed an elaborated comparison between different state-of-the-art methods on a variety of datasets. This comparison showed that our method, although it trains much faster than other methods, is competitive with state-of-the-art methods for predicting the next activity, while requiring much less resources. Our comparisons also show that most of the state-of-the-art methods perform equally well. The runtime experiments showed that the efficiency of both training and testing, especially suffix prediction, can still be improved. Reducing these runtimes will make all methods more feasible for real-life problems.

We also discovered some interesting avenues for further research. One of them is an in-depth analysis of the used datasets for evaluating predictions. A better understanding of why methods do perform well on some datasets but worse on others is important when trying to improve prediction techniques in BP. To further improve our DBN method we want to investigate the combination of Neural Networks with probabilistic methods, in order to combine the benefit of both methods. Combining the explainability and efficiency of our probabilistic method with the predicting power of the Neural Networks. Last it would be interesting to investigate how DBNs or other classification methods could be used for outcome prediction. We could use the same mechanics as explained in this paper, only changing the target the model learns from events to outcome.

Acknowledgments. The computational resources and services used in this work were provided by the VSC (Flemish Supercomputer Center), funded by the Research Foundation - Flanders (FWO) and the Flemish Government – department EWI.

References

1. van der Aalst, W.M.P.: Process Mining: Discovery, Conformance and Enhancement of Business Processes, 1st edn. Springer, Heidelberg (2011). https://doi.org/10.1007/978-3-642-19345-3
2. Becker, J., Breuker, D., Delfmann, P., Matzner, M.: Designing and implementing a framework for event-based predictive modelling of business processes. In: Enterprise Modelling and Information Systems Architectures-EMISA 2014 (2014)
3. Breuker, D., Matzner, M., Delfmann, P., Becker, J.: Comprehensible predictive models for business processes. Mis Q. **40**(4), 1009–1034 (2016)
4. Camargo, M., Dumas, M., González-Rojas, O.: Learning accurate LSTM models of business processes. In: Hildebrandt, T., van Dongen, B.F., Röglinger, M., Mendling, J. (eds.) BPM 2019. LNCS, vol. 11675, pp. 286–302. Springer, Cham (2019). https://doi.org/10.1007/978-3-030-26619-6_19
5. Damerau, F.J.: A technique for computer detection and correction of spelling errors. Commun. ACM **7**(3), 171–176 (1964)
6. Di Francescomarino, C., Ghidini, C., Maggi, F.M., Milani, F.: Predictive process monitoring methods: which one suits me best? In: Weske, M., Montali, M., Weber, I., vom Brocke, J. (eds.) BPM 2018. LNCS, vol. 11080, pp. 462–479. Springer, Cham (2018). https://doi.org/10.1007/978-3-319-98648-7_27
7. Di Mauro, N., Appice, A., Basile, T.M.A.: Activity prediction of business process instances with inception CNN models. In: Alviano, M., Greco, G., Scarcello, F. (eds.) AI*IA 2019. LNCS (LNAI), vol. 11946, pp. 348–361. Springer, Cham (2019). https://doi.org/10.1007/978-3-030-35166-3_25
8. van Dongen, B., Borchert, F.: Bpi challenge 2018. Eindhoven university of technology (2018). https://doi.org/10.4121/uuid:3301445f-95e8-4ff0-98a4-901f1f204972
9. van Dongen, B.: Bpi challenge 2012. Eindhoven university of technology. https://data.4tu.nl/repository/uuid:3926db30-f712-4394-aebc-75976070e91f

10. van Dongen, B.: Bpi challenge 2015. Eindhoven university of technology. https:// doi.org/10.4121/uuid:31a308ef-c844-48da-948c-305d167a0ec1
11. Evermann, J., Rehse, J.R., Fettke, P.: Predicting process behaviour using deep learning. Decis. Support Syst. **100**, 129–140 (2017)
12. Hinkka, M., Lehto, T., Heljanko, K.: Exploiting event log event attributes in RNN based prediction. In: Welzer, T., et al. (eds.) ADBIS 2019. CCIS, vol. 1064, pp. 405–416. Springer, Cham (2019). https://doi.org/10.1007/978-3-030-30278-8_40
13. Lakshmanan, G.T., Shamsi, D., Doganata, Y.N., Unuvar, M., Khalaf, R.: A Markov prediction model for data-driven semi-structured business processes. Knowl. Inf. Syst. **42**(1), 97–126 (2013). https://doi.org/10.1007/s10115-013-0697-8
14. Lin, L., Wen, L., Wang, J.: MM-Pred: a deep predictive model for multi-attribute event sequence. In: Proceedings of the 2019 SIAM International Conference on Data Mining, pp. 118–126. SIAM (2019)
15. Pauwels, S., Calders, T.: Detecting and explaining drifts in yearly grant applications. arXiv preprint arXiv:1809.05650 (2018)
16. Pauwels, S., Calders, T.: Detecting anomalies in hybrid business process logs. ACM SIGAPP Appl. Comput. Rev. **19**(2), 18–30 (2019)
17. Pearl, J.: Probabilistic reasoning in intelligent systems: networks of plausible inference. Elsevier (2014)
18. Russell, S.J., Norvig, P.: Artificial intelligence: a modern approach (2009)
19. Tax, N., Verenich, I., La Rosa, M., Dumas, M.: Predictive business process monitoring with LSTM neural networks. In: Dubois, E., Pohl, K. (eds.) CAiSE 2017. LNCS, vol. 10253, pp. 477–492. Springer, Cham (2017). https://doi.org/10.1007/978-3-319-59536-8_30
20. Verenich, I.: Helpdesk, Mendeley data, v1 (2016). https://doi.org/10.17632/39bp3vv62t.1

Predictive Process Mining Meets Computer Vision

Vincenzo Pasquadibisceglie[1(✉)], Annalisa Appice[1,2], Giovanna Castellano[1,2], and Donato Malerba[1,2]

[1] Department of Informatics, Università degli Studi di Bari Aldo Moro,
via Orabona, 4, 70125 Bari, Italy
{vincenzo.pasquadibisceglie,annalisa.appice,
giovanna.castellano,donato.malerba}@uniba.it
[2] Consorzio Interuniversitario Nazionale per l'Informatica - CINI, Bari, Italy

Abstract. Nowadays predictive process mining is playing a fundamental role in the business scenario as it is emerging as an effective means to monitor the execution of any business running process. In particular, knowing in advance the next activity of a running process instance may foster an optimal management of resources and promptly trigger remedial operations to be carried out. The problem of next activity prediction has been already tackled in the literature by formulating several machine learning and process mining approaches. In particular, the successful milestones achieved in computer vision by deep artificial neural networks have recently inspired the application of such architectures in several fields. The original contribution of this work consists of paving the way for relating computer vision to process mining via deep neural networks. To this aim, the paper pioneers the use of an RGB encoding of process instances useful to train a 2-D Convolutional Neural Network based on Inception block. The empirical study proves the effectiveness of the proposed approach for next-activity prediction on different real-world event logs.

Keywords: Predictive Process Mining · Computer vision · Deep learning · Convolutional Neural Networks · Inception blocks

1 Introduction

Integrating effective analysis and control procedures into the ecosystem of modern companies covers a crucial role to monitor the good performance of production processes. Process monitoring requires proper methodologies and tools that facilitate and support the role of the company's managers. Thanks to the large amount of data produced by large companies logging their production processes, we are now able to monitor the running instances of these processes (traces) by resorting to predictive models acquired from event log data through process mining algorithms. In particular, process mining coupled with machine

D. Fahland et al. (Eds.): BPM Forum 2020, LNBIP 392, pp. 176–192, 2020.
https://doi.org/10.1007/978-3-030-58638-6_11

learning has paved the way for research developments in Predictive Process Mining (PPM) attracting attention of industry and academy. These developments promise new opportunities and multiple benefits to ensure that the production activities will run in a desired manner by avoiding predicted failures and deviations from designed process structures at runtime. Indeed, the ability to accurately predict different factors of running traces, such as the next activity, the completion cycle time until a trace is resolved or the outcome of the trace [10], may support decision making tasks such as resource allocation, alert of compliance issues and recommendations to take different actions [3]. There are at least two different kinds of predictive monitoring: one focused on predicting the next activity and the other one focused on predicting an outcome or feature of the whole process instance. In this work the evaluation is focused on the next activity only. Predicting the next activity in a running trace is a fundamental problem in productive process monitoring as such predictive information may allow analysts to intervene proactively and prevent undesired behaviors [23,32].

In the last decade, several studies have investigated the application of predictive analytic approaches in process management [8,11]. In any case, accounting for the results achieved with deep artificial neural networks [15], especially in computer vision [20,29], significant interest has recently arisen in applying deep learning to analyze event logs and gain accurate insights into the future of the logged processes. Following this research direction, a few deep neural networks have been recently designed to address the problem of predicting the next process activity. These studies are mainly based on the idea of elaborating sequences of events. To this aim, they adopt LSTM architectures [30,31] and more recently CNN architectures [12,24]. We note that, as these approaches basically focus on the sequence of activities stored in the event log, they mainly consider the information in the control-flow perspective. However, various studies [4,5,28] rooted in [1] have been proved that information embedded in multiple perspectives (e.g. activity perspective, resource perspective, performance perspective and so on) may contribute to increase the predictive power of a machine learning method.

Based on these premises, we propose a novel PPM approach, called PRE-MIERE (PREdictive process MIning basEd on tRace imagEs), to predict the next activity of running traces. The proposed approach takes advantage from representing traces over multiple perspectives and transforms these data into RGB-like form. In this way, we are able to fully realize the power of a Convolution Neural Network for the next activity prediction. The decision of investing on a CNN architecture, that accepts a sample as an image (i.e. a matrix of size $m \times n$), is motivated by the fact that this architecture, due to the ability of processing image data with the kernel sliding on two dimensions, has gained very high accuracy in computed vision for tasks of image classification [19,26].

In short, the advantage of our proposal is twofold. Firstly, information collected along any process perspective can be, in principle, taken into account for encoding the traces into the image format. Secondly, robust CNN architectures, designed to process data in the image format, can be properly used to process trace data. This allows us to transfer the achievements of computer vision to the goals of process mining. As an additional contribution, we evaluate the advantage of Inception blocks when training our CNN architecture. Preliminary experiments performed on benchmark event logs prove that our attempt of applying a computer vision approach to process trace data produces promising results in terms of predictive accuracy paving the way for new investigations in this direction.

To our knowledge, this is the first study where computer vision properly reinforces process mining. Even authors of [12,24], who have firstly trained CNN architectures in PPM, did not consider a proper image representation of traces. They have represented traces as sequence of activities [12] or sequence of activity frequency and elapsed time [24]. Although these studies have mapped traces data as matrix data, they have associated one characteristic (activity, timestamp) to each column of the matrix. From this point of view, the trace representations adopted in [12,24] resemble the multi-variate time series format instead of the image format. Differently, in this study, we propose an image representation of traces with pixels associated with color values – each pixel capturing a characteristic of the trace scene.

The paper is organized as follows. An overview of related works is provided in the next Section. The proposed approach is described in Sect. 3. Section 4 describes the experimental setup and discusses the results on different real-world event logs. Finally, Sect. 5 draws conclusions and outlines future work.

2   Related Work

The large amount of data derived from running production processes allows process mining techniques to extract useful knowledge from event logs. This knowledge enables the managers of a company to predict the future behavior of a trace promoting an optimal management of the entire process [1].

The first studies addressing the task of next activity prediction in PPM literature are mainly based on process discovery algorithms, which resort to formal language models (e.g. Petri networks) to describe how recorded processes are performed (see [10] for a survey). More recently, predictive accuracy has been gained by using machine learning algorithms to learn predictive process models [6,14,25]. On the other hand, the recent success of deep learning in different application fields (e.g. computer vision [19,26,29] or cybersecurity [3, 17]) have also conditioned research in PPM inspiring the employment of deep neural networks in process mining to build predictive systems useful for decision support at strategic levels.

In this Section, we focus on deep learning related studies as this is the approach investigated in our study. In fact, a few deep neural networks have been recently formulated for task of the next activity prediction. They are mainly based on representations of running traces in the form of sequences (possibly obtained by some embedding technique). In particular, Tax et al. also [30] present a study comparing different neural network architectures (RNN, LSTM, GRU) for the next activity prediction. Di Mauro et al. [12] illustrate a deep learning approach based on either an LSTM architecture or a 1D CNN architecture trained with Inception Blocks. They elaborate an embedding representation of activities enriched with a vector of temporal features. Evermann et al. [13] adopt an approach based on embedding representation of activities combined with resources and the life cycle, in order to train a recurrent neural networks (RNN) with long short-term memory (LSTM). Camargo et al. [9] train RNNs with LSTMs layers, in order to predict sequences of next events, their timestamps and their associated resource pools. Similarly to [13] and [12], they use an embedding representation of the categorical information. In any case, even if they account for the resource information, they do not use the raw resource data. In fact, they resort to the algorithm described in [27] to group the resources in roles based on information enclosed in the resource activity profiles. Specifically, they process the roles, instead of the resources, in the input data by handling an activity-based representation of the resource view. Finally, Pasquadibisceglie et al. [24] adopt a 2-channel representation of traces that represent the activity and temporal characteristics as two columns of a matrix and train a 2D CNN architecture from these data.

The above mentioned approaches leverage basically on rough activities and timestamps. The approach proposed in this work makes use of complex characteristics that are engineered on the trace events. On the other hand, as these characteristics are encoded in RGB-like images, trace data can be elaborated with a 2D CNN having a kernel properly sliding on the two dimension of a scene. There are a few studies, e.g. [17,21], that already introduce the idea of resorting to an encoding algorithm, in order to transform rough data in image-like data. However, they are conducted in cyber security. So, to the best of our knowledge, this study is the first attempt of developing this idea in PPM research and address the task of next activity prediction as a computer vision task.

3   Proposed Approach

In this section, we introduce PREMIERE that is a novel deep learning approach to solve the next-activity prediction problem by resorting to a computer vision solution. In general, predicting the next activity can be defined as the prediction of the executing activity of the next event in a running trace. The prediction is commonly performed by considering the prefix trace, i.e. the sequence of past events recorded from the beginning of the current trace. In PREMIERE, prefix traces are encoded in the RGB image-like format so that the next activity prediction can be dealt as an image classification problem.

To derive a predictive model by learning from data, we assume the availability of an event log containing a list of process traces and we need to create a labeled dataset from the given event log. Each process trace under consideration identifies the execution of a process instance and consists of a finite sequence of events $\sigma_i = \langle e_1, e_2, \ldots, e_n \rangle$ such that each event appears at most ones in the event log. From each trace in the event log, we can derive several prefix traces σ^k with $1 \leq k < |\sigma^k|$. Hence, a trace is a complete process instance (started and ended), while a prefix trace is an instance in execution (running trace).

Table 1. An excerpt of a trace stored in an event log assuming that the activity alphabet of the log comprises 27 distinct activities, while the resource log comprises 48 distinct resources.

Trace ID	Activity	Resource	Timestamp
10102	Act1	11	2011/10/20 11:06:16.081
10102	Act2	11	2011/10/27 13:46:27.098
10102	Act3	11	2011/10/27 13:46:49.205
10102	Act4	12	2011/10/27 15:11:16.295
10102	Act5	11	2011/11/17 08:13:20.068
10102	Act3	11	2011/10/24 08:00:30.100
10102	Act4	12	2011/10/24 08:12:06.123
10102	Act6	11	2011/11/17 08:14:21.708

In PREMIERE, the labeled dataset includes all the prefixes of all the traces in the input event log, labelled with their next-activity. Prefix traces are first mapped onto a vector of characteristics extracted over multiple process perspectives (see Sect. 3.1). These data are then encoded in the RGB-like form (see Sect. 3.2). A 2D CNN architecture with inception blocks is finally trained on the collected prefix trace images (as they are labeled with the next activity), in order to learn an image classification model (see details in Sect. 3.3). We note that the learned image classifier can be used for predicting the next activity of any running trace based on the trace RGB image-like representation.

3.1 Feature Extraction

The feature extraction stage is based on the work described in [4], where various feature engineers have been introduced, in order to build features describing the activities, control-flows, resources and performances of traces. Authors of [4] describe two feature engineering schema: the former computing aggregate information on the various perspectives of the prefix traces, while the latter modeling the sequence hidden in the prefix traces using a windowing mechanism. Experimented reported in [4] show that the higher accuracy is achieved when the classification model to predict the next activity is learned using the aggregate schema. Although, this conclusion has been drawn by using traditional classification algorithms to train the predictive models, we assume that it can be reasonably applied to predictive modeling with deep learning also.

In particular, the feature engineering schema, that deploys the aggregate operators, includes feature engineers defined to summarize information on the control-flow, activity, resource and performance, respectively.

- The *aggregate control-flow feature engineer* evaluates the presence of direct relationship between two activities within a prefix trace. In particular, for any possible combination of two activities, the number of times these combinations occurs in the prefix trace is calculated. Let us consider the example of trace in Table 1, transitions $Act1 \rightarrow Act2$, $Act2 \rightarrow Act3$, $Act4 \rightarrow Act5$ and $Act4 \rightarrow Act6$ occur once defining one-valued transition features. Transition $Act3 \rightarrow Act4$ occurs twice defining 2-valued transition feature. Remaining potential transitions (e.g. $Act2 \rightarrow Act1$) never observed in the prefix trace are zero valued.

– The *aggregate activity feature engineer* measures the number of occurrences of each activity within a prefix trace. Let us consider the example of trace in Table 1, activities $Act1$, $Act2$, $Act5$ and $Act6$ define one-valued activity features. Activities $Act3$ and $Act4$ define 2-valued activity features. Remaining potential activities, belonging to the activity alphabet of the log (i.e. the set of distinct activities that may be, in principle, executed), but never observed in the prefix trace, are zero valued.

– The *aggregate resource feature engineers* measures the number of times that each resource triggers an activity in the prefix trace. Let us consider the example of trace in Table 1, resource 11 triggers activities of six events, while resource 12 triggers activities of 2 events. As for the activity features, remaining resources, belonging to the resource alphabet of the log (i.e. the set of distinct resources that may, in principle, trigger an activity), but never observed in the prefix trace, are zero valued.

– The *aggregate time feature engineer* is based on the temporal perspective. It constructs five features that represent the length of the prefix trace (i.e. the number of events in the prefix trace, the duration of the prefix trace (i.e. the time difference between the last event in the prefix trace and the first event in the prefix trace), as well as the calculated minimum, maximum, average and median time difference between consecutive events in the prefix trace.

3.2   RGB Encoding

Once the above-described features have been extracted by log pre-processing, we obtain a vector identifying different characteristics for each prefix trace. Inspired by the work of Taejoon Kim et al. [17], we convert the one-dimensional vector of numerical features to a 3-channel (e.g. RGB) image. The basic idea is to map each feature value to a single pixel of the image. To this aim, we consider a RGB-like encoding function with 24 bits for each pixel that is, in this way, assigned to a single feature. Firstly, in order to assign the same weight to the independent features, min-max normalization is applied so that the value v_f of each feature ranges in $[0.0, 1.0]$. Then starting from the value v_f, we calculate a new value x_f for that feature as follows:

$$x_f = v_f \times (2^{24} - 1).$$

The encoded value x_f, that ranges in $[0, 2^{24} - 1]$, can then be considered as the value of a single 24-bit pixel. Having pixels represented with 24 bits, the resulting image can be regarded as a 3-channel image, namely a RGB image. As an example, Fig. 1(a) shows RGB images derived by applying the encoding

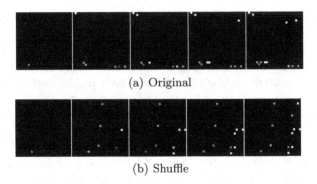

(a) Original

(b) Shuffle

Fig. 1. Example of RGB images extracted from consecutive prefix traces of one trace stored in the Receipt Phase Log (see the log description reported in Sect. 4.1): Original in (a) and Shuffle (b)

scheme to consecutive prefix traces extracted from a trace. We note that richer information is plotted in the image as new events are enclosed in the prefix trace.

It could be argued that the RGB values come from an arbitrary encoding of the activities as numbers. Actually, changing the order of the features in the encoding would lead to different RGB values, as well as to a different distribution of coloured pixels in the image, as it can be seen from Fig. 1(b). However, a shuffling of features influences all resulting images in the same way, i.e. it produces a transformation that is applied to all training images simultaneously, thus the predictive accuracy of PREMIERE does not depend on the order of features. This is confirmed by experimental results shown in Sect. 4.3.

3.3 Neural Network Architecture

The RGB images extracted from traces of an event log are used as training set to learn a classification model for next activity prediction. To learn the model we use a deep neural network architecture based on the Inception module. The concept of Inception module comes from the GoogLeNet architecture that was proposed in [29] to reduce the overall number of network parameters, thus reducing overfitting and requiring less computational resources for the network training in comparison to traditional Convolutional Neural Networks. This is obtained by replacing the fully connected layers of traditional CNNs by sparse layers. An Inception architecture is then obtained by repeating spatially the same convolutional building block, called Inception module. The basic structure of an Inception module combines one max pooling layer with three parallel 2D convolutional layers with different kernel size that simultaneously apply to the same input. The outputs of all layers are concatenated through a concatenation layer.

In this work we consider a very light neural network that takes advantage of the Inception module as building block to achieve computational efficiency enabling the run of the model even with limited computational and memory

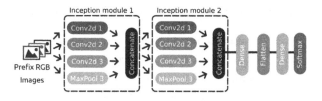

Fig. 2. The adopted deep neural network architecture based on Inception modules.

resources. Using other deep learning architectures typically adopted in computer vision, such as VGG16 [26], ResNet [16] and other standard CNNs [2,19,22,26], would require more computational resources. The proposed architecture leads to better results when compared to such architectures both in terms of computational efficiency and prediction accuracy.

The proposed architecture stacks two Inception modules on top of each other. The outcome from the concatenation layer of the second Inception module is forwarded to a dense layer and then passed to a flatten layer that provides the feature vector to be fed in the final fully connected layer. This final layer of non-linear neurons produces the final classification outcome in form of class probabilities. To compute the output probabilities the Softmax function is used. As activation function we use the ReLU (Rectifier Linear Unit) non-linear function $f(x) = max(0; x)$ because it limits the gradient vanishing downside as its by-product is usually one once x is positive. Figure 2 shows the architecture of the adopted deep neural network. The training of the neural network is accomplished through the Back Propagation learning algorithm.

3.4   Implementation Details

PREMIERE has been implemented in Python 3.6.9 – 64 bit version – using Keras 2.3.1[1] library that is a high-level neural network API using TensorFlow 1.15.0[2] as the back-end. As in [12], the optimization of the hyper-parameters was conducted by considering the 20% of the training set as validation set. In particular, the tree-structured Parzen estimator (TPE) [7] was adopted for hyper-parameter optimization. Table 2 reports the hyper-parameters optimized and the corresponding range of possible values explored with TPE. The Backpropagation training was applied with early stopping to avoid overfitting. Specifically, the training phase was stopped when there is no improvement of the loss on the validation set for 20 consecutive epochs. To minimize the loss function we used the Adam (Adaptive moment estimation) optimizer [18]. The maximum number of epochs was set to 200. The source code of the proposed approach is available on the GitHub[3] repository.

---

[1] https://keras.io/.

[2] https://www.tensorflow.org/.

[3] https://github.com/vinspdb/PREMIERE.

Table 2. Configuration of neural network hyperparameters.

Parameters	Value
Learning rate	{0.00001, 0.01}
Dropout rate	{0.0, 1.0}
Dense layer units	[32, 64, 128]
Batch size	[32, 64, 128]

Table 3. Event log description: number of distinct activities (column 2), number of traces (column 3), number of events (column 4), perspectives available in the event description (column 5).

Event log	#distinct activities	#traces	#events
BPI2012Complete	23	13087	164506
BPI2012WComplete	6	9658	72413
Receipt phase	27	1434	8577
BPI13Problems	7	2306	9011

4   Experiments

To provide a compelling evaluation of the effectiveness of our approach, we have conducted a range of experiments on several event logs. The main objective of these experiments is to investigate the accuracy performance of PREMIERE compared to that of the recent state-of-the-art deep learning approaches that address the task of predicting the next activity of a running trace. A description of the event logs is reported in the followings.

4.1   Event Logs

The experiments have been performed on four real-life event logs, whose characteristics are summarized in Table 3.

BPI Challenge 2012.[4] This log originally comprises events collected by a Dutch financial institute. It pertains to an application process for a personal loan or overdraft within a global financing organization. The activities related to a loan application process are categorized into three sub-processes: activities that track the state of the application (A), activities that track the state of work items associated with the application (W) and activities that track the state of the offer (O). The events of sub-processes A and O contain only the completion life-cycle transition, while sub-process W includes both scheduled started and completed life-cycle transitions. In this work, to enable the comparison with the competitor results, we have considered two event logs extracted from BPI

---

[4] http://www.win.tue.nl/bpi/2012/challenge.

Challenge 2012, that is: BPI2012Complete (that contain all traces, but retain the completed events of such traces) and BPI2012WComplete (that contains traces of sub-process W, but retain the completed events of such traces).

Receipt Phase.[5] This log originates from the CoSeLoG project executed under NWO project number 638.001.211. The CoSeLoG project aimed at investigating (dis)similarities between several processes of different municipalities in the Netherlands. This event log contains the records of the execution of the receiving phase of the building permit application process in an anonymous municipality.

BPI13 Problems.[6] This log contains events collected from problem management system of Volvo IT in Belgium. In particular, dataset BPI13 Problems consists of 2 subsets (Open Problems and Closed Problems). Open Problems' subset contains 819 traces with 2351 events on 5 distinct activities. Closed Problems' subset contains 1487 traces with 6660 events on 7 activities. The two subgroups were merged, in order to obtain a single log containing 9011 events.

4.2   Experimental Setup

For each event log, we have evaluated the accuracy of each compared approach by partitioning the event log in training and testing data. We have learned the predictive model from the training data and measured the accuracy performance of such model when it is used to predict the next activity on unseen testing data.

Competitors. For the comparison, we have considered five, recent, state-of-the-art competitors that use various deep neural network architectures, that is:

– Embedding + 1D CNNs + Inception blocks that are based on the implementation described in [12];
– LSTMs that are described in [30].[7]
– Embedding + RNNs with LSTMs that are described in [13];
– 2D CNNs that are based on the implementation described in [24];
– LSTMs that are described in [9].

As the authors of the competitors made the code available, we have run all the compared algorithms in the same experimental setting to perform a safe comparison. Parameters of the competitors are set according to the code provided by the authors.

Experimental Setting. We have performed a 3-fold cross validation of each event log. For each trial, we have trained the predictive model on the training set (two folds) and evaluated the accuracy of the model on the testing set (the hold-out fold). We analyze the mean performance achieved across all three testing trials.

---

[5] https://data.4tu.nl/repository/uuid:a07386a5-7be3-4367-9535-70bc9e77dbe6.

[6] https://www.win.tue.nl/bpi/doku.php?id=2013:challenge.

[7] For this deep neural network we have used the implementation provided in [12];.

Evaluation Metrics. To evaluate the effectiveness of the proposed approach and compare it to the competitors, we used standard machine learning metrics that are commonly used to evaluate the predictive ability of classification models. Specifically, we computed the following metrics to evaluate predictions on testing traces:

- *Accuracy* that is defined as the proportion of correctly predicted instances of all testing instances, that is, $ACC = \dfrac{\sum\limits_{i=1}^{k} n_i tp_i}{n}$, where tp_i is the number of events with activity i that are correctly predicted as executing activity i; n_i is the true size of activity i (i.e. the number of events with activity i); $n = \sum\limits_{i} n_i$ is the total size of predictions; k is the number of distinct activity types.

- *Precision* that determines how many activities are correctly classified for a particular activity type, given all predictions of that activity, that is, $PREC = \dfrac{\sum\limits_{i=1}^{k} n_i \dfrac{tp_i}{tp_i + fp_i}}{n}$, where fp_i is the number of events not executing activity i that are wrongly predicted as executing activity i;

- *REC* that determines how many activities are correctly predicted for a particular activity type given all occurrences of that activity type, that is, $R = \dfrac{\sum\limits_{i=1}^{k} n_i \dfrac{tp_i}{tp_i + fn_i}}{n}$, where fn_i that is the number of events executing activity i that are wrongly predicted as not executing activity i.

- *F-measure* that is the harmonic weighted mean of precision and recall, that is, $F = \dfrac{\sum\limits_{i=1}^{k} n_i 2 \dfrac{P_i \times R_i}{P_i + R_i}}{n}$. The higher the F-measure, the better the balance between precision and recall achieved by the evaluated approach.

4.3   Results and Discussion

We start analysing the effectiveness of the inception blocks in the proposed architecture. To this aim, we compare the accuracy of PREMIERE to that of its baseline, denoted as PREMIERE-NoInception, which is derived by removing the inception blocks from the adopted deep learning architecture. The results reported in Table 4 show that inception blocks actually allow us to gain in accuracy in all the considered datasets.

We proceed by comparing the performance of PREMIERE to that of competitors. The results reported in Table 5 highlight that PREMIERE outperforms its competitors along the considered accuracy metrics. As the competitors experiment various deep learning architectures (e.g. 1D CNNs with inception blocks in [12], LSTMs in [30] and [9], RNNs with LSTMs in [13] and 2D CNNs in [24]),

Table 4. Accuracy, precision, recall and F-score (mean ± standard deviation on 3 folds) for both PREMIERE (Inc) and its baseline without Inception – PREMIERENoInc – on all the event logs. The best results are in bold.

LOG	Approach	ACC	PREC	REC	F
Receipt Phase	Inc	**0.848 ± 0.021**	**0.829 ± 0.031**	**0.848 ± 0.021**	**0.828 ± 0.027**
	NoInc	0.785 ± 0.091	0.794 ± 0.052	0.785 ± 0.091	0.771 ± 0.096
BPI12WComp	Inc	**0.822 ± 0.013**	**0.831 ± 0.016**	**0.822 ± 0.013**	**0.815 ± 0.012**
	NoInc	0.786 ± 0.032	0.789 ± 0.036	0.786 ± 0.032	0.779 ± 0.031
BPI12Comp	Inc	**0.789 ± 0.001**	**0.799 ± 0.010**	**0.789 ± 0.001**	**0.744 ± 0.003**
	NoInc	0.763 ± 0.008	0.752 ± 0.013	0.763 ± 0.008	0.729 ± 0.003
BPI13Prob	Inc	**0.595 ± 0.006**	**0.591 ± 0.016**	**0.595 ± 0.006**	**0.579 ± 0.006**
	NoInc	0.554 ± 0.013	0.561 ± 0.027	0.554 ± 0.013	0.534 ± 0.005

this analysis allows us to confirm that the combination of 2D convolutions and inception blocks allow us to design a more effective deep learning architecture for this task.

In addition, we point out that we are able of using 2D convolutions in PRE-MIERE thank to the idea of transforming traces in RGB images. 2D convolutions have been already adopted in the architecture defined in [24], where they have been used without the RGB information introduced in this paper and without the inception blocks. By considering that PREMIERE systematically outperforms [24], the collected results provide the empirical evidence of the stronger effectiveness of the imaging trace representation adopted here. To further support this conclusion, we focus the attention on the results achieved with PREMIERENoInc reported in Table 4. We note that, the 2D CNN architecture of PREMIERE outperforms the 2D CNN of [24] even when trained without inception blocks.

In general, the algorithm described in [12] is the runner-up of this comparative analysis except for dataset BPI13Prob, where the runner-up is the algorithm presented in [13]. To explain this runner-up, we recall that both PREMIERE and [12] use a deep neural network architecture with convolutions and inception blocks. So, from this point of view, our analysis also assesses the effectiveness of using convolutions and inception blocks in deep learning for next activity prediction. On the other hand, PREMIERE processes a 2D representation of the trace data, while [12] considers a 1D representation of the same data. Therefore, we can again conclude that the accuracy that PREMIERE commonly gains on [12] can be due to the imagery encoding representation introduced in PREMIERE. In general, PREMIERE achieves an increase of the accuracy that is of 1.65% with respect to its runner-up in Receipt Phase, 4.13 % in BPI12WComp, 0.12% in BPI12Comp and 2.18% in BPI13Prob, respectively.

Table 5. Accuracy, precision, recall and F-score (mean ± standard deviation on 3 folds) for both PREMIERE and its competitors [9, 12, 13, 24, 30] on all the event logs. The best results are in bold.

LOG	Approach	ACC	PREC	REC	F
Receipt Phase	PREMIERE	**0.848 ± 0.021**	0.829 ± 0.031	**0.848 ± 0.021**	**0.828 ± 0.027**
	[12]	0.834 ± 0.010	0.791 ± 0.026	0.834 ± 0.010	0.803 ± 0.015
	[30]	0.830 ± 0.051	0.805 ± 0.065	0.830 ± 0.051	0.805 ± 0.062
	[13]	0.812 ± 0.009	0.793 ± 0.014	0.812 ± 0.009	0.798 ± 0.012
	[24]	0.782 ± 0.082	0.796 ± 0.038	0.782 ± 0.082	0.757 ± 0.095
	[9]	0.841 ± 0.013	**0.836 ± 0.025**	0.841 ± 0.013	0.825 ± 0.019
BPI12WComp	PREMIERE	**0.822 ± 0.013**	**0.831 ± 0.016**	**0.822 ± 0.013**	**0.815 ± 0.012**
	[12]	0.788 ± 0.006	0.773 ± 0.006	0.788 ± 0.006	0.766 ± 0.004
	[30]	0.787±0.005	0.778±0.006	0.787±0.005	0.767±0.007
	[13]	0.711 ± 0.005	0.676 ± 0.003	0.711 ± 0.005	0.676 ± 0.007
	[24]	0.776 ± 0.009	0.759 ± 0.005	0.776 ± 0.009	0.748 ± 0.006
	[9]	0.780 ± 0.009	0.757 ± 0.009	0.780 ± 0.009	0.748 ± 0.008
BPI12Comp	PREMIERE	**0.789 ± 0.001**	**0.799 ± 0.010**	**0.789 ± 0.001**	0.744 ± 0.003
	[12]	0.788 ± 0.002	0.782 ± 0.017	0.788 ± 0.002	**0.746 ± 0.002**
	[30]	0.759 ± 0.038	0.710 ± 0.061	0.759 ± 0.038	0.706 ± 0.053
	[13]	0.720 ± 0.002	0.687 ± 0.003	0.720 ± 0.002	0.672 ± 0.003
	[24]	0.763 ± 0.007	0.758 ± 0.011	0.763 ± 0.007	0.727 ± 0.002
	[9]	0.768 ± 0.002	0.771 ± 0.020	0.768 ± 0.002	0.718 ± 0.002
BPI13Prob	PREMIERE	**0.595 ± 0.006**	**0.591 ± 0.016**	**0.595 ± 0.006**	**0.579 ± 0.006**
	[12]	0.529 ± 0.013	0.520 ± 0.005	0.529 ± 0.013	0.478 ± 0.008
	[30]	0.530 ± 0.013	0.520 ± 0.006	0.530 ± 0.013	0.478 ± 0.009
	[13]	0.582 ± 0.004	0.550 ± 0.004	0.582 ± 0.004	0.524 ± 0.004
	[24]	0.502 ± 0.005	0.471 ± 0.008	0.502 ± 0.005	0.439 ± 0.008
	[9]	0.505 ± 0.007	0.455 ± 0.545	0.505 ± 0.007	0.435 ± 0.017

At the completion of this study, we explored the sensitivity of the performance of PREMIERE to the order according to the aggregate features have been processed to be transformed into imagery RGB pixels. To this aim, we considered the dataset Receipt Phase and run PREMIERE on ten trials that have been defined by shuffling the vector of the aggregate features. The results are reported in Table 6. It can be seen that accuracy metrics do not change significantly along the various trials. This confirms that the adopted imaging encoding strategy (that considers the features in the same order on all the traces) produces performances, which are stable with respect to the features' order.

Table 6. Mean and standard deviation of accuracy, precision, recall, F-score of PRE-MIERE measured on 10 trials of the 3-fold CV on Receipt Phase.

N. Shuffle	ACC	PREC	REC	F
Shuffle 1	0.845 ± 0.019	0.831 ± 0.024	0.845 ± 0.019	0.826 ± 0.024
Shuffle 2	0.849 ± 0.022	0.827 ± 0.035	0.849 ± 0.022	0.831 ± 0.027
Shuffle 3	0.847 ± 0.018	0.834 ± 0.021	0.847 ± 0.018	0.827 ± 0.022
Shuffle 4	0.844 ± 0.021	0.828 ± 0.031	0.844 ± 0.021	0.828 ± 0.026
Shuffle 5	0.845 ± 0.018	0.827 ± 0.025	0.845 ± 0.018	0.827 ± 0.025
Shuffle 6	0.850 ± 0.020	0.831 ± 0.028	0.850 ± 0.020	0.833 ± 0.025
Shuffle 7	0.847 ± 0.022	0.829 ± 0.032	0.847 ± 0.022	0.828 ± 0.022
Shuffle 8	0.847 ± 0.023	0.830 ± 0.032	0.847 ± 0.023	0.829 ± 0.023
Shuffle 9	0.848 ± 0.018	0.829 ± 0.031	0.848 ± 0.018	0.832 ± 0.018
Shuffle 10	0.846 ± 0.018	0.828 ± 0.032	0.846 ± 0.018	0.830 ± 0.015

5 Conclusion

In this paper we have proposed PREMIERE, a novel deep learning approach for next activity prediction, which tries to transfer achievements of computer vision to process mining. In particular, we take advantage of the power of 2D CNN architectures, widely used in computer vision tasks, by converting the problem of next activity prediction in a task of image classification. To this aim, we adopt an encoding scheme that transforms characteristics of event data into RGB images. This RGB encoding function can be applied to any characteristic engineered from trace data. Once the trace data are transformed in RGB images, they become the natural input of a 2D CNN, that is trained for image classification and used to predict the next activity of a running trace. The Inception is used to reinforce the accuracy of the architecture. Experimental results on several real-world event logs confirm the effectiveness of PREMIERE in terms of gain of accuracy achieved on various deep learning competitors using LSTMS, RNNs with LSTMs, 1D CNNs and 2D CNNs.

As future work, we plan to experiment standard 2D CNN architectures (e.g. ResNet) in the considered task. In addition, we intend to extend this research by accounting for knowledge hidden in how the traces evolve as new events are logged. By extending the proposed holistic combination of computer vision and process mining, we plan to model the trace development as a video and investigate the performance of 3D-CNN architectures in the prediction of the next activity based on a video representation of the trace. In addition, as stakeholders may be interested to know what is the eventual, final outcome of the process, in addition to just one step ahead, we plan to extend the investigation of the proposed computer vision approach to a larger list Key Performance Indicators (KPIs) of interest, such as remaining time, total costs, final customer (un)satisfaction. Finally, we plan to experiment the effectiveness of the proposed approach in a streaming setting with the training performed on event collected until the current time point and testing performed on future events of running traces. This will require the definition of a streaming framework to update the predictive model as new events are collected and deal with concept drift happening as new events (unobserved before) appear in the process.

Acknowledgement. Authors wish to thank unknown reviewers for the useful suggestions provided to improve the final quality of the paper. The research of Vincenzo Pasquadibisceglie is funded by PON RI 2014-2020 - Big Data Analytics for Process Improvement in Organizational Development - CUP H94F18000260006. The work is partially supported by the POR Puglia FESR-FSE 2014-2020 - Asse prioritario 1 - Ricerca, sviluppo tecnologico, innovazione - Sub Azione 1.4.b bando Innolabs - Research project KOMETA (Knowledge Community for Efficient Training through Virtual Technologies), funded by Regione Puglia.

References

1. van der Aalst, W.M.P.: Process Mining - Data Science in Action, 2nd edn. Springer, Heidelberg (2016). https://doi.org/10.1007/978-3-662-49851-4
2. Abdel-Hamid, O., Mohamed, A., Jiang, H., Deng, L., Penn, G., Yu, D.: Convolutional neural networks for speech recognition. IEEE/ACM Trans. Audio, Speech Lang. Process. **22**(10), 1533–1545 (2014)
3. Andresini, G., Appice, A., Di Mauro, N., Loglisci, C., Malerba, D.: Exploiting the auto-encoder residual error for intrusion detection. In: 2019 IEEE European Symposium on Security and Privacy Workshops (EuroS PW), pp. 281–290 (2019)
4. Appice, A., Di Mauro, N., Malerba, D.: Leveraging shallow machine learning to predict business process behavior. In: 2019 IEEE International Conference on Services Computing (SCC), pp. 184–188 (2019)
5. Appice, A., Malerba, D.: A co-training strategy for multiple view clustering in process mining. IEEE Trans. Serv. Comput. **9**(6), 832–845 (2016)
6. Appice, A., Malerba, D., Morreale, V., Vella, G.: Business event forecasting. In: Spender, J., Schiuma, G., Albino, V. (eds.) 10th International Forum on Knowledge Asset Dynamics, IFKAD 2015, pp. 1442–1453. Inst Knowledge Asset Management (2015)
7. Bergstra, J., Bardenet, R., Bengio, Y., Kégl, B.: Algorithms for hyper-parameter optimization. In: Proceedings of the 24th International Conference on Neural Information Processing Systems, NIPS 2011, pp. 2546–2554. Curran Associates Inc., USA (2011)
8. Breuker, D., Matzner, M., Delfmann, P., Becker, J.: Comprehensible predictive models for business processes. J. MIS Q. **40**, 1009–1034 (2016)
9. Camargo, M., Dumas, M., González-Rojas, O.: Learning accurate LSTM models of business processes. In: Hildebrandt, T., van Dongen, B.F., Röglinger, M., Mendling, J. (eds.) BPM 2019. LNCS, vol. 11675, pp. 286–302. Springer, Cham (2019). https://doi.org/10.1007/978-3-030-26619-6_19
10. Di Francescomarino, C., Ghidini, C., Maggi, F.M., Milani, F.: Predictive process monitoring methods: which one suits me best? In: Weske, M., Montali, M., Weber, I., vom Brocke, J. (eds.) BPM 2018. LNCS, vol. 11080, pp. 462–479. Springer, Cham (2018). https://doi.org/10.1007/978-3-319-98648-7_27

11. Di Francescomarino, C., Ghidini, C., Maggi, F.M., Petrucci, G., Yeshchenko, A.: An eye into the future: leveraging a-priori knowledge in predictive business process monitoring. In: Carmona, J., Engels, G., Kumar, A. (eds.) BPM 2017. LNCS, vol. 10445, pp. 252–268. Springer, Cham (2017). https://doi.org/10.1007/978-3-319-65000-5_15

12. Di Mauro, N., Appice, A., Basile, T.M.A.: Activity prediction of business process instances with inception CNN models. In: Alviano, M., Greco, G., Scarcello, F. (eds.) AI*IA 2019. LNCS (LNAI), vol. 11946, pp. 348–361. Springer, Cham (2019). https://doi.org/10.1007/978-3-030-35166-3_25

13. Evermann, J., Rehse, J.R., Fettke, P.: Predicting process behaviour using deep learning. Decis. Support Syst. **100**, 129–140 (2017)

14. Di Francescomarino, C., Dumas, M., Federici, M., Ghidini, C., Maggi, F.M., Rizzi, W.: Predictive business process monitoring framework with hyperparameter optimization. In: Nurcan, S., Soffer, P., Bajec, M., Eder, J. (eds.) CAiSE 2016. LNCS, vol. 9694, pp. 361–376. Springer, Cham (2016). https://doi.org/10.1007/978-3-319-39696-5_22

15. Goodfellow, I., Bengio, Y., Courville, A.: Deep Learning. MIT Press, Cambridge (2016)

16. He, K., Zhang, X., Ren, S., Sun, J.: Deep residual learning for image recognition. In: 2016 IEEE Conference on Computer Vision and Pattern Recognition (CVPR) (2016)

17. Kim, T., Suh, S.C., Kim, H., Kim, J., Kim, J.: An encoding technique for CNN-based network anomaly detection. In: 2018 IEEE International Conference on Big Data (Big Data), pp. 2960–2965, December 2018

18. Kingma, D.P., Ba, J.: Adam: a method for stochastic optimization (2014)

19. Krizhevsky, A., Sutskever, I., Hinton, G.E.: Imagenet classification with deep convolutional neural networks. In: Pereira, F., Burges, C.J.C., Bottou, L., Weinberger, K.Q. (eds.) Advances in Neural Information Processing Systems, vol. 25, pp. 1097–1105. Curran Associates, Inc. (2012)

20. Krizhevsky, A., Sutskever, I., Hinton, G.E.: Imagenet classification with deep convolutional neural networks. CACM **60**(6), 84–90 (2017)

21. Li, Z., Qin, Z., Huang, K., Yang, X., Ye, S.: Intrusion detection using convolutional neural networks for representation learning. In: Liu, D., Xie, S., Li, Y., Zhao, D., El-Alfy, E.-S.M. (eds.) ICONIP 2017. LNCS, vol. 10638, pp. 858–866. Springer, Cham (2017). https://doi.org/10.1007/978-3-319-70139-4_87

22. Long, J., Shelhamer, E., Darrell, T.: Fully convolutional networks for semantic segmentation. In: Proceedings of the IEEE Conference on Computer Vision and Pattern Recognition, pp. 3431–3440 (2015)

23. Márquez-Chamorro, A.E., Resinas, M., Ruiz-Cortés, A.: Predictive monitoring of business processes: a survey. IEEE Trans. Serv. Comp. **11**(6), 962–977 (2018)

24. Pasquadibisceglie, V., Appice, A., Castellano, G., Malerba, D.: Using convolutional neural networks for predictive process analytics. In: 2019 International Conference on Process Mining (ICPM), pp. 129–136 (2019)

25. Pravilovic, S., Appice, A., Malerba, D.: Process mining to forecast the future of running cases. In: Appice, A., Ceci, M., Loglisci, C., Manco, G., Masciari, E., Ras, Z.W. (eds.) NFMCP 2013. LNCS (LNAI), vol. 8399, pp. 67–81. Springer, Cham (2014). https://doi.org/10.1007/978-3-319-08407-7_5

26. Simonyan, K., Zisserman, A.: Very deep convolutional networks for large-scale image recognition. arXiv:1409.1556 (2014)

27. Song, M., van der Aalst, W.M.: Towards comprehensive support for organizational mining. Decis. Support Syst. **46**(1), 300–317 (2008)

28. Song, M., Günther, C.W., van der Aalst, W.M.P.: Trace clustering in process mining. In: Ardagna, D., Mecella, M., Yang, J. (eds.) BPM 2008. LNBIP, vol. 17, pp. 109–120. Springer, Heidelberg (2009). https://doi.org/10.1007/978-3-642-00328-8_11

29. Szegedy, C., et al.: Going deeper with convolutions. In: The IEEE Conference on Computer Vision and Pattern Recognition (CVPR), June 2015

30. Tax, N., Teinemaa, I., van Zelst, S.J.: An interdisciplinary comparison of sequence modeling methods for next-element prediction. Softw. Syst. Model. 1–21 (2020). https://doi.org/10.1007/s10270-020-00789-3

31. Tax, N., Verenich, I., La Rosa, M., Dumas, M.: Predictive business process monitoring with LSTM neural networks. In: Dubois, E., Pohl, K. (eds.) CAiSE 2017. LNCS, vol. 10253, pp. 477–492. Springer, Cham (2017). https://doi.org/10.1007/978-3-319-59536-8_30

32. Teinemaa, I., Dumas, M., Rosa, M.L., Maggi, F.M.: Outcome-oriented predictive process monitoring: review and benchmark. TKDD **13**(2), 17:1–17:57 (2019)

Prescriptive Business Process Monitoring for Recommending Next Best Actions

Sven Weinzierl[(✉)] ⓘ, Sebastian Dunzer[(✉)] ⓘ, Sandra Zilker[(✉)] ⓘ,
and Martin Matzner[(✉)] ⓘ

Friedrich-Alexander-Universität Erlangen-Nürnberg,
Fürther Straße 248, Nürnberg , Germany
{sven.weinzierl,sebastian.dunzer,sandra.zilker,martin.matzner}@fau.de

Abstract. Predictive business process monitoring (PBPM) techniques predict future process behaviour based on historical event log data to improve operational business processes. Concerning the next activity prediction, recent PBPM techniques use state-of-the-art deep neural networks (DNNs) to learn predictive models for producing more accurate predictions in running process instances. Even though organisations measure process performance by key performance indicators (KPIs), the DNN's learning procedure is not directly affected by them. Therefore, the resulting next most likely activity predictions can be less beneficial in practice. Prescriptive business process monitoring (PrBPM) approaches assess predictions regarding their impact on the process performance (typically measured by KPIs) to prevent undesired process activities by raising alarms or recommending actions. However, none of these approaches recommends actual process activities as actions that are optimised according to a given KPI. We present a PrBPM technique that transforms the next most likely activities into the next best actions regarding a given KPI. Thereby, our technique uses business process simulation to ensure the control-flow conformance of the recommended actions. Based on our evaluation with two real-life event logs, we show that our technique's next best actions can outperform next activity predictions regarding the optimisation of a KPI and the distance from the actual process instances.

Keywords: Prescriptive business process monitoring · Predictive business process monitoring · Business process management

1 Introduction

Predictive business process monitoring (PBPM) techniques predict future process behaviour to improve operational business processes [9]. A PBPM technique constructs predictive models from historical event log data [10] to tackle different prediction tasks like predicting next activities, process outcomes or remaining time [4]. Concerning the next activity prediction, recent PBPM techniques use

© Springer Nature Switzerland AG 2020
D. Fahland et al. (Eds.): BPM Forum 2020, LNBIP 392, pp. 193–209, 2020.
https://doi.org/10.1007/978-3-030-58638-6_12

state-of-the-art deep neural networks (DNNs) to learn predictive models for producing more accurate predictions in running process instances [25]. DNNs belong to the class of deep-learning (DL) algorithms. DL is a subarea of machine learning (ML) that identifies intricate structures in high-dimensional data through multi-representation learning [8]. After learning, models can predict the next most likely activity of running process instances.

However, providing the next most likely activity does not necessarily support process stakeholders in process executions [21]. Organisations measure the performance of processes through key performance indicators (KPIs) in regard to three dimensions: time, cost and quality [23]. Recent PBPM techniques rely on state-of-the-art DNNs that can only learn predictive models from event log data. Even though an event log can include KPI information, it does not directly affect such an algorithm's learning procedure unless a KPI is the (single) learning target itself. As a consequence, the learned models can output next activity predictions, which are less beneficial for process stakeholders.

Some works tackled this problem with prescriptive business process monitoring (PrBPM) approaches. PrBPM approaches assess predictions regarding their impact on the process performance – typically measured by KPIs – to prevent undesired activities [21]. To achieve that, existing approaches generate alarms [5,11,13,21] or recommend actions [3,6].

However, none of these approaches recommends next best actions in the form of process activities that are optimised regarding a given KPI for running processes. In our case, *best* refers to a KPI's optimal value regarding the future course of a process instance. Additionally, the next best actions, which depend on next activity predictions and prediction of a particular KPI, might obscure the actual business process. Therefore, transforming methods should check whether a recommended action is conform regarding a process description.

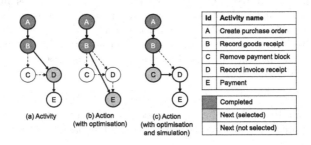

Fig. 1. A next activity prediction vs. a next best action recommendation.

Given a running process instance of a purchase order handling process after finishing the first two activities, a DNN model predicts the next most likely activity D (cf. (a) in Fig. 1). Additionally, the KPI *time* is of interest and the first two activities A and B take each 1 h, the predicted activity D takes 2 h and the last activity E takes 2 h. In sum, the complete process instance takes 6 h and a deadline of 5 h – that exists due to a general agreement – is exceeded.

In contrast, the recommended action with optimisation can be the activity E (cf. (b) in Fig. 1). Even though the complete process instance takes 4 h, the mandatory activity D is skipped. With an additional simulation, the activity "Remove payment block" can be recommended (cf. (c) in Fig. 1) taking 1 h. Afterwards, the activities D and E are followed with a duration of 2 h and 1 h. Here, the activity E takes 1 h instead of 2 h since the payment block is already removed. Thus, the complete process instance takes 5 h, and the deadline is met.

In this paper, we provide a PrBPM technique for recommending the next best actions depending on a KPI. Thereby, it conducts a business process simulation (BPS) to remain within the allowed control-flow. To reach our research objective, we develop a PrBPM technique and evaluate its actions regarding the optimisation of a KPI and the distance from ground truth process instances with two real-life event logs.

This paper is an extended and revised version of a research-and-progress paper [26]. Additionally, it includes a BPS and an evaluation with two real-life logs. The paper is structured as follows: Sect. 2 presents the required background for our PrBPM technique. Section 3 introduces the design of our PrBPM technique for recommending next best actions. Further, we evaluate our technique in Sect. 4. Section 5 provides a discussion. The paper concludes with a summary and an outlook on future work in Sect. 7.

2   Background

2.1   Preliminaries

PBPM or PrBPM techniques require event log data. We adapt definitions by Polato et al. [15] to formally describe the terms *event, trace, event log, prefix* and *suffix*. In the following, $\mathcal{A}$ is the set of process activities, $\mathcal{C}$ is the set of process instances (cases), and $\mathbb{C}$ is the set of case ids with the bijective projection $id : \mathbb{C} \rightarrow \mathcal{C}$, and $\mathcal{T}$ is the set of timestamps. To address time, a process instance $c \in \mathcal{C}$ contains all past and future events, while events in a trace σ_c of c contain all events up to the currently available time instant. $\mathcal{E} = \mathcal{A} \times \mathbb{C} \times \mathcal{T}$ is the event universe.

Definition 1 (Event). *An event $e \in \mathcal{E}$ is a tuple $e = (a, c, t)$, where $a \in \mathcal{A}$ is the process activity, $c \in \mathbb{C}$ is the case id, and $t \in \mathcal{T}$ is its timestamp. Given an event e, we define the projection functions $F_p = \{f_a, f_c, f_t\}$: $f_a : e \rightarrow a$, $f_c : e \rightarrow c$, and $f_t : e \rightarrow t$.*

Definition 2 (Trace). *A trace is a sequence $\sigma_c = \langle e_1, \dots, e_{|\sigma_c|} \rangle \in \mathcal{E}^*$ of events, such that $f_c(e_i) = f_c(e_j) \wedge f_t(e_i) \leq f_t(e_j)$ for $1 \leq i < j \leq |\sigma_c|$. Note a trace σ_c of process instance c can be considered as a process instance σ_c.*

Definition 3 (Event log). *An event log $\mathcal{L}_\tau$ for a time instant τ is a set of traces, such that $\forall \sigma_c \in \mathcal{L}_\tau . \exists c \in \mathcal{C} . (\forall e \in \sigma_c . id(f_c(e)) = c) \wedge (\forall e \in \sigma_c . f_t(e) \leq \tau)$, i. e. all events of the observed cases that already happened.*

Definition 4 (Prefix, suffix of a trace). *Given a trace* $\sigma_c = \langle e_1, .., e_k, .., e_n \rangle$, *the prefix of length* k *is* $hd^k(\sigma_c) = \langle e_1, .., e_k \rangle$, *and the suffix of length* k *is* $tl^k(\sigma_c) = \langle e_{k+1}, .., e_n \rangle$, *with* $1 \leq k < n$.

2.2 Long Short-Term Memory Neural Networks

Our PrBPM technique transforms next activity predictions into the next best actions. Thus, next activity predictions are the basis for our PrBPM technique. To predict next activities, we use a "vanilla", i.e. basic, long short-term memory network (LSTM) [7] because most of the PBPM techniques for predicting next activities rely on this DNN architecture [24]. LSTMs belong to the class of recurrent neural networks (RNNs) [8] and are designed to handle temporal dependencies in sequential prediction problems [1]. In general, an LSTM consists of three layers: an input layer (receiving data input), a hidden layer (i.e. an LSTM layer with an LSTM cell) and an output layer (providing predictions).

An LSTM cell uses four gates to manage its memory over time to avoid the problem of gradient exploding/vanishing in the case of longer sequences [1]. First, a forget gate that determines how much of the previous memory is kept. Second, an input gate controls how much new information is stored into memory. Third, a gate gate or candidate memory that defines how much information is stored into memory. Fourth, an output gate that determines how much information is read out of the memory.

To learn an LSTM's parameters, a loss function (e.g. the cross-entropy loss for classification) is defined on a data point (i.e. prediction and label) and measures the penalty. Additionally, a cost function in its basic form calculates the sum of loss functions over the training set. The LSTM's parameters are updated iteratively via a gradient descent algorithm (e.g. stochastic gradient descent), in that, the gradient of the cost function is computed by backpropagation through time [19]. After learning the parameters, an LSTM model with adjusted parameter values exists.

2.3 Business Process Simulation

Actions optimised according to a KPI can be not conform to the process control-flow. Thus, suggesting process-conform actions to process stakeholders is essential. Consequently, we add control-flow knowledge to our PrBPM technique with formal process models.

A well-known approach to assess the quality of process executions is business process simulation (BPS). Several approaches examine processes and their variants regarding compliance or performance with BPS [2,16]. We refer to discrete-event-driven BPS [22]. Here, simulation models formally contain discrete events which are interrelated via process semantics.

BPS usually delivers its insights to users [17]. Unlike existing approaches, such as [18,27], we use the simulation results to process the predictions of an LSTM. Thus, we use discrete-event-driven [22] short-term simulation [18] as a boundary measure to ensure that the DNN-based next best action makes sense from a control-flow perspective. Alike Rozinat et al. [18], our simulation starts from a non-empty process state to aid in recommending the next best action from the current state on.

3   A PrBPM Technique for Recommending Next Best Actions

Our PrBPM technique transforms next activity predictions into the next best actions. The technique consists of an *offline* and an *online component*. In the offline component, it learns a DNN for predicting next activities and values of a KPI. In the online component, the next best actions are recommended based on the next activity and KPI value predictions.

3.1   Offline Component

The offline component receives as input an event log $\mathcal{L}_\tau$, and outputs the two ML models m_{pp} and m_{cs}. While m_{pp} (process prediction model) predicts next activities and a KPI value related to next activities, m_{cs} (candidate selection model) selects a fix set of suffix candidates. The technique learns both models from individually pre-processed versions of $\mathcal{L}_\tau$. Figure 2 visualises the steps of the offline component. In the following, we describe the steps of the offline component based on the exemplary finished process instance σ_1^f, as represented in (1). The last attribute per event is the KPI; here the defined costs for executing an activity.

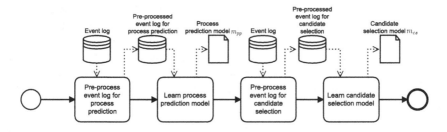

Fig. 2. Four-step offline component scheme with the two models m_{pp} and m_{cs}.

$$\begin{aligned}
\sigma_1^f = \langle &(1, \text{``Create Application''}, 2011\text{-}09\text{-}30\ 16\text{:}20\text{:}00, 0), \\
&(1, \text{``Concept''}, 2011\text{-}09\text{-}30\ 17\text{:}30\text{:}00, 10), \\
&(1, \text{``Accepted''}, 2011\text{-}09\text{-}30\ 18\text{:}50\text{:}00, 20), \\
&(1, \text{``Validating''}, 2011\text{-}09\text{-}30\ 19\text{:}10\text{:}00, 40)\rangle.
\end{aligned} \tag{1}$$

Pre-process Event Log for Process Prediction. m_{pp} is a multi-task DNN for predicting next activities and a KPI value at each time step of a running process instance. For m_{pp}, the pre-processing of $\mathcal{L}_\tau$ comprises four steps. First, to determine the end of each process instance in $\mathcal{L}_\tau$, it adds a termination event to the end of each process instance. So, for σ_1^f, as represented in (1), we add the event (1, "End", 2011-09-30 19:10:00, 0) after the fourth event with the activity name "Validating". Additionally, for termination events, we always overtake the timestamp value of the previous event and set the value of the KPI to 0. Second, it onehot-encodes all activity names in the process instances as numeric values (cf. (2) for σ_1^f including the termination event's activity)[1].

$$\sigma_1^f = \langle (0,0,0,0,1),(0,0,0,1,0),(\dots),(0,1,0,0,0),(1,0,0,0,0) \rangle. \qquad (2)$$

This step is necessary since LSTMs, as used in this paper, use a gradient descent optimisation algorithm to learn the network's parameters. Third, it crops prefixes out of process instances by using the function $hd^k()$. For instance, a prefix with size three of σ_1^f is:

$$hk^3(\sigma_1^f) = \langle (0,\ 0,\ 0,\ 0,\ 1),(0,0,0,1,0),(0,0,1,0,0) \rangle. \qquad (3)$$

Lastly, it transforms the cropped input data into a three-order tensor (prefixes, time steps and attributes). Additionally, m_{pp} needs two label structures for parameter learning. First, for the onehot-encoded next activities, a two-dimensional label matrix is required. Second, if the KPI values related to the next activities are scaled numerically, a one-dimensional label vector is required. If the values are scaled categorically, a two-dimensional label matrix is needed.

Create Process Prediction Model. m_{pp} is a multi-task DNN. The model's architecture follows the work of Tax et al. [20]. The input layer of m_{pp} receives the data and transfers it to the first hidden layer. The first hidden layer is followed by two branches. Each branch refers to a prediction task and consists of two layers, a hidden layer and an output layer. The output layer of the upper branch realises next activity predictions, whereas the lower creates KPI value predictions. Depending on the KPI value's scaling (i.e. numerical or categorical), the lower branch solves either a regression or classification problem. Each hidden layer of m_{pp} is an LSTM layer with an LSTM cell.

Pre-process Event Log for Candidate Selection. m_{cs} is a nearest-neighbour-based ML algorithm for finding suffixes "similar" to predicted suffixes. For m_{cs}, the pre-processing of $\mathcal{L}_\tau$ consists of three steps. First, it ordinal-encodes all activity names in numerical values. For example, the ordinal-encoded representation of σ_1^f, as depicted in (1) including the termination event's activity, is $\langle (1),(2),(3),(4),(5) \rangle$. Second, it crops suffixes out of process instances through the function $tl^k(\cdot)$. For instance, the suffix with size three of σ_1^f ($lt^3(\sigma_1^f)$) is $\langle (4),(5) \rangle$. Lastly, the cropped input data is transformed into a two-dimensional matrix (suffixes and attributes).

---

[1] We include the temporal control-flow attributes of Tax et al. [20] in the implementation, which are not described in this paper for better understanding.

Create Candidate Selection Model. m_{cs} is a nearest-neighbour-based ML algorithm. It retrieves k suffixes "nearest" to a suffix predicted for a given prefix (i.e. a running process instance at a certain time step). The technique learns the model m_{cs} based on all suffixes cropped out of $\mathcal{L}_\tau$.

3.2   Online Component

The online component receives as input a new process instance, and the two trained predictive models m_{pp} and m_{cs}. It consists of five steps (see Fig. 3), and outputs next best actions. After pre-processing (first step) of the running process instance, a suffix of next activities and its KPI values are predicted (second step) by applying m_{pp}. The second step is followed by the condition, whether the sum of the KPI values of the suffix and the respective prefix exceeds a threshold or not. If the threshold is exceeded, the predicted suffix of activities is transferred from the second to the third step (i.e. find candidates) and the procedure for generating next best actions starts. Otherwise, it provides the next most likely activity. To find a set of suffix candidates, the technique loads m_{cs} from the offline component. Subsequently, it selects the best candidate from this set depending on the KPI and concerning BPS. Finally, the first activity of the selected suffix represents the best action and is concatenated to the prefix of activities (i.e. running process instance at a certain time step). If the best action is the end of the process instance, the procedure ends. Otherwise, the procedure continues and predicts the suffix of the new prefix. In the following, we detail the online component's five steps. Thereby, we refer to the running process instance σ_2^r, for that the second event has just finished.

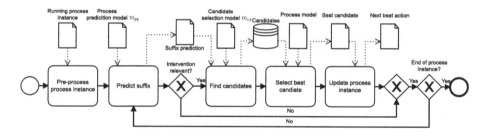

Fig. 3. Five-step online component scheme with the two models m_{pp} and m_{cs}.

$$\sigma_2^r = \langle (2, \text{"Create Application"}, \text{2012-09-30 18:00:00}, 20), \\ (2, \text{"Concept"}, \text{2012-09-30 18:30:00}, 20) \rangle. \tag{4}$$

Pre-process Process Instance. To predict the suffix of the running process instance with m_{pp}, we onehot-encode all activity names in numerical values and transform the output into a third-order tensor.

Predict Suffix. Based on a running process instance (prefix), m_{pp} predicts the next sequence of activities and the KPI values. To get the complete suffix, we apply m_{pp} repeatedly. Afterwards, the technique calculates the sum of the KPI values over the activities of the complete process instance consisting of the prefix and its predicted suffix. For instance, if the prefix of a process instance is σ_2^r, one potential suffix is:

$$s_{\sigma_2^r} = \langle(\text{"Accepted"}, 20),(\text{"Validating"}, 40),(\text{"End"}, 10)\rangle. \tag{5}$$

For a better intuition, we omit the suffixes' encoding in the online component. The values 20, 40 and 10 assigned to the events are KPI values (e.g. cost values) predicted by m_{pp}. In line with Tax et al. [20], we do not perform the suffix prediction for prefixes with size ≤ 1 since the amount of activity values is insufficient. After predicting the suffix, the total costs of σ_2^r are 110. To start the procedure for recommending the next best actions, the total KPI value of an instance has to exceed a threshold value t. The value of t can be defined by domain experts or derived from the event log (e.g. average costs of process instances). Regarding σ_2^r, the procedure starts because we assume $t = 100$ ($110 > t$).

Find Candidates. Second, for the predicted suffix, m_{cs} from the offline component reveals a set of alternatives with a meaningful control-flow.

For example, m_{cs} ($k = 3$) selects based on $s_{\sigma_2^r}$ the following three suffix alternatives:

$$m_{cs}(s_{\sigma_2^r}) = [\langle(\text{"End"}, 10)\rangle,$$
$$\langle(\text{"Accepted"}, 20),(\text{"Validating"}, 40),(\text{"End"}, 10)\rangle,$$
$$\langle(\text{"Validating"}, 20),(\text{"Accepted"}, 10),(\text{"Validating"}, 10),$$
$$(\text{"End"}, 10)\rangle]. \tag{6}$$

In (6), the first and the third suffix result in total costs (50 and 90) falling below t.

Select the Best Candidate. In the third step, we select the next best action from the set of possible suffix candidates. We sort the suffixes by the KPI value. Thus, the first suffix is the best one in regard to the KPI. To incorporate control-flow knowledge, a simulation model checks the resulting instance. Thereby, we reduce the risk of prescribing nonsensical actions. The simulation uses a formal process model to retrieve specific process semantics. The simulation produces the current process state from the prefix and the process model. If the prefix does not comply with the process model, the simulation aborts the suffix selection for the prefix and immediately recommends an intervention. Otherwise, we check the k selected suffixes whether they comply with the process model in the simulation from the current process state on. If a candidate suffix fails the simulation, our technique omits it in the selection. However, when all suffix candidates infringe the simulation, the technique assumes the predicted next activity as the best action candidate. Concerning the candidate set from (6), the best candidate is suffix three since it does not infringe the simulation model.

Update Process Instance. To evaluate our technique, we assume that a process stakeholder performs in each case the recommended action. Thus, if the best suffix candidate exists, the activity (representing the next best action) and the KPI value of the first event are concatenated to the running process instance (i.e. prefix). After the update, σ_2^r comprises three events, as depicted in (7).

$$\begin{aligned} \sigma_2^r = \langle & (2, \text{``Create Application''}, 2012\text{-}09\text{-}30\ 18\text{:}00\text{:}00, 20), \\ & (2, \text{``Concept''}, 2012\text{-}09\text{-}30\ 18\text{:}30\text{:}00, 20), \hspace{2cm} (7) \\ & (2, \textbf{``Validating''}, -, \textbf{20})\rangle. \end{aligned}$$

The technique repeats the complete procedure until the termination event is reached.

4 Evaluation

We provide an evaluation regarding our PrPBM technique's optimisation of a KPI and the distance from ground truth process instances. For that, we developed a prototype that recommends next best actions depending on the KPI *throughput time* and concerning a process simulation realised with DCR graphs. We compare our results to a representative baseline [20] for two event logs.

4.1 Event Logs

First, we use the *helpdesk*[2] event log containing data from an Italian software company's ticketing management process. It includes $21,348$ events, $4,580$ process instances, 226 process instance variants and 14 activities. Second, we include the *bpi2019*[3] event log from the BPI challenge 2019, provided by a company for paints and coatings. It depicts a purchase order handling processes. For this event log, we only considered a random 10%-sampling with sequences of 30 events or shorter, due to the high computation effort. It includes $101,714$ events, $24,900$ process instances, $3,255$ process instance variants and 32 activities.

4.2 Process Models

We used DCR graphs as models for the BPS in our technique's best candidate selection. In Fig. 4, we present the DCR graph for the *helpdesk* event log. The three most important constraints are the following. First, after "Closed" the other activities should not happen. Second, if "Assign seriousness" occurs, someone must take over the responsibility. Third, before a ticket is closed, "Resolve ticket" must occur.

Figure 5 shows the DCR graph for the *bpi2019* event log. The three most essential constraints are the following. First, "Create Purchase Order Item" may only happen once per order. Second, After the goods were received, "Change Quantity" and "Change price" should not occur. Third, "Record Goods Receipt", "Record Invoice Receipt" and "Clear Invoice" must eventually follow each other.

---

[2] https://data.mendeley.com/datasets/39bp3vv62t/1.

[3] https://data.4tu.nl/repository/uuid:a7ce5c55-03a7-4583-b855-98b86e1a2b07.

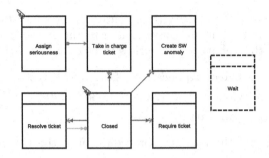

Fig. 4. DCR graph for the *helpdesk* event log.

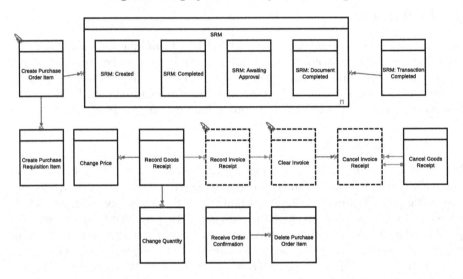

Fig. 5. DCR graph for the *bpi2019* event log.

4.3 Procedure

We split both event logs in a 2/3 training and 1/3 test set with a random process-instance-based sampling. As a baseline, we use the most cited next event PBPM technique from Tax et al. [20]. We evaluate the technique in two ways. First, we evaluate the optimisation of the KPI *throughput time* (*in-time* value) by the percentage of process instances that could comply with the temporal threshold for different prefix sizes. The temporal threshold is the average throughput time of a process instance in an event log. Second, we evaluate the *distance* from the ground truth process instance through the average Damerau-Levenshtein distance. This metric determines the distance between two strings or sequences through the minimum number of operations (consisting of insertions, deletions or substitutions of a single character, or transposition of two adjacent characters), i.e. the lower the value, the more similar the strings are.

To train the multi-task LSTM m_{pp}, we apply the *Nadam* optimisation algorithm with a *categorical cross-entropy loss* for next activity predictions and a *mean squared error* for *throughput time* (KPI) predictions. Moreover, we set the batch size to 256, i.e. gradients update after every 256th sample of the training set. We set the default values for the other optimisation parameters. For training the candidate selection model m_{cs}, we apply the nearest-neighbour-based ML algorithm ball tree [14]. Ball tree utilises a binary tree data structure for maintaining spatial data hierarchically. We choose a spatial-based algorithm to consider the semantic similarity between the suffixes of activities and KPI values. Moreover, we set the hyperparameter k (number of "nearest" neighbours) of m_{cs} to 5, 10 and 15. Thereby, we check different sizes of the suffix candidate set.

Finally, technical details and the source code are available on GitHub[4].

4.4 Results

Figure 6 shows the results for the *helpdesk* event log. For most of the prefixes in the *helpdesk* event log, our technique's next best actions are more *in time* than next activity predictions. While for $k = 10$ next best actions have the lowest *in-time* values compared to next activity predictions, *in-time* values of next best

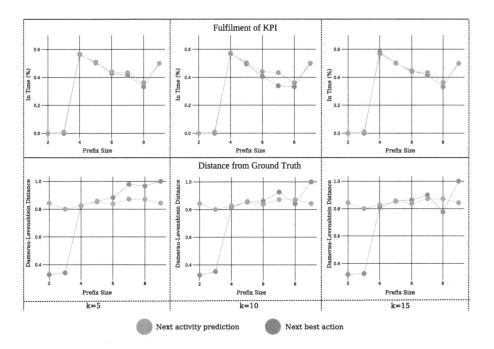

Fig. 6. Results for the *helpdesk* event log.

---

[4] https://github.com/fau-is/next-best-action.

actions with $k = 5$ and $k = 15$ are rather similar to each other. Furthermore, the higher the k, the lower is the *distance* of the next best actions from the actual process instances. Up to prefix size 4, the *distance* of the next best actions is lower compared to next activity predictions.

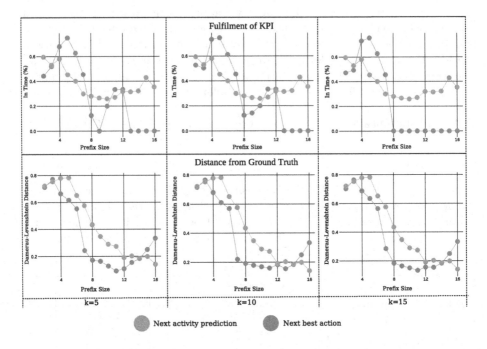

Fig. 7. Results for the *bpi2019* event log.

Figure 7 shows the results for the *bpi2019* event log. For most of the prefixes with a size ≥ 8, the next best actions of our technique are more *in time* than next activity predictions. For $k = 15$ and prefixes ≥ 8, next best actions have an *in-time* value of 0. With an increasing k, the *in-time* values of the next best actions vary less from prefix size 2 to 12. In contrast to next activity predictions, the *distance* of next best actions is lower for prefixes with a size > 3 and < 15. Over the three k values, the *distance* of next best actions is rather similar.

5   Discussion

Our contribution to academia and practice is a PrBPM technique that recommends the next best actions depending on a KPI while concerning BPS. Moreover, the evaluation presents an instantiation of our PrBPM technique. The KPI is the *throughput time*, and a DCR graph realises BPS via the event log.

Based on our results, our PrBPM technique can provide actions with lower *in-time* values and less *distance* from the ground truth compared to the next most likely activities for both event logs. However, the *in-time* values (i.e. the percentage of process instances that could comply with the temporal threshold) of next best actions differs more from the baseline's next activity prediction for the *bpi2019* event log than for the *helpdesk* event log. The *helpdesk* event log has a lower instance variability than the *bpi2019* event log. Therefore, fewer process paths exist from which our technique can recommend actions with lower *in-time* values. Further, the number of candidates k has an effect on the KPI's optimisation. While we get the actions with the lowest *in-time* values with $k = 10$ for the *helpdesk* event log, the KPI values with $k = 5$ and $k = 15$ are similar to each other. For the *bpi2019* event log, our technique provides actions with the lowest *in-time* value if k is set to 15. The results with $k = 5$ are similar to those of $k = 10$. A higher k value leads to lower *in-time* values in the *bpi2019* event log because of a higher instance variability. On the contrary, the *helpdesk* event log needs a lower k value. Regarding the *distance* from ground truth process instances, most of the next best actions (especially those for the *bpi2019* event log) reach a better result than next activity predictions. A reason for that could be the limited predictive quality of the underlying DNN model for predicting next activities. However, our technique integrates control-flow knowledge and therefore overcomes this deficit to a certain degree. Moreover, for the *bpi2019* event log, our technique provides actions with lower *in-time* values for prefixes with a size ≥ 8. In terms of the *helpdesk* event log, we get actions with lower *in-time* values for shorter prefixes. We suppose that our technique requires a longer prefix for event logs with higher instance variability to recommend next best actions. Finally, even though our technique provides actions with lower *in-time* values, it seems that it does not terminate before the baseline. Our results show the aggregated values over different prefix sizes. Thus, we assume that few sequences, for which the termination can not be determined, distort the results.

Despite all our efforts, our technique bears three shortcomings. First, we did not optimise the hyperparameters of the DNN model m_{pp}, e.g. via random search. Instead, we set the hyperparameters for m_{pp} according to the work of Tax et al. [20]. We used the same setting since we compare our technique's next best actions to their next activity predictions. Second, even though our technique is process-modelling-notation agnostic, we argue that declarative modelling is an appropriate approach for the process simulation. Due to its freedoms, declarative modelling facilitates the partial definition of the control-flow. As a consequence, we have a more flexible definition of a process's control-flow than by using a restricted procedural process model. While our DNN-based technique copes well with rather flexible processes, other techniques using *traditional* ML algorithms (e.g. a decision tree) might handle restricted processes faster and with a higher predictive quality. Third, for our PrBPM technique's design, we neither consider cross-instance nor cross-business-process dependencies. In an organisational environment, additional effects like direct and indirect rebound effects can hinder our technique.

6   Related Work

A variety of PBPM techniques were proposed by researchers as summarised by, e.g. Márquez-Chamorro et al. [10] or Di Francescomarino et al. [4]. Many of these techniques are geared to address the next activity prediction task. For that, most of the recent techniques rely on LSTMs [24] such as Weinzierl et al. [25]. To predict not only the next activities with a single predictive model, Tax et al. [20] suggest a multi-task LSTM-based DNN architecture. With this architecture, they predict the next activities and their timestamps. Metzger et al. [12] extend their architecture by another LSTM layer to additionally predict the binary process outcome whether a delay occurs in the process or not. These techniques output predictions and do not recommend next best actions.

Furthermore, researchers suggested PrBPM approaches that raise alarms or recommend actions to prevent undesired activities. Metzger et al. [11] investigate the effect of reliability estimates on (1) intervention costs (called adaption cost) and (2) the rate of non-violation of process instances by performing a simulation of a parameterised cost model. Thereby, they determine the reliability estimates based on the predictions of an ensemble of multi-layer perceptron classifiers at a pre-defined point in the process. In a later work [13], reliability estimates were determined based on an ensemble of LSTM classifiers at different points in the process. The recommendation of actions is not part of these works. Teinemaa et al. [21] propose a concept of an alarm-based PrBPM framework. They suggest a cost function for generating alarms that trigger interventions to prevent an undesired outcome or mitigate its effect. In a later work [5], a multi-perspective extension of this framework was presented. In both versions, the framework focuses on alarms. Gröger et al. [6] present a PrBPM technique that provides action recommendations for the next process step during the execution of a business process to avoid a predicted performance deviation. Performance deviation is interpreted as a binary outcome prediction, i.e. exists a deviation or not. In detail, an action recommendation comprises several action items and is represented by a rule extracted from a learned decision tree. An action item consists of the name and value of a process attribute. Even though this approach recommends actions in the form of process attribute values of the next process step which are optimised according to a KPI (e.g. lead time), process steps as next best actions are not recommended. Conforti et al. [3] propose a PrBPM technique that predicts risks depending on the deviation of metrics during process execution. The technique's purpose is to provide decision support for certain actions such as the next process activity which minimises process risks. However, this technique can only recommend actions which are optimised regarding the KPI risk. Thus, with the best of our knowledge, there is no PrBPM approach that transforms next most likely activity predictions into the next best actions (represented by activities) depending on a given KPI.

7 Conclusion

Next activity predictions provided by PBPM techniques can be less beneficial for process stakeholders. Based on our motivation and the identified research gap, we argue that there is a crucial need for a PrBPM technique that recommends the next best actions in running processes. We reached our research goal with the evaluation of our developed PrBPM technique in Sec. 5. Thereby, we show that our technique can outperform the baseline regarding KPI fulfilment and distance from ground truth process instances. Further research might concern different directions. First, we plan to adapt existing loss functions for LSTMs predicting next most likely activities. Such a loss function can enable an LSTM to directly consider information on KPIs in the learning procedure. Second, future research should further develop existing PrBPM approaches. More advanced multi-tasking DNN architectures can facilitate the recommendation of more sophisticated next best actions. For instance, next best actions that optimise more than one KPI. Finally, we call for PrBPM techniques that are aware of concept evolution. Our technique is not able to recommend an activity as the best action if it was not observed in the training phase of the ML models.

Acknowledgments. This project is funded by the German Federal Ministry of Education and Research (BMBF) within the framework programme *Software Campus* under the number 01IS17045.

References

1. Bengio, Y., Simard, P., Frasconi, P., et al.: Learning long-term dependencies with gradient descent is difficult. Trans. Neural Netw. **5**(2), 157–166 (1994)
2. Centobelli, P., Converso, G., Gallo, M., Murino, T., Santillo, L.C.: From process mining to process design: a simulation model to reduce conformance risk. Eng. Lett. **23**(3), 145–155 (2015)
3. Conforti, R., de Leoni, M., La Rosa, M., van der Aalst, W.M.P.: Supporting risk-informed decisions during business process execution. In: Salinesi, C., Norrie, M.C., Pastor, Ó. (eds.) CAiSE 2013. LNCS, vol. 7908, pp. 116–132. Springer, Heidelberg (2013). https://doi.org/10.1007/978-3-642-38709-8_8
4. Di Francescomarino, C., Ghidini, C., Maggi, F.M., Milani, F.: Predictive process monitoring methods: which one suits me best? In: Weske, M., Montali, M., Weber, I., vom Brocke, J. (eds.) BPM 2018. LNCS, vol. 11080, pp. 462–479. Springer, Cham (2018). https://doi.org/10.1007/978-3-319-98648-7_27
5. Fahrenkrog-Petersen, S.A., et al.: Fire now, fire later: alarm-based systems for prescriptive process monitoring. arXiv preprint arXiv:1905.09568 (2019)
6. Gröger, C., Schwarz, H., Mitschang, B.: Prescriptive analytics for recommendation-based business process optimization. In: Abramowicz, W., Kokkinaki, A. (eds.) BIS 2014. LNBIP, vol. 176, pp. 25–37. Springer, Cham (2014). https://doi.org/10.1007/978-3-319-06695-0_3
7. Hochreiter, S., Schmidhuber, J.: Long short-term memory. Neural Comput. **9**(8), 1735–1780 (1997)
8. LeCun, Y., Bengio, Y., Hinton, G.: Deep learning. Nature **521**(7553), 436 (2015)

9. Maggi, F.M., Di Francescomarino, C., Dumas, M., Ghidini, C.: Predictive monitoring of business processes. In: Jarke, M., et al. (eds.) CAiSE 2014. LNCS, vol. 8484, pp. 457–472. Springer, Cham (2014). https://doi.org/10.1007/978-3-319-07881-6_31

10. Márquez-Chamorro, A., Resinas, M., Ruiz-Cortás, A.: Predictive monitoring of business processes: a survey. IEEE Trans. Serv. Comput. (TSC) **11**(6), 962–977 (2017). https://ieeexplore.ieee.org/document/8103817

11. Metzger, A., Föcker, F.: Predictive business process monitoring considering reliability estimates. In: Dubois, E., Pohl, K. (eds.) CAiSE 2017. LNCS, vol. 10253, pp. 445–460. Springer, Cham (2017). https://doi.org/10.1007/978-3-319-59536-8_28

12. Metzger, A., Franke, J., Jansen, T.: Data-driven deep learning for proactive terminal process management. In: Proceedings of the 17th International Conference on Business Process Management (BPM), pp. 196–211 (2019)

13. Metzger, A., Neubauer, A., Bohn, P., Pohl, K.: Proactive process adaptation using deep learning ensembles. In: Giorgini, P., Weber, B. (eds.) CAiSE 2019. LNCS, vol. 11483, pp. 547–562. Springer, Cham (2019). https://doi.org/10.1007/978-3-030-21290-2_34

14. Omohundro, S.M.: Five Balltree Construction Algorithms. International Computer Science Institute, Berkeley (1989)

15. Polato, M., Sperduti, A., Burattin, A., de Leoni, M.: Data-aware remaining time prediction of business process instances. In: Proceeding of the International Joint Conference on Neural Networks (IJCNN), pp. 816–823. IEEE (2014)

16. Redlich, D., Gilani, W.: Event-driven process-centric performance prediction via simulation. In: Daniel, F., Barkaoui, K., Dustdar, S. (eds.) BPM 2011. LNBIP, vol. 99, pp. 473–478. Springer, Heidelberg (2012). https://doi.org/10.1007/978-3-642-28108-2_46

17. Rosenthal, K., Ternes, B., Strecker, S.: Business process simulation: a systematic literature review. In: Proceedings of the 26th European Conference on Information Systems (ECIS) (2018)

18. Rozinat, A., Wynn, M.T., van der Aalst, W.M., ter Hofstede, A.H., Fidge, C.J.: Workflow simulation for operational decision support. Data Knowl. Eng. **68**(9), 834–850 (2009)

19. Rumelhart, D.E., Hinton, G.E., Williams, R.J.: Learning representations by back-propagating errors. Nature **323**(6088), 533–536 (1986)

20. Tax, N., Verenich, I., La Rosa, M., Dumas, M.: Predictive business process monitoring with LSTM neural networks. In: Dubois, E., Pohl, K. (eds.) CAiSE 2017. LNCS, vol. 10253, pp. 477–492. Springer, Cham (2017). https://doi.org/10.1007/978-3-319-59536-8_30

21. Teinemaa, I., Tax, N., de Leoni, M., Dumas, M., Maggi, F.M.: Alarm-based prescriptive process monitoring. In: Weske, M., Montali, M., Weber, I., vom Brocke, J. (eds.) BPM 2018. LNBIP, vol. 329, pp. 91–107. Springer, Cham (2018). https://doi.org/10.1007/978-3-319-98651-7_6

22. Tumay, K.: Business process simulation. In: Proceedings of the Winter Simulation Conference, pp. 93–98. ACM (1996)

23. van der Aalst, W.M.P.: Process Mining: Data Science in Action, 2nd edn. Springer, Heidelberg (2016). https://doi.org/10.1007/978-3-662-49851-4

24. Weinzierl, S., et al.: An empirical comparison of deep-neural-network architectures for next activity prediction using context-enriched process event logs. arXiv:2005.01194 (2020b)

25. Weinzierl, S., Stierle, M., Zilker, S., Matzner, M.: A next click recommender system for web-based service analytics with context-aware LSTMs. In: Proceedings of the 53rd Hawaii International Conference on System Sciences (HICSS) (2020)

26. Weinzierl, S., Zilker, S., Stierle, M., Park, G., Matzner, M.: From predictive to prescriptive process monitoring: recommending the next best actions instead of calculating the next most likely events. In: Proceedings of the 15th International Conference on Wirtschaftsinformatik. AISeL (2020c)

27. Wynn, M.T., Dumas, M., Fidge, C.J., ter Hofstede, A.H.M., van der Aalst, W.M.P.: Business process simulation for operational decision support. In: ter Hofstede, A., Benatallah, B., Paik, H.-Y. (eds.) BPM 2007. LNCS, vol. 4928, pp. 66–77. Springer, Heidelberg (2008). https://doi.org/10.1007/978-3-540-78238-4_8

BPM Adoption and Maturity

IT Culture and BPM Adoption in Organizations

Brian Letts[✉] and Vu Tran[✉]

Capella University, Minneapolis, MN 55402, USA
bjletts@protonmail.com, vu.tran@capella.edu

Abstract. The present study investigates the relationship between IT organizational culture type and Business Process Management adoption in organizations implementing IT solutions. Specifically, the study investigates how the success of BPM adoption varies by organizational culture type. The target population consisted of IT resources who work in the United States at organizations with at least 50 employees and who have participated in the development and implementation of a BPM initiative involving an IT solution within the last two years. A survey was conducted with 157 anonymous participants representing the target population. The study found the highest level of BPM adoption success was with the adhocracy culture type compared to the market and hierarchy culture types. There is a significant positive correlation between the adhocracy culture type and BPM adoption as measured by BPO. There is also a significant negative correlation between the market culture type and BPM adoption as measured by BPO and PPI. The insights gained by this study can help practitioners make informed decisions on their BPM adoption approach within their IT community and scholars in future research on the relationship between organizational culture and BPM.

Keywords: BPM · BPM culture · Organizational culture · BPM adoption · IT culture

1 Introduction

The purpose of implementing business process management (BPM) initiatives is to improve the competitiveness and performance of organizations by focusing on making end-to-end business processes more effective and efficient [1] and adding value for the customer [2]. Recent studies reported that BPM is one of the top concerns of information technology (IT) executives [3] and the vast majority of the most prominent organizations worldwide had implemented some form of BPM initiative or planned to within the next three to five years [4]. Other studies found the primary driver enabling the success of BPM initiatives is the strategic use of information technology (IT) [5, 6]. Some found the most significant BPM initiatives involve the use of an IT solution [5, 7]. The purpose of implementing IT projects is to change some business process aspect of the impacted organization for the better [7, 8].

With IT project failure rates of 70% [9], it is critically important IT leaders at all levels of an organization understand the factors that lead to successful BPM initiatives involving IT solutions [5]. Much of the early research and practice has concentrated

© Springer Nature Switzerland AG 2020
D. Fahland et al. (Eds.): BPM Forum 2020, LNBIP 392, pp. 213–228, 2020.
https://doi.org/10.1007/978-3-030-58638-6_13

on the role of IT in support of BPM, but recent research defines BPM from a more holistic perspective to include methods, governance, people, strategic alignment, and organizational culture [10]. Organizational culture is an important factor in the success of IT-driven changes to business processes [11]. Many BPM initiatives fail in large part due to the organization's current culture is not supportive of the change required [12, 13].

BPM has a distinctive and recognizable culture [14]. Within an organization, IT has been found to have its own unique organizational culture separate and distinct from the organization as a whole [15]. IT organizational culture can have more influence on BPM success than the national culture or the culture of the organization [16]. The level of success of BPM projects, specifically BPM initiatives involving IT solutions, is dependent upon the cultural fit of the IT organizational culture and the culture required for the BPM initiative [11, 13, 17].

The aim of this research is to provide insights into the IT organizational culture type most conducive to the BPM adoption with organizations located in the United States. The organization of the remainder of the present study is an overview of BPM, BPM adoption, organizational culture, BPM culture, IT organizational culture, the research design, and the results of the study including interpretation of findings, conclusions and recommendations for future research.

2 Background

The organizational benefits of successfully adopting BPM as a management approach are improved performance, lower costs, higher quality, increased customer satisfaction, and achievement of organizational goals and strategy [18]. A study finds that organizations fall short of achieving all of the above benefits if they fail to adopt BPM [19]. Understanding the factors influencing the success of BPM initiatives that involve the implementation of IT solutions is critical to improving the overall success of BPM initiatives and the organization's performance [5]. The low success rate of BPM initiatives may be due to the incongruence between the organizational culture type and the BPM culture needed for the success of the BPM initiative [20]. Many BPM initiatives fail due to ignoring the cultural conflicts of the resources involved in the effort [21].

Researchers have proposed, and studies have found support for the position that cultural fit is a key factor to successful implementation of BPM in organizations [17, 20]. A recent study found different types of organizational cultures influence the BPM adoption rate and the success of BPM initiatives for organizations located in Slovenia [22]. In conjunction with cultural fit, researchers found the strategic use of IT to support BPM initiatives is another key to BPM success [5, 11, 13]. This study uses the Hribar and Mendling model [22] to investigate the relationship between IT organization culture and BPM initiatives in organizations within the United States. Specifically, this study analyzes the relationship between the types of IT organizational cultures and BPM initiatives. This present study follows the assertion that the type of IT organization culture (rather than national or organizational culture) has the greatest influence on BPM adoption within an organization implementing BPM initiatives involving IT solutions [13, 16, 22].

2.1 BPM Adoption

For the present study, BPM adoption is "the use and deployment of BPM concepts in organizations" [23] and the extent the resources of the organization use BPM concepts as they execute business processes reflects the organization's level of BPM adoption [24]. Many scholarly researchers used proxies to measure the level of BPM adoption [22, 25, 26]. The present study used the instrument described in [22], which combines the Business Process Orientation (BPO) maturity model developed by McCormack and Johnson [27] and the Process Performance Index (PPI) instrument developed by Rummler and Brache [28] together to operationalize and measure the level of BPM adoption. As the scores of the BPO and PPI instruments increase, so should the success of the BPM adoption [26]. A higher level of BPM adoption should lead to a higher level of BPM initiative success [24].

2.2 Organizational Culture and BPM Culture

For the present study, organizational culture is a concatenation of values, beliefs, historical precedents, knowledge and understanding, and learned modes of acceptable behavior of an organization [29]. The predominant theoretical approach by research scholars studying organizational cultures is conceptualizing it based on values (Schein, 2010). Once an organizational culture is measured based on values, a comparison between the organizational culture variable with other variables is possible [30]. A BPM culture is an organizational culture whose values support the objectives of BPM (i.e., possessing processes that are efficient and effective) [8, 14]. The success of BPM initiatives depends on upon identifying and discerning the characteristics of an organization's culture, i.e., its BMP culture, that either promote or impede an approach to an organization's adoption of BPM [12]. Based on research by [14], Customer orientation, Excellence, Responsibility, and Teamwork (CERT) are cultural values considered critical for BPM success.

2.3 IT Organizational Culture

Of all the different definitions of culture (i.e. national, organizational, occupational), Schein [16] posits occupational culture, such as an IT culture, may have the largest influence on the behavior of individuals in an organization. Individuals within an occupational culture work within the organization's culture while accentuating their unique occupational culture characteristics at work [16]. IT has a unique culture that is different from the rest of the organization [15]. Research by [15] resulted in identifying a set of five core IT cultural values (i.e., self-rule, organization and precision, technological innovation, admiration for technology, and enjoyable work environment). The present study follows the assertion that the type of IT organization's culture, i.e., IT organization's BPM culture, (rather than national or organizational culture) has the greatest influence on BPM adoption within an organization implementing BPM initiatives involving IT solutions. The organization's IT culture supporting BMP initiatives and the BMP culture necessary for BPM success are the same.

The present study operationalized IT organizational culture using the organizational culture assessment instrument (OCAI) which is based on the Competing Values Framework (CVF) [31]. The OCAI identifies four unique types of organizational culture:

hierarchy, market, clan, and adhocracy [31]. An internal focus and a controlling mindset dominate the hierarchy culture type [32]. The market culture focuses on results, setting and reaching goals, competition, hard-driving, and winning [38] and has an orientation to the external environment, which concentrates mainly on customers [31]. The clan culture focuses on team cohesion, unity, support for others, loyalty, development and a sense of family [33]. An adhocracy culture type represents innovativeness, risk-taking, individualism, growth, and a dynamic environment [32].

3   Research Design

3.1   Research Method

This study replicated the Hribar and Mendling's study [22] investigating the relationship between organizations located in the United States engaging in BPM adoption implementing IT initiatives and the IT organizational culture supporting these efforts. Following the research question from the [22] paper, the specific research question for the present study was how does the success of BPM adoption with organizations located in the United States implementing BPM initiatives involving IT solutions vary between different types of organizational culture within IT? The specific sub-questions for the present study are listed below followed by Table 1 presenting the associated hypotheses, which are identical to those in the [22] study except organizational culture is more specifically defined (i.e. IT).

- Research sub-question 1: What is the relationship, if any, between the dominant IT organizational culture type and business process orientation (BPO)?
- Research sub-question 2: What is the relationship, if any, between the dominant IT organizational culture type and process performance index (PPI)?
- Research sub-question 3: What is the relationship, if any, between BPO and PPI?

Table 1. Sub-questions and associated hypotheses

Sub-Question	Hypothesis
1	H_0^1 The dominant IT organizational culture type will have no impact on BPO
1	H_a^1 The dominant IT organizational culture type will have a significant impact on BPO
2	H_0^2 The dominant IT organizational culture type will have no impact on PPI
2	H_a^2 The dominant IT organizational culture type will have a significant impact on PPI
3	H_0^3 BPO and PPI are not correlated with each other
3	H_a^3 BPO and PPI are positively correlated with each other

We adopt the position taken by [13] and [22] that the relationship between an organizational culture and BPM success is a causal relationship, which was subsequently supported by the work of [22]. We recommend additional studies such as the present study to confirm this position. We expect to find a negative correlation between the hierarchy organizational culture type and BPM adoption as [12] posit many of the policies and procedures of hierarchical organizations conflict with BPM concepts. We also expect to find a positive correlation between the clan, market, and adhocracy organizational culture types and BPM adoption as all of these have at least two of the four CERT values that are imperative to BPM success.

The purpose of the present quantitative study was to determine the relationship between the organizational culture type, operationalized using the OCAI, and the level of BPM adoption, operationalized using the combined business process orientation and process performance index instruments, with organizations located in the United States implementing BPM initiatives involving IT solutions.

3.2 Survey Design

Three separate, previously empirically validated, well-established instruments to measure organizational culture and BPM comprise the Hribar and Mendling's instrument [22] used in the present study. In our exchanges with one of the authors of the [22] study, he recommended we use the version of the instrument where the OCAI instrument section used a 1–5 scale and the results of which were converted to percentages following his guidance. Table 2 lists the details of the types of variables used in the present study as originally defined in [22].

Table 2. Description of variables

Variable name	Variable Type	Independent/Dependent	Measurement Scale Range
OCAI	Continuous	Independent	1-5
BPO	Continuous	Dependent	1.0-5.0
PPI	Continuous	Dependent	10-50

The underpinning of the OCAI is the CVF, which identifies four distinct and competing culture types: hierarchy, clan, adhocracy, and market [32]. Schmiedel, vom Brocke, and Recker's study [14] posit the CVF culture types align well with the four BPM (i.e. CERT) culture values making the use of the OCAI ideal to validate the framework in the present study. Detail information on the OCAI instrument can be found in [33].

Fifteen questions separated into three distinct areas: process view, process jobs, and process management and measurement systems comprise the BPO maturity model. The BPO maturity model is an average for all of the questions, and the higher the average, the higher the process orientated maturity level of the organization. There are four BPO levels, and their value ranges are 1.0–2.0 representing ad hoc, 2-0-3.0 representing defined, 3.0–4.0 representing linked, and 4.–5.0 representing integrated [27]. Detail information on the BPO instrument can be found in [27].

The PPI is an instrument that measures the level in which organizational resources manage the organization as a system and is process-centered [25]. Ten questions regarding BPM process-centric critical success factors grouped into three broad dimensions:

(1) process view; (2) process jobs; and (3) process management and measurement systems comprise the PPI descriptive maturity model [25]. Detail information on the PPI instrument can be found in [28].

3.3 Population and Sampling

The target population for the present study was IT resources who work in the United States at organizations with at least 50 employees and who have participated in the development and implementation of a BPM initiative involving an IT solution within the last two years. Koh Tas and Low Sui [34] posit the business process practices of small organizations with less than 50 employees are likely to be unsophisticated or highly variable. Table 3 lists the differences in the populations between the present study and the [22] study.

Table 3. Differences in population

Present Study	Hribar and Mendling (2014)
IT resources only	Top managers or process owners
Resources working for an organization in the United States	Organizations located in Slovenia
Implementation of a BPM initiative involving IT solution within the last two years	Experience with any BPM initiative (not limited to just BPM initiatives involving an IT solution)

Kasim, Haračić, and Haračić [4] noted that 94% of the most prominent organizations worldwide had implemented some form of BPM initiative or planned to within the next three to five years and the main driver enabling the success of BPM initiatives is the strategic use of IT [5]. There are nearly 475,000 private organizations in the United States with 50 or more employees [35]. With 94% of organizations who have implemented some degree of a BPM initiative, the population would be approximately 446,000.

According to [36], it is necessary to conduct an a priori analysis prior to obtaining research data to determine the necessary sample size N which "is computed as a function of the required power level $(1 - \beta)$, the pre-specified significance level α, and the population effect size to be detected with probability $1 - \beta$" (p. 176). The free online sample calculator, G*Power© determined the appropriate sample size needed for the present study. Using an a priori power analysis for the F-test with a medium effect size of 0.30, a significance level of 0.05, a power level of 0.80 for four groups resulted in G*Power© calculating a minimum sample size of 128 or 32 for each organizational culture type [37].

Survey data was obtained from the target large population using an online survey company (Qualtrics.com©) with very large marketing panels that include IT resources at all levels and in every business sector. Anyone self-identifying as a member of IT (e.g. software engineer, system analyst, business system analyst, system architect, IT project manager and IT managers) would be included in the Qualtrics IT-focused marketing panel.

Qualtrics used simple random sampling to send an email inviting participants within their marketing panels to participate in the survey. Survey participants must have responded in the affirmative on all of the following inclusion questions:

- Are you an information technology professional?
- Are you a member of IT in your organization?
- Do you work for an organization located in the United States?
- Does your organization have at least 50 employees?
- Do you have experience developing and implementing a business process management (BPM) initiative involving an IT solution within the last two years?

3.4 Data Collection and Analysis

Qualtrics randomly selected individuals within their marketing panels who met the criteria to take the survey (i.e. individual self-identified as a member of IT) and sent them an email with a link to participate in the survey. Qualtrics initially released the survey to a predetermined number of panel members, and then Qualtrics periodically released additional emails to panel members. Qualtrics ceased sending out emails to panel members after they received the requested number of completed surveys. Qualtrics only provided completed surveys that met all of the inclusion criteria. Microsoft Excel 2010 and IBM Statistical Package for the Social Sciences (SPSS©) Statistics Grad Pack version 23.0 was used to analyze the data obtained from Qualtrics.

4 Results

4.1 Descriptive Analysis

Qualtrics provided a total of 230 cases, of which a total of 157 useful cases were used for data analysis, after removing invalid cases and outliers. Invalid cases included cases that provided the same or alternating answer on the items within one or more of the instrument sections, had more than one dominant organizational culture type, or answered questions within the survey that contradicted what the participant indicated on the inclusion questions. The researchers followed [38] and identified and removed outliers greater than three standard deviations from the mean of the variable. Twenty-six cases had at least one question of the 15 questions used to calculate the BPO variable in which the respondent selected "X – Cannot Judge." To enable the ability to use these cases, the mean for each question from the other respondents from which there is a value was determined and then assigned the mean to the cases missing a value for that question. This technique is in accordance to recommendations from [39].

Table 4 presents the descriptive statistics of the final data set. An analysis of the means for each culture type found adhocracy had the highest level of BPM adoption success as measured by both BPO (M = 40.47) and PPI (M = 4.44). The market culture type had the lowest level of BPM adoption success as measured by both BPO (M = 37.07) and PPI (M = 3.99). Thus, the comparison of the means for BPO and PPI indicates the adhocracy culture type is the most favorable to BPM adoption success, while the market culture type is least favorable. Whether the differences between the means are significant was determined through inferential statistics.

Table 4. Descriptive statistics of dependent variables

Dominant Culture Type	Dependent Variable	N	Min	Max	Mean	Median	Std. Dev.
Clan	PPI	51	26.00	46.00	39.67	40.00	4.36
	BPO	51	2.67	4.73	4.18	4.29	0.46
Adhocracy	PPI	17	29.00	47.00	40.47	42.00	4.91
	BPO	17	3.62	4.80	4.44	4.53	0.35
Market	PPI	41	25.00	46.00	37.07	36.00	5.83
	BPO	41	3.00	4.67	3.99	4.07	0.48
Hierarchy	PPI	48	29.00	46.00	39.42	40.00	5.00
	BPO	48	2.60	4.87	4.07	4.13	0.47

4.2 Scale Reliability

Cronbach's α was calculated to ascertain if items on each of the three separate instruments (i.e., BPO, PPI, and OCAI) reliably measured the intended construct. [40] posit values of the Cronbach's α between 0.7–0.8 indicate acceptable reliability, between 0.8–0.9 good reliability, and greater than 0.9 excellent reliability. PPI and BPO have good reliabilities with Cronbach's $\alpha > 0.8$ (i.e., PPI $\alpha = .86$ and BPO $\alpha = .81$). Three out of the four culture types had excellent reliability (i.e., clan $\alpha = .94$, market $\alpha = .93$, and hierarchy $\alpha = .93$), while the adhocracy culture type had good reliability ($\alpha = .86$). All of the instruments used in the present study had sufficient reliability.

4.3 Analysis of Hypotheses

The assumptions for the ability to use parametric tests are the data need to have a normal distribution, independence, and there is homogeneity of variance [41]. Table 5 presents the results of the analysis of the data indicating the assumptions for the ability to use parametric tests were satisfied.

Table 5. Analysis of data for parametric tests assumptions

Normal Distribution	Independence	Homogeneity of Variance	
Q-Q Plots Indicated Approximate Normality	Participants Randomly Selected	Levene's Test Non-Significant ($p > .05$)	
		PPI - $F(3, 153) = 2.20$, $p = 0.091$	BPO - $F(3, 153) = 1.04$, $p = 0.377$

The ANOVA test was used to determine if there were any statistical differences between the means from the BPO scores for each of the four culture types as well as between the means from the PPI scores for each of the four culture types. Based on the analysis, there is a statistically significant difference ($\alpha = .05$) between the means from the PPI scores of the four culture types ($F(3, 153) = 2.90$, $p = .037$, $\eta_p^2 = .054$) and from the BPO scores of the four culture types ($F(3, 153) = 4.37$, $p = .006$, $\eta_p^2 = .079$). There is a significant effect between the independent variable, IT organizational culture type on the dependent variables, PPI and BPO. The researcher used the Tukey-Kramer post-hoc test to determine which culture types were different within the BPO and PPI variables [42]. Table 6 presents the results of the Tukey-Kramer post hoc test for BPO. Table 7 presents the results of the Tukey-Kramer post hoc test for PPI.

Table 6. Tukey-Kramer post hoc results for BPO

Dominant Culture Type	Dominant Culture Type	Mean Difference	Std. Error	Sig.	95% Confidence Interval	
					Lower Bound	Upper Bound
Clan	Adhocracy	-0.26	0.11	0.100	-0.56	0.03
	Market	0.19	0.10	0.306	-0.08	0.46
	Hierarchy	0.11	0.09	0.794	-0.14	0.36
Adhocracy	Clan	0.26	0.11	0.100	-0.03	0.56
	Market	0.45*	0.11	0.002	0.14	0.76
	Hierarchy	0.37*	0.11	0.008	0.08	0.67
Market	Clan	-0.19	0.10	0.306	-0.46	0.08
	Adhocracy	-0.45*	0.11	0.002	-0.76	-0.14
	Hierarchy	-0.08	0.10	0.970	-0.35	0.20
Hierarchy	Clan	0.11	0.09	0.794	-0.36	0.14
	Adhocracy	-0.37*	0.11	0.008	-0.67	-0.08
	Market	0.08	0.10	0.970	-0.20	0.35

* The mean difference is significant at the .05 level.

Table 7. Tukey-Kramer post hoc results for PPI

Dominant Culture Type	Dominant Culture Type	Mean Difference	Std. Error	Sig.	95% Confidence Interval	
					Lower Bound	Upper Bound
Clan	Adhocracy	-0.80	1.34	0.990	-4.61	3.00
	Market	2.59	1.10	0.116	-0.37	5.55
	Hierarchy	0.25	0.94	1.000	-2.29	2.79
Adhocracy	Clan	0.80	1.34	0.990	-3.00	4.61
	Market	3.40	1.50	0.160	-0.77	7.56
	Hierarchy	1.05	1.39	0.969	-2.87	4.97
Market	Clan	-2.59	1.10	0.116	-5.56	0.37
	Adhocracy	-3.40	1.50	0.160	-7.56	0.77
	Hierarchy	-2.34	1.16	0.247	-5.47	0.79
Hierarchy	Clan	-0.25	0.94	1.000	-2.79	2.29
	Adhocracy	-1.05	1.39	0.969	-4.97	2.87
	Market	2.34	1.16	0.247	-0.79	5.47

Note. Based on observed means. The error term is Mean Square (Error) = 25.294.

The analysis of correlation using Pearson's r correlation coefficient calculated the degree of strength and direction of the relationship between all of the four organizational culture types and the level of BPM adoption as measured by BPO and PPI. The Pearson's r correlation coefficient calculation between BPO and PPI addressed the third sub-research question and associated hypothesis. The range of values for the Pearson's r is from -1.00 to +1.00 and Cohen [43] provides the following guidelines to interpret the Pearson's r

values: $r = 0.10$ to 0.29 or -0.10 to -0.29 (small); $r = 0.30$ to 0.49 or -0.30 to -0.49 (medium); and $r = 0.50$ to 1.0 or -0.50 to -1.00 (large). Table 8 provides the results of the Pearson's r correlation coefficient test on the variables.

Table 8. Pearson's r correlation matrix

Variable	Test	PPI	BPO	Clan	Adhocracy	Market	Hierarchy
PPI	Pearson's r	1					
	Sig. (2-tailed)						
BPO	Pearson's r	0.68**	1				
	Sig. (2-tailed)	0.000					
Clan	Pearson's r	0.09	0.08	1			
	Sig. (2-tailed)	0.260	0.314				
Adhocracy	Pearson's r	0.10	0.24**	-0.24**	1		
	Sig. (2-tailed)	0.210	0.003	0.002			
Market	Pearson's r	-0.22**	-0.17*	-0.41**	-0.21**	1	
	Sig. (2-tailed)	0.005	0.035	0.000	0.009		
Hierarchy	Pearson's r	0.05	-0.08	-0.46**	-0.23**	-0.40**	1
	Sig. (2-tailed)	0.500	0.322	0.000	0.004	0.000	

* Correlation is significant at the 0.05 level (2-tailed).
** Correlation is significant at the 0.01 level (2-tailed).

The results of the Pearson's r test show a significant ($\alpha = .05$) large positive correlation ($r = .68$) between PPI and BPO ($p < .001$) independent of culture type indicating organizations with a high PPI score also have a high BPO score. The test results indicate the adhocracy culture type has a significantly ($\alpha = .05$) positive correlation with BPO ($r = .24$). The market culture type has a significantly ($\alpha = .05$) negative correlation with both BPO ($r = -.17$) and PPI ($r = -.22$) indicating the stronger the market culture, the less likely the organization adopts BPM. Each of the relationships between the four culture types has a significantly ($\alpha = .05$) negative correlation with the other culture types and evenly divided between a small or medium strength, all significant at the 0.01 level (2-tailed).

4.4 Summary of Hypothesis Testing

The specific research question for the present study was how does the success of BPM adoption with organizations located in the United States implementing BPM initiatives involving IT solutions vary between different types of organizational culture within IT? Table 9 presents the results of our study indicating all of the hypotheses were supported.

Table 9. Summary of hypotheses tests

Hypothesis	Result
H_a^1 The dominant IT organizational culture type will have a significant impact on BPO	Supported
H_a^2 The dominant IT organizational culture type will have a significant impact on PPI	Supported
H_a^3 BPO and PPI are positively correlated with each other	Supported

Based on the results of our study, the adhocracy organizational culture type achieved the highest level of BPM adoption and conversely, the market organizational culture type achieved the lowest level of BPM adoption.

4.5 Interpretations of Findings

The results indicate an IT organizational culture type of adhocracy has a greater degree of BPM adoption success as measured by BPO when compared to the market and hierarchy organizational culture types. Organizations attained the highest level of BPM adoption success with the adhocracy culture type, while organizations attained the lowest with the market culture type. Practitioners seeking to improve factors that influence BPM adoption success within their organization can use this information to tailor their IT culture to more suit the adhocracy culture type, particularly if their current dominant IT organizational culture type is market or hierarchy.

The characteristics of the adhocracy culture type include individualism, risk-taking, innovativeness, growth, and a dynamic environment [32]. The adhocracy culture exemplifies the belief that an organization's competitive advantage with the organization's existing products and service is temporary and to maintain a competitive advantage, new products and services must be continually provided to the organization's customers [31, 33]. The development of innovative IT products that provide improvements in effectiveness and efficiencies are examples of BPM IT initiatives [7], which may explain why the adhocracy culture type within the IT community was found to be the most favorable for BPM adoption.

The five IT culture values (i.e., self-rule, organization and accuracy, technological innovation, admiration for technology, and enjoyable work environment) align well with the values associated with the adhocracy culture type [15]. [15] argue there are less cultural conflicts when cultural values align and may explain why the adhocracy culture type achieved the highest level of BPM adoption success.

According to [32], the values associated with the market culture type are a focus on results, setting and reaching goals, hard-driving, winning, and competition. A potential reason why the results found a negative correlation with the market culture type and BPM adoption is the cultural aspects of IT do not align well with any of the values associated with the market culture type. Practitioners who determine their organization's IT community has a dominant market culture type may search for ways to incrementally tailor their IT culture towards characteristics more closely aligned with the adhocracy culture type.

The [22] study found a causal relationship where the clan culture type had the highest support for BPM adoption success and the hierarchy culture type had the lowest level of support for BPM adoption success. The differences between this study and [22] may be due to national differences between the United States and Slovenia. According to [44], the culture of the local society affects an organization's culture and argues national culture is a factor in the performance of an organization, decision-making, behavior, and how people react to the types of transformational changes that come from BPM IT initiatives.

Comparing to the [22] results, the differences in the results may have less to do with location and more to do with the type of resources. IT resources were targeted in the present study while participants in the [22] study were top managers and process owners, who may be more representative of the culture of the organization. [15] and [16] argue occupational cultures, such as an IT culture, may be a more significant factor influencing behavior than national culture or the organization's culture. The IT community is an integral component of many of the most significant BPM initiatives [5, 7] and the high failure rate of BPM IT initiatives may be a result of a cultural conflict between the IT culture and the culture necessary for BPM success [14].

According to [45], the success of an IT project depends on the cultural fit between the IT culture and the culture of those to whom the IT product impacts. When the IT product is a BPM IT initiative, the cultural fit is between the IT culture and the BPM culture (i.e., CERT values) necessary for the success of the BPM IT initiative [14]. The CERT value related to "create" aligns well with the adhocracy culture type. IT's role in most organizations is to create new, innovative products, which may explain why the results found the adhocracy culture type to be the most favorable to BPM adoption success whereas the [22] study found the clan type most favorable.

4.6 Limitations

A limitation of the present study is that participants were from a cross-section of industries throughout the United States rather than a single industry. Rather than provide a narrow view of the relationship between IT organizational culture to the success of BPM initiatives involving IT solutions, the present study intended to provide a broad view. The present study only analyzes the dominant cultural type of the IT organization and does not consider the level of other non-dominant culture types present relative to the dominant culture type and the influence this might have on the BPM adoption success. [14] indicate successful BPM initiatives need the presence of all four CVF culture types. The sample size of the adhocracy culture type was small ($n = 17$), which may have impacted the results. Changes in the IT organizational culture could also be the consequence of the implementation of BPM IT initiatives or other factors not considered in this study. Most BPM IT initiatives effect the business processes of how organizations deliver value to their customers which would affect the organizational culture, but there would be less effect on the IT organizational culture as IT business processes likely did not change.

4.7 Summary, Conclusions and Future Work

An organizational culture not supportive of the changes required to implement BPM initiatives is a significant factor in organizations failing to adopt BPM [13]. This study found a positive correlation between BPM adoption and the adhocracy culture type. The more an organization identifies with the adhocracy culture type, the more the organization achieves BPM adoption success as measured by BPO. The values of the adhocracy culture type align well with the values of the IT culture identified by [15], which, in turn, may be a better cultural fit with the BPM CERT values. The results of the present study argue the success of BPM IT initiatives should account for the IT organizational culture.

This study found a negative correlation between BPM adoption and the market culture type. The more an organization identifies with the market culture type, the less likely that organization can achieve BPM adoption success. In contrast to the adhocracy culture type, there is a lack of alignment between the IT culture values identified by [15] and the characteristics that reflect the market culture type.

For organizations implementing BPM IT initiatives, an organizational cultural analysis should be conducted to determine the dominant culture type within their IT community. If the dominant culture type of the IT community is adhocracy, then the organization should implement policies and practices that enhance and support the characteristics of the adhocracy culture type. If the dominant culture type of the IT community is not adhocracy, then the organization should begin an incremental process of adapting the IT culture to more reflect the adhocracy culture type. Examples of policies and practices that support the adhocracy culture type and are aligned with the IT cultural values [15] and CERT values [14] are rewarding employees for innovativeness and risk-taking, providing opportunities to learn new technologies beneficial to the organization, decentralizing decision-making, and encouraging and inspiring employees to help each other to achieve organizational goals specifically tied to BPM IT initiatives.

BPM initiatives challenge organizations to provide increased value to their customers by implementing innovative ideas. Company executives should leverage the core IT cultural value of technological innovation by challenging their IT team to deliver industry-leading IT solutions. Company executives should leverage the IT cultural value of a dynamic, changing environment by insisting on a very aggressive release schedule of IT products that consistently brings value to the customer. To support the IT cultural value of growth, organizations should have policies to cover the costs of continuing education, as well as undergraduate and graduate level degrees related to IT and then be supportive of implementing innovative concepts and designs employees have learned into the organization.

We recommend conducting future quantitative research increasing the sample size of the adhocracy culture type. We recommend conducting future quantitative research to narrow down the industry type (e.g., aerospace, IT, manufacturing). Follow-on studies should use a larger total sample size, which would provide a more precise mean and a smaller margin of error. We recommend conducting this research in a different location (e.g. Germany or the European Union) to determine if similar results are found. Finally, we recommend a qualitative study to explore the underlying factors of the alignment between an IT organizational culture type of adhocracy and BPM adoption success.

References

1. Harmon, P.: The scope and evolution of business process management. In: Brocke, J, Rosemann, M. (eds.) Handbook on Business Process Management 1. International Handbooks on Information Systems, pp. 37–80. Springer, Heidelberg (2015). https://doi.org/10.1007/978-3-642-00416-2_3
2. Zairi, M.: Business process management: a boundaryless approach to modern competitiveness. Bus. Process Manag. J. **3**, 64–80 (1997). https://doi.org/10.1108/14637159710161585
3. Luftman, J., Derksen, B.: European key IT and management issues & trends for 2014. CIONET Eur. Bus. IT Trend Inst. **1**, 36 (2014)

4. Kasim, T., Haračić, M., Haračić, M.: The improvement of business efficiency through business process management. Econ. Rev.: J. Econ. Bus./Ekonomska Revija: Casopis za Ekonomiju i Biznis **16**(1), 31–43 (2018)
5. Rahimi, F., Moller, C., Hvam, L.: Business process management and IT management: the missing integration. Int. J. Inf. Manag. **36**(1), 142–154 (2015)
6. vom Brocke, J., Rosemann, M.: Handbook on Business Process Management: Strategic Alignment, Governance, People and Culture. International Handbooks on Information Systems, vol. 2. Springer, Berlin (2010)
7. Polakovič, P., Šilerová, E., Hennyeyová, K., Slováková, I.: Business process management in linking enterprise information technology in companies of agricultural sector. AGRIS on-Line Pap. Econ. Inform. **10**(3), 119–126 (2018). https://doi.org/10.7160/aol.2018.100310
8. Tallon, P.P., Queiroz, M., Coltman, T.R., Sharma, R.: Business process and information technology alignment: construct conceptualization, empirical illustration, and directions for future research. J. Assoc. Inf. Syst. **17**(9), 563 (2016). https://doi.org/10.17705/1jais.00438
9. Macdougall, C., Michaelides, R.: Combatting IT failure rates through IT program executive sponsorship. Paper presented at Project Management Institute Research and Education Conference, Newtown Square, PA, Project Management Institute, Phoenix 29 July 2014
10. Rosemann, M., vom Brocke, J.: The six core elements of business process management. In: vom Brocke, J., Rosemann, M. (eds.) Handbook on Business Process Management 1. IHIS, pp. 105–122. Springer, Heidelberg (2015). https://doi.org/10.1007/978-3-642-45100-3_5
11. Trkman, P.: The critical success factors of business process management. Int. J. Inf. Manag. **30**(2), 125–134 (2010)
12. Alibabaei, A., Aghdasi, M., Zarei, B., Stewart, G.: The role of culture in business process management initiatives. Aust. J. Basic Appl. Sci. **4**(7), 2143–2154 (2010)
13. Schmiedel, T., vom Brocke, J., Recker, J.: Culture in business process management: how cultural values determine BPM success. In: vom Brocke, J., Rosemann, M. (eds.) Handbook on Business Process Management 2. IHIS, pp. 649–663. Springer, Heidelberg (2015). https://doi.org/10.1007/978-3-642-45103-4_27
14. Schmiedel, T., vom Brocke, J., Recker, J.: Which cultural values matter to business process management? Bus. Process Manag. J. **19**(2), 292–317 (2013). https://doi.org/10.1108/14637151311308321
15. Jacks, T., Palvia, P., Iyer, L., Sarala, R., Daynes, S.: An ideology of IT occupational culture: the ASPIRE values. ACM SIGMIS DATABASE: Database Adv. Inf. Syst. **49**(1), 93–117 (2018)
16. Schein, E.H.: Some thoughts about the uses and misuses of the concept of culture. J. Bus. Anthropol. **4**(1), 106–113 (2015)
17. Tumbas, S., Schmiedel, T.: Developing an organizational culture supportive of business process management. In: Proceedings of the 11th International Conference on Wirtschaftsinformatik. Leipzig, Germany (2013)
18. Hammer, M.: What is business process management? In: vom Brocke, J., Rosemann, M. (eds.) Handbook on Business Process Management 1. IHIS, pp. 3–16. Springer, Heidelberg (2015). https://doi.org/10.1007/978-3-642-45100-3_1
19. Beckett, C., Myers, M.D.: Organizational culture in business process management: the challenge of balancing disciplinary and pastoral power. Pac. Asia J. Assoc. Inf. Syst. **10**(1) (2018). https://doi.org/10.17705/1pais.10102
20. Indihar Štemberger, M.I., Buh, B., Glavan, L.M., Mendling, J.: Propositions on the interaction of organizational culture with other factors in the context of BPM adoption. Bus. Process Manag. J. **24**(2), 425–445 (2018). https://doi.org/10.1108/BPMJ-02-2017-0023
21. Aparecida da Silva, L., Pelogia Martins Damian, I., Inês Dallavalle de Pádua, S.: Process management tasks and barriers: functional to processes approach. Bus. Process Manag. J. **18**, 762–776 (2012). https://doi.org/10.1108/14637151211270144

22. Hribar, B., Mendling, J.: The correlation of organizational culture and success of BPM adoption. In: Proceedings of the 2014 European Conference on Information Systems, Tel Aviv, Israel (2014). ISBN 978-0-9915567-0-0

23. Reijers, H.A., van Wijk, S., Mutschler, B., Leurs, M.: BPM in practice: who is doing what? In: Hull, R., Mendling, J., Tai, S. (eds.) BPM 2010. LNCS, vol. 6336, pp. 45–60. Springer, Heidelberg (2010). https://doi.org/10.1007/978-3-642-15618-2_6

24. Vukšić, V.B., Glavan, L.M., Susa, D.: The role of process performance measurement in BPM outcomes in Croatia. Econ. Bus. Rev. Central S.-E. Eur.**17**(1), 117–143/149–150 (2015)

25. Bandara, W., Gable, G.G., Rosemann, M.: Business process modeling success: an empirically tested measurement model. In: Proceedings of International Conference on Information Systems, pp. 1–20. University of Wisconsin, Milwaukee (2006)

26. Skrinjar, R., Trkman, P.: Increasing process orientation with business process management: critical practices. Int. J. Inf. Manag. **33**(1), 48–60 (2013)

27. McCormack, K., Johnson, W.: Business Process Orientation: Gaining the E-Business Competitive Advantage. St. Lucie Press, Florida (2001)

28. Rummler, G.A., Brache, A.P.: Improving Performance: How to Manage the White Space in the Organizational Chart, 2nd edn. Wiley, San Francisco (1995)

29. Allaire, Y., Firsirotu, M.E.: Theories of organizational culture. Organ. Stud. **5**, 193–226 (1984). https://doi.org/10.1177/017084068400500301

30. Napitupulu, I.H.: Organizational culture in management accounting information system: survey on state-owned enterprises (SOEs) Indonesia. Glob. Bus. Rev. **19**, 556–571 (2018). https://doi.org/10.1177/0972150917713842

31. Cameron, K.S., Quinn, R.: Diagnosing and Changing Organizational Culture: Based on the Competing Values Framework, 2nd edn. Wiley, San Francisco (2006)

32. Chatterjee, A., Pereira, A., Bates, R.: Impact of individual perception of organizational culture on the learning transfer environment. Int. J. Train. Dev. **22**(1), 15–33 (2018). https://doi.org/10.1111/ijtd.12116

33. Lizbetinová, L., Lorincová, S., Caha, Z.: The application of the organizational culture assessment instrument (OCAI) to logistics. Enterprises/Primjena instrumenta procjene organizacione kulture (OCAI) na logisticke tvrtke. Nase More, **63**(3), 170–176 (2016)

34. Koh Tas, Y., Low Sui, P.: Organizational culture and TQM implementation in construction firms in Singapore. Const. Manag. Econ. **26**, 237–248 (2008). https://doi.org/10.1080/01446190701874397

35. Counts by Company Size (n.d.). https://www.naics.com/business-lists/counts-by-company-size/. Accessed 13 Feb 2018

36. Faul, F., Erdfelder, E., Lang, A.G., Buchner, A.: G* power 3: a flexible statistical power analysis program for the social, behavioral, and biomedical sciences. Behavi. Res. Methods **39**, 175–191 (2007)

37. Kang, K., Kim, S., Kim, S., Oh, J., Lee, M.: Comparison of knowledge, confidence in skill performance (CSP) and satisfaction in problem-based learning (PBL) and simulation with PBL educational modalities in caring for children with bronchiolitis. Nurse Educ. Today **35**, 315–321 (2015). https://doi.org/10.1016/j.nedt.2014.10.006

38. Leys, C., Ley, C., Klein, O., Bernard, P., Licata, L.: Detecting outliers: do not use standard deviation around the mean, use absolute deviation around the median. J. Exp. Soc. Psychol. **49**, 764–766 (2013). https://doi.org/10.1016/j.jesp.2013.03.013

39. Downey, R.G., King, C.V.: Missing data in Likert ratings: a comparison of replacement methods. J. Gen. Psychol. **125**(2), 175–191 (1998). https://doi.org/10.1080/00221309809595542

40. Gushta, M.M., Rupp, A.A.: In: Salkind, N.J. Encyclopedia of Research Design (2010). SAGE Publications, Thousand Oaks (2010)

41. Field, A.: Discovering Statistics Using IBM SPSS Statistics: North American Edition, 5th edn. SAGE Publications, Thousand Oaks (2017)
42. Amani, M., Ghadimi, N., Aslani, M.R., Ghobadi, H.: Correlation of serum vascular adhesion protein-1 with airflow limitation and quality of life in stable chronic obstructive pulmonary disease. Respir. Med. **132**, 149–153 (2017). https://doi.org/10.1016/j.rmed.2017.10.011
43. Cohen, J.: Statistical Power Analysis for the Behavioral Sciences, 2nd edn. Erlbaum, Hillsdale (1988). https://doi.org/10.4324/9780203771587
44. Hofstede, G.: Culture's Consequences: International Differences in Work-related Values. SAGE Publications, Beverly Hills (1980)
45. Rao, V., Ramachandran, S.: Occupational cultures of information systems personnel and managerial personnel: potential conflicts. Commun. Assoc. Inf. Syst. **29**, 581–604 (2011). https://doi.org/10.17705/1CAIS.02931

Process Mining Adoption

A Technology Continuity Versus Discontinuity Perspective

Rehan Syed[1]([✉]), Sander J. J. Leemans[1], Rebekah Eden[1],
and Joos A. C. M. Buijs[2]

[1] Queensland University of Technology, Brisbane, Australia
r.syed@qut.edu.au
[2] Algemene Pensioen Groep, Heerlen, Netherlands

Abstract. Process mining is proffered to bring substantial benefits to adopting organisations. Nevertheless, the uptake of process mining in organisations has not been as extensive as predicted. In-depth analysis of how organisations can successfully adopt process mining is seldom explored, yet much needed. We report our findings on an exploratory case study of the early stages of the adoption of process mining at a large pension fund in the Netherlands. Through inductive analysis of interview data, we identified that successful adoption of process mining requires overcoming tensions arising from discontinuing old practices while putting actions into place to promote continuity of new practices. Without targeted strategies implemented to transition users away from old practices, data quality is jeopardised, decision-making is impeded, and the adoption of process mining is ultimately hampered.

1 Introduction

Using Business Process Management (BPM) principles, organisations can improve and optimise their business processes [15]. A key contributor to BPM initiatives is process mining, which involves the data-driven analysis of the historical behaviour of business processes. Process mining techniques provide organisations with the ability, amongst other things, to monitor performance indicators, automatically discover process models, identify resource constraints and bottlenecks, and determine the extent of regulatory performance [1]. Through applying these techniques, process mining is proffered to improve organisation's decision making practices [2]. Recently, process mining has seen a large uptake by organisations across many fields, including healthcare processes [12,32], shared services [31], financial services [4,10], software development [11], and insurance [38]. Further evidence of the surge of process mining is the recent entry of many vendors into the market [24].

Despite the recent uptake of process mining, several issues have been experienced. Moreover, to the best of our knowledge, there has been limited research attention into how process mining has been used *within* organisations and how organisations adapt (or not) to the new technology is uncharted territory.

© Springer Nature Switzerland AG 2020
D. Fahland et al. (Eds.): BPM Forum 2020, LNBIP 392, pp. 229–245, 2020.
https://doi.org/10.1007/978-3-030-58638-6_14

In this paper, we examine the ongoing adoption of process mining within an organisation using the theory of technology discontinuity [34], which states that technology is a central force affecting environmental conditions, and that populations within organisational communities may appear or disappear based on the rise and fall of technology. That is, technology is an important source of variances in the environment and thus is a critical factor in population dynamics. New technologies aim to 'discontinue' legacy technologies, whereas users' familiarity with the legacy systems pushes for the 'continuity', which creates tension paradigms (for more detail, see Sect. 2). As such, our **Research Question** is *What are the factors that influence process mining continuity in organisations?*

To provide insights into this question, we conducted an exploratory, inductive case study of the early stages of the adoption of a process mining tool at a large pension fund in the Netherlands. We followed an inductive approach; thus, we did not set out to study process mining adoption through the theory of technology discontinuity [34]. Rather, its importance emerged throughout our data analysis, which involved constantly iterating between data and literature. Nevertheless, for simplicity, we structure the paper sequentially. Next, we provide the theoretical background. Subsequently, we detail our methodology and the background of our case organisation. Following, we present the key challenges and enablers to consider in the adoption of process mining. Then, we discuss the implications of our findings through the theory of technology discontinuity and conclude by outlining our theoretical and practical contributions.

2   Theoretical Background

In the past, a common misconception was held that implementing new technologies in organisations would directly and automatically result in benefits. Following the widely reported IT productivity paradox, where technologies had been implemented yet benefits not obtained for long periods of time [7–9], this technology deterministic attitude has been largely rebuked [30]. Many argued that technologies must be accepted [14,35] and used [30] if benefits would be obtained. Although not a necessary and sufficient condition, technology acceptance is considered to be a precursor to use and benefits. This spurred cumulative research into the technology acceptance [36], which focused on factors related to performance and effort expectations, social norms, facilitators, as well as demographic variables that predicts an individuals intention to use a technology. However, in the context of our case study rather than these individual factors being critical to the acceptance and ongoing use of process mining, it was the tensions between legacy practices and new practices, as discussed later, which was salient. As such, below we focus on literature related to discontinuing the use of existing practices and technologies.

The notion of the discontinuity of technology emerged in the Management discipline, with the development of the theory of technology discontinuity [3,34]. This theory sort to explain, at a macro level, how technology change influences the organisational landscape. The authors [34] explained that process discontinuities occur in the form of process substitution, or in process innovation that results in major breakthroughs in any given industry. These process discontinuities can be categorised as competence-destroying or competence-enhancing, as described below:

- Competence-destroying: refers to new ways of making a product or completing tasks that require new skills and abilities. It also requires new technologies as the resultant practice is fundamentally different to the existing practice.
- Competence-enhancing: refers to improvements in existing ways of making products or completing tasks which do not make existing skills and abilities obsolete, rather it is an incremental improvement to the technology [34].

Competence-destroying and competence enhancing discontinuities typically occur due to changes in the competitive environment, which requires organisations to shift and for management to put new initiatives and technologies in place. Yet, The introduction of a new technology in an organisation often leads to organisational change and adoption issues [13]. For technological change to seed, organisational members are required to discontinue their legacy practices and technologies in favour of new technologies and practices, which can be marked by resistance. The aim of the discontinuation of old technologies is to give way to the new concepts, processes, and systems. The legacy technologies that initially bring the innovation and build the foundation of new business models themselves become a blockade for new technologies due to their familiarity and institutionalisation [25]. Our analysis, as later described demonstrated that with competence-enhancing practices, tensions arose as individuals have the potential to revert back to legacy practices. As [3] notes, "older technological orders seldom vanish quietly; competition between old and new technologies is fierce".

3 Methodology

We performed a qualitative, inductive, case study [16] to explore the early stages of the adoption of process mining. In this section, we first describe the case organisation, followed by the data collection and analysis techniques used.

3.1 Case Organisation

The case organisation, APG (Algemene Pensioen Groep), is a large provider of services to pension funds in the Netherlands. APG's direct customers are pension fund providers who outsource some of their end-user focused processes to APG. As these customers have many different processes, rules and regulations, and cater to the needs of 4.6 million pension-fund participants[1], there is a vast potential for process improvement of more than $2,230$ pension-related processes in APG.

To optimise its processes, APG has been collecting data from several process-related systems for over 8 years. Recently, APG commenced collecting this data

---

[1] See https://www.apg.nl/en.

in a single centralised data warehouse ("data core") bringing data under central management using the DAMA-DMBOK data management approach [23]. However, several analyses are still performed on data directly from separate source systems. To leverage the data and to assist with process optimisation efforts, APG had been using business intelligence tools. Two years ago, APG switched its focus to process mining and implemented and commenced using Celonis Process Mining[2]. The process mining initiative was largely driven in a bottom-up manner, whereby Celonis was piloted in a dozen cases, of which a few led to bigger projects: for instance, a large centralised customer journey analysis [10].

In the early phases, governance was less strict, as a 'launch and learn' approach was adopted. This lead to some departments within APG to use Celonis on their own. At the same time, they were missing specific expert guidance, data delivery and governance support. Recently, more governance, expertise, and business user guidance has been set-up in the organisation.

At the time of data collection, APG was on the verge of rolling out the use of Celonis to a large number of users through a generic process analysis dashboard, covering all 2,230 pension related processes. Dashboards were being built, training was being performed and the first users were starting to use Celonis on a daily basis.

3.2   Data Collection and Analysis

Data was primarily collected from APG through performing semi-structured interviews (see Table 1 for interview protocol summary) with participants during the early stages of adoption (December 2019).

Nine interviews were performed, which lasted between 30 and 45 min on average. Purposeful sampling [18] was used to identify interview participants to ensure different perspectives were garnered to facilitate constant comparison. Interviews were, therefore, conducted with data/process intelligence experts, and the Celonis dashboard end-users. All interviews were recorded and transcribed. nVivo (v12) was used as a data repository to collate our interview coding.

To analyse the interviews, we first performed open coding [17] to inductively identify the key themes related to the adoption of Celonis at APG. As such, we did not have a framework to deductively analyse the interviews; rather, we let the themes emerge [17]. As a result, 500 reference nodes inductively emerged from the interview data. The tensions experienced between discontinuing old practice in favour of continuing new practices emerged as a central theme.

Subsequently, as per [19] we performed on-coding, which involved constantly comparing the different themes together, resulting in a refined list of challenges and enablers related to the tensions associated with the adoption of process mining. As discussed in Sect. 4, seven challenges and four enablers of process mining have been identified.

We then performed theoretical coding [37] to identify the key relationships between the challenges and enablers (see: Sect. 5.1). Throughout this process, we

---

[2] See https://www.celonis.com/.

Table 1. Interview protocol.

Question type	Purpose	Example interview questions
Introductory Questions	To understand the processes the participants are involved with and how they interact with Celonis (e.g., process mining expert, end user)	1. How do you use Celonis to perform your work? 2. Why do you use Celonis to perform your work?
Overview of using the system	To understand how the system is used to attain the user goal and the factors that alter how the system used	3. How do you use Celonis in a way that helps you attain your goals? 4. How does your use of Celonis vary depending on different factors?
Dashboard building process	To understand the process and factors behind the design and build of the dashboard	5. When you build dashboards how do you select the elements (e.g., features) to use? 6. What is the most difficult question you have attempted to answer using the dashboard elements?
Validation of dashboard output	To understand how the users assess the validity of the information provided by the dashboard	7. Has Celonis ever provided inaccurate information? How did you discover this? 8. How do you validate your conclusions from Celonis?
Factors influencing use	To understand the facilitators and constraints for users to effectively adopt these dashboards	9. What facilitates the effective use of Celonis? 10. What constrains the effective use of Celonis?
Outcomes	To understand the impact of using Celonis	11. What are the impacts of using Celonis? 12. Do you have any stories that highlight the effect of Celonis?

constantly compared our findings to literature and recognised the importance of technology continuity theory and other relevant literature (see Sect. 5). This iterative process of open coding, on-coding, and theoretical coding continued until theoretical saturation was reached [21]. We determined that theoretical saturation was attained when the challenges, enablers, and relationships between challenges and enablers were stable (i.e., no new themes related to these themes or relationships emerged from the analysis).

Throughout this iterative inductive analysis process, to maintain reliability of the coding coder-corroboration was used [33]. Following this approach three researchers coded the interview data and then discussed any differences until consensus was reached. While our coding was manually performed [20], we also supplemented our findings with additional analyses performed in nVivo, including cross-tab and matrix-coding queries to discover interrelationships between different factors [5]. Near and AND operators were used to analyse relevant concepts discussed in the interview data.

4   Findings

In this section, we provide an account of the challenges (Table 2) and enablers (Table 3) experienced by the APG in the early stage of process mining adoption.

4.1   Challenges of Process Mining

Overall, we identified seven process mining adoption challenges which are summarised in Table 2 and detailed below

Table 2. Identified process mining challenges.

Challenge	Definition	Reference interview quotes
Governance	Organisational level challenges concerning the policies, regulations, roles and responsibilities, and accountability	*You have a question who do you go to, do you go to me, do you go to the owner of the dashboard? ... We are still arguing about that and it is still not clear; ... So the first issue we have, which is not a Celonis issue, it is ... the way we govern our data and structure our data (A8)*
Collaborative tensions	The inter-dependency, information collaboration, and communication tensions between teams in a process mining initiative	*The pension dept. and the change dept. don't have a role in the community ... using the dashboard. Report comes from the communication department. ... They are serving other departments, which is what we have to overcome (A1)*
Data & information quality	Users' misunderstanding of process mining outcomes due to inaccurate information, inconsistent interpretation of data	*... It's quite difficult sometimes to get the right data from all the systems. [It's] the biggest limitation inside APG, to get the data and build the view, on which you can build the dashboard that shows what you want to see or analyse (A2)*
Technical	Users' resistance towards dashboard and process mining systems due to their perceptions of the technical aptitude of the system to provide accurate information	*systems are currently limited, because of history, because the systems have been built on for ten years or something, and you have this like little sub-process, I guess you can call them (A9)*
Process complexities	The unclear process boundaries, interfaces and emergent complexities due to several interrelated processes	*[In] pension administration, most processes are handled in one system, and that was quite hard already, and here they have 7 different teams systems, and one process can span across ... all 7 systems. ... You need to connect and extract the data from all the systems, model views on that data, so that's quite an extensive process (A1)*

(continued)

Table 2. (*continued*)

Challenge	Definition	Reference interview quotes
User tailoredness	User resistance towards standardised features of a process mining system thus limiting the flexibility for user-customisation	*... They have all kinds of different tools they are used to. Human habit is to grab what they know. This is new, so, every time we show them more and more, they get more and more interested. But, there is a saying in Dutch "what I don't know I don't eat", they have to get used to it, it's not that they are avoiding it, no, they are not used to it yet (A3)*
User-related	User challenges related to training, confidence in their ability to use the PM tools, and sunk costs	*There are always challenges, if you are using a new tool. You have to learn the buttons: where is what, how does it interact when I do something here, how does it work? The biggest challenge is to start with the first dashboard (A5)*

Governance Issues. The participants mentioned the absence of appropriate governance mechanisms early on as a challenge for process mining initiatives. The need for a well-defined structure, policies and regulations, and clearly defined separation of responsibilities were mentioned as vital needs to enable employees to use process mining. The dependency on receiving the required technical or context expert advice led to delays in determining the quality and accuracy of the process mining outcomes. The support requests from users who were not familiar with process mining tools faced delays due to unwritten and ad-hoc practices in the absence of a sound governance mechanism.

Collaborative Tensions. The initial split between the design team and the end-users did not always work out well. The initial artefacts were designed and implemented by the Celonis experts, but the end-users were contacted too late. Furthermore, the lack of coordination between the technical implementation team and the end-users led to increased confusions on how to use the process mining features effectively. The differences in the definitions between different collaborating departments should be rectified, aligned, and incorporated in technical feature design. Additionally, the absence of clearly defined roles and responsibilities and the final ownership of dashboard has created confusion amongst the staff. The ambiguity related to who would provide the post-implementation support, who should address technical queries, and who should take the final approval and decisions on data access and quality was yet to be addressed.

Data & Information Quality Challenges. The users' confidence on process mining outcomes were significantly influenced by the quality of data and information. The data visualisation is generated by using different data sources; hence, consistency and accuracy mismatches resulted in users lacking confidence in the generated output. Furthermore, the data source itself was mentioned as a primary criterion to build user confidence in the process mining outcomes. The inconsistent naming conventions and redundancies that existed in the legacy data sources contributed significantly to incorrect insights. The bureaucratic inter-dependencies between different organisational functions slow-downed the ability of the technical team to take corrective actions. Access to the right data sources was considered as the primary reason that hampered the technical staff's ability to provide relevant insights to the users. The respondents also confirmed the significant loss of development time, due to communication issues, that was needed to overcome the stakeholders' differences on naming attributes and process definitions. For example, a process was considered to be a straight-through-process when the automation rate reaches 100% by technical staff, whereas, 90% was considered sufficient by the beneficiary department. The respondents also confirmed the difficulties in data interpretation were not caused by Celonis, but it was a result of incorrect data input or combination of different data-sets. The long-term view of system expansion and future requirements of data quality were not taken into consideration during initial design phases. The efforts to prioritise and improve the input data quality via the data-core to Celonis were negatively affected by the technical limitations of the old system, which did not have the option to update.

Technical Challenges. The nature of technical issues ranges from Celonis design to the legacy nature of existing systems. The respondents have mentioned that a few requirements had not been developed because Celonis process explorer did not support the functionality. The existing confusions on post-implementation maintenance aspects of Celonis were linked with the absence of governance frameworks and policies. The respondents appreciated the modular development approach adopted by the implementation team, since it supported users' familiarity and expectations of Celonis. The real-time data availability was mentioned as a vital element to address stakeholders' demands for on-time information.

Process Complexities. The extensive amount of process exponentially adds to the analytical complexities. There are 2, 230 processes used in the organisation, which complicated the ability to perform deeper analysis of data. The nature of user requirements depends on data from a variety of processes from different departments, and hence, users were not able to explain exactly the type of analytical output suitable for their needs. Sub-dashboards were developed as a workaround for performing multi-level analysis to address the inter-dependencies between processes and sub-processes, which led to production of ambiguous interpretations of analysis.

User-Tailordness. The users were quite familiar with their previous systems for process and data analysis. Different tools were used by different users based on their familiarity and experience with the tool. The respondent mentioned Celonis is user-friendly; however, they also mentioned that there exists resistance to use Celonis because it is designed to provide standard features without addressing specific needs.

User Related Challenge. Users' familiarity with old tools and techniques caused the risk of users creating their own dashboards by spending significant amount of time on personalisation. Furthermore, with the customisation, the alignment with the available or required data also posed a major risk. Respondents mentioned that the development of useful technical features was not an issue; however, whether or not these features would be used by the end-user to create useful insights was yet a concern. Training and development had been recognised as a major challenge by the respondents. Staff with different level of technical capabilities required a wide variety of training interventions.

4.2 Enablers of Process Mining

Four process mining enablers were identified, which are summarised in Table 3 and described below.

Table 3. Identified process mining enablers.

Enabler	Definition	Reference interview quotes
Actionable insights	Users' ability to take meaningful actions resulting from process mining analysis	*With excel you were testing your hypotheses and seeing what the outcomes were and making changes based on that. Whereas now you are able to actually explore and find new areas to target (A7)*
Confidence in process mining	User's trust and confidence in the accuracy, reliability, and applicability of process mining	*I think it was because they had large data-sets and that you run into the limits of using just excel and I think they didn't always know of other possibilities well, we showed them what else was possible but I guess the reason they came to us was the limitations with their current methods (A9)*

(continued)

Table 3. (*continued*)

Enabler	Definition	Reference interview quotes
Perceived benefits	Expected individual and organisational benefits associated with process mining	*We saw possibilities in Celonis beyond the process mining itself. In the way we wanted our reports, we used Celonis for it. Another way of using Celonis is to set up a dashboard that will help us analyse processes within APG but more thoroughly (A3)*
Training & development	Actions and activities performed to improve awareness, familiarity, and users skills to use process mining tools in the organisation	*First ... we demoed the dashboard ... we just show an impression of the dashboard, this dashboard has this and this ... no more in depth questions about how do I see this. Just this is what you see in the dashboard. [Then] we will plan more sessions for just a few users and we will go more in depth with them (A8)*

Actionable Insights. The participants explained the ability to generate action-able insights is a key factor driving the Celonis adoption. Users with analytical mindsets and above average technical competencies are increasingly using the system to address their operational intelligence needs. The capabilities to per-form deeper analysis are well-appreciated by the stakeholders. By using the insights provided by Celonis, the staff can now visually see the actual progress and bottlenecks that restrict achievement of their key performance indicators. *... it was like the tool for the time, they had to finish their target within 180 days, we showed them that by far they didn't reach that goal. And they didn't know where to improve it, waiting times, and how to improve it. So we, after two week, we said this is (exactly) what the process is doing, you have wrong date, days, we see only you have achieved only 55% of the goal of finishing the process within 180 days, but even in those cases we showed them how much time it would take and where the bottlenecks were, and what the waiting time was, we showed them with very good clarity (A6).*

Confidence in Process Mining. The participants acknowledged dashboards were instrumental in maintaining self-service capabilities for users. There was a strong consensus on the effectiveness of Celonis in assisting users to perform complex analysis in an easy-to-use manner. The confidence in Celonis' ability to provide evidence-based information has resulted in signs of increasing use in APG. *That one [dashboard] doesn't lie. That's what I like about it, the system is proven by itself, it's developing, Celonis itself is developing... the management saw more and more possibilities in the way Celonis provides a view on it, so they* *wanted more and more information out of Celonis, or into Celonis to make it better visible for them (A3).*

Perceived Benefits. The use of Celonis is gradually increasing at APG, and the participants have already started to see the future benefits of Celonis for different stakeholder groups. One such area was attributed to conformance checking capabilities of the tools used. The ability to provide a holistic process overview by incorporating the complex process dependencies and inter-dependencies was well-perceived by respondents. *I guess it would be interesting for an auditor to look at a large amounts, large transactions and if large transactions need to have a certain signature, then maybe you could build it in a rule. Ok, the process needs to follow this for large amounts and the signature needs have to be checked otherwise it's non-conformation and if it's non-conformance then we look into why it happened. (A9).*

Training and Development. Various training and development activities were introduced to assist staff. Workshops were used to introduce concepts and develop users' skills. Users were also given hands-on demonstrations of the key features of Celonis. The development team took responsibility to provide training support to the end-users. *We had a workshop of an hour, to get to know it. We looked at it: how should you start? But it was based on an existing dashboard. When I started a new one for myself, I just clicked everywhere until I was ready. Learning by clicking, just do it, you cannot break it (A5).*

5   Discussion

In this section, we discuss our the relationships between the challenges and enablers of process mining adoption at APG. These relationships spoke to the tension that results from competence-enhancing discontinuity whereby there exists the pull to the old legacy processes and systems and a pull towards novel practices and processes. In doing so, we integrate relevant literature to present consensuses and contradictions. Following, we also provide insights into how APG perceived our findings.

5.1   Interrelationships Between Challenges and Enablers

In order to explore the links between challenges and enablers of process mining, we used NVivo's cross-tab query with a 'Near' operator. A 'Near' operator is used to identify the words within a specified word distance from each other. The results of the Near query are illustrated in Table 4.

"Actionable Insights" were found to be associated with collaborative tensions, data and information quality challenges, and process complexities. The value of process mining capabilities to provide interesting, valid, and useful insights was acknowledged by the respondents as a key enabler that counters the users'

Table 4. Interrelationships between challenges and enablers.

	Actionable insights	Confidence in PM	Perceived benefits	Training & development
Governance issues	**0**	**0**	0	1
Collaborative tensions	2	0	0	2
Data & information quality	3	8	2	1
Technical challenges	0	0	0	0
Process complexity	1	0	1	0
User tailoredness	0	1	0	0
User issues	0	1	0	0

intentions of continued use of old practices and tools, *I was quite happy with the dashboard, and these guys here at asset management used it not for the process, because the processes here are handled differently, but to show the people what they can do with Celonis, so actually I used it as an example (A2).* The respondents also referred to process mining as a viable technique to overcome the process complexities and reduce the information overload on users by generating meaningful insights: *Our processes are kind of different so one process can have 7 sub-processes or 7 ways to flow into other processes, have like one happy path so its all different per process. We adapt our dashboard to the process. So this one process has like seven or eight ways of possible happy flows, so how many have go through happy flows, 1, 2, 3, 4, 5, 6, to 8. From that point we analyse, so whatever question the business has we try to implement it into the dashboard and try to analyse what are the bottlenecks (A8).* The case details explain that the trust and confidence in process mining systems are dependent on the accessibility to, and reliability of, data sources. [28] recommended the use of ontologies to define the scope and cases from the data sources, depending on the nature of data requirements by diverse users.

The next relationships relate to the "Confidence in Process Mining" and Data & Information Quality challenges, User Tailoredness, and User Issues. As respondents mentioned, their confidence in process mining developed because of the completeness and consistency of information that Celonis provides. The end users' confidence in the data-core used by the organisation for the dashboard operations paved the way to maintain their trust in the process mining accuracy: *Sometimes, you just have to state that clearly the hypothesis was not correct, because the data states otherwise, and the data is 100% correct. If you find out that really isn't possible, then you have to go back and see if the data is correct (A2).*

On the other hand, the standardised nature of dashboards was mentioned as a potential risk for users' experiences with their old and specifically customised dashboards; hence, the standardised features may act as a factor contributing to their continued use of legacy systems, leading to process-continuity of old practices: *Those dashboards [legacy] really have the features they like and they need, and nothing else. And now there will be a dashboard that is not specially made for them personally, so there might be a risk that they keep using their own information base (A1).* Our observations are aligned with the findings discussed by [29], which states that information quality is a key determinant with an indirect effect on user trust and risk reduction.

"Perceived benefits" of process mining superseded the data & information quality challenges. The dashboards' capability to provide information in a modular, focused, and yet integrated overview of associated processes positively influenced the end-users' ability to comprehend the complexities involved. *About the dashboards, so like I said, the process knowledge is at a higher level within APG. Processes can be grouped into customer journey, so whenever someone retires they first get for instance, we have like process A, then process B, then process C, then process D, and all of these processes contribute to one customer journey, and this dashboard shows information about the customer journey itself (A8).* Our findings confirm the recommendation by [24] that users' awareness of the benefits of process mining can help overcome the challenges associated with enterprise adoption.

The "Training & Development" activities were mentioned as a viable option to overcome governance issues, collaborative tensions, and data & information quality challenges. The participants appreciated the clearly-defined training responsibilities that have helped them to understand the process mining tool and to troubleshoot problems: *The tool Celonis has been released by [development] team, so basically what I would usually say if you don't know Celonis at all, they [users] will come ask me about the dashboard, I don't understand, how does this work, how can I filter stuff, but we maintain the dashboard not the tool. So if they have questions about the tool they should get training by [development] team. That's what I proposed and that's what hopefully we will do (A8).* Along similar lines, the interaction with the development team for training purposes helped ease the collaborative tensions between different departments: *We are working together with [development team], who had more knowledge than I have about Celonis (A2).* The vital importance of training was also acknowledged by previous studies [6,26,27].

The insights gained from the above analysis reflect the two sets of practices that create the competence-enhancing tensions in organisations embarking in process mining initiatives. We did not identify an enabling counter factor to reduce the impact of technical challenges. Our findings resonate with [22], which states that technical system quality does not have a direct or indirect organisational impact. We concur that the technical challenges were not perceived as a barrier, because most technical issues are hidden from the end-users. The confidence in process mining was mostly observed as an enabler for the new practices (i.e. in this case process mining); however, it may contribute to user frustration towards unfamiliar dashboards/features as well. Figure 1 illustrates the interrelationships between the identified challenges and process mining enablers.

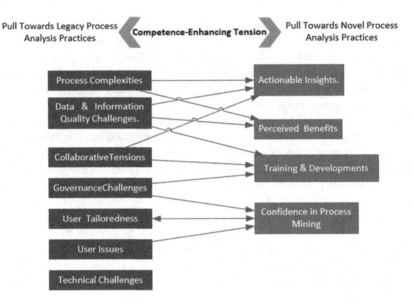

Fig. 1. Competence-enhancing tensions.

5.2   Reaction of APG

We provided the findings of this paper to APG. Overall, our findings align with APG's perception of the situation, and also mention some issues which are already addressed. For example, at the time of the interviews, distinction of roles (data owner(s), dashboard owner, dashboard maintainer, dashboard builder, etc.) was under development, and have been implemented and communicated since. This already makes the way-of-working and expectations and responsibilities clear. In the meantime, APG has also set-up a 'self-service' team next to the teams that build the dashboards. The self-service team is the go-to point for end-users and maintainers of the dashboards built; hence, it streamlines support. Some other findings mentioned by respondents were already addressed in the weeks and months before the interviews; for instance, balancing between generic and custom dashboards. Furthermore, because of the 'data-core' (the central data warehouse) concept and the application of the DAMA-DMBOK data management approach [23] APG is actually fostering discussions on aligning terminology within APG, providing insights in the data quality in the source systems, and thus enabling combining data from different (legacy) systems.

6   Conclusion

Process mining technologies aim to provide data-driven support to business process management initiatives and as such are being introduced in many organisations across many industries worldwide. To obtain tangible ongoing benefits from process mining, organisations need to adapt and incorporate process mining tools into business process management initiatives. In this paper we performed an exploratory, inductive case study of the factors that influence process mining adoption in organisations. To this end, we conducted interviews at a large Dutch pension fund in the early stages of adoption of process mining. Through repeated analysis and literature study, the importance of the continuity lens emerged. Through thematic analysis, we identified 7 challenges related to the adoption of process mining, and 4 enablers to overcome the challenges. Using the insights from the data, we present an initial framework explaining how challenges and enablers of process mining interact. The case is limited to a single organisation's experience. The findings of this study will be beneficial for the organisation embarking into their process mining journey. We propose future research to further explore the relationships identified by focusing on organisations with mature use of process mining.

References

1. van der Aalst, W.M.P.: Process Mining - Data Science in Action, 2nd edn. Springer, Cham (2016). https://doi.org/10.1007/978-3-662-49851-4
2. van der Aalst, W.M.P., Pesic, M., Song, M.: Beyond process mining: from the past to present and future. In: Pernici, B. (ed.) CAiSE 2010. LNCS, vol. 6051, pp. 38–52. Springer, Heidelberg (2010). https://doi.org/10.1007/978-3-642-13094-6_5
3. Anderson, P., Tushman, M.L.: Technological discontinuities and dominant designs: a cyclical model of technological change. Adm. Sci. Q. 35(4), 604–633 (1990)
4. Azemi, E., Bala, S.: Exploring BPM adoption and strategic alignment of processes at Raiffeisen Bank Kosovo. In: BPM Forum, vol. 2428, pp. 37–48 (2019)
5. Bazeley, P., Jackson, K.: Qualitative Data Analysis with NVivo. SAGE Publications Limited, Thousand Oaks (2013)
6. Bostrom, R.P., Olfman, L., Sein, M.K.: The importance of learning style in end-user training. MIS Q. 14(1), 101–119 (1990)
7. Brynjolfsson, E.: The productivity paradox of information technology. Commun. ACM 36(12), 66–77 (1993)
8. Brynjolfsson, E., Hitt, L.M.: Beyond the productivity paradox. Commun. ACM 41(8), 49–55 (1998)
9. Brynjolfsson, E., Rock, D., Syverson, C.: Artificial intelligence and the modern productivity paradox: a clash of expectations and statistics. Technical report, National Bureau of Economic Research (2017)
10. Buijs, J.C.A.M., Bergmans, R.F.M., Hasnaoui, R.E.: Customer journey analysis at a financial services provider using self service and data hub concepts. In: BPM, vol. 2428, pp. 25–36 (2019)
11. Caldeira, J., e Abreu, F.B., Reis, J., Cardoso, J.: Assessing software development teams' efficiency using process mining. In: ICPM, pp. 65–72. IEEE (2019)
12. Canjels, K.F., Imkamp, M.S.V., Boymans, T.A.E.J., Vanwersch, R.J.B.: Unraveling and improving the interorganizational arthrosis care process at Maastricht UMC+: an illustration of an innovative, combined application of data and process mining. In: BPM Industry Forum, vol. 2428, pp. 178–189 (2019)
13. Christensen, C.M., Overdorf, M.: Meeting the challenge of disruptive change. Harv. Bus. Rev. 78(2), 66–77 (2000)

14. Davis, F.D.: Perceived usefulness, perceived ease of use, and user acceptance of information technology. MIS Q. **13**(3), 319–340 (1989)
15. Dumas, M., Rosa, M.L., Mendling, J., Reijers, H.A.: Fundamentals of Business Process Management, 2nd edn. Springer, Heidelberg (2018). https://doi.org/10.1007/978-3-662-56509-4
16. Eisenhardt, K.M.: Building theories from case study research. Acad. Manag. Rev. **14**(4), 532–550 (1989)
17. Fernández, W.D., et al.: The grounded theory method and case study data in IS research: issues and design. In: ISFW: CC, vol. 1, pp. 43–59 (2004)
18. Flick, U.: An Introduction to Qualitative Research. Sage Publications Limited, Thousand Oaks (2018)
19. Glaser, B.: Theoretical Sensitivity: Advances in the Methodology of Grounded Theory. Sociology Press, Mill Valley (1978)
20. Glaser, B.: Doing Grounded Theory: Issues and Discussions. Sociology Press, Mill Valley (1998)
21. Glaser, B.G., Strauss, A.L.: Discovery of Grounded Theory: Strategies for Qualitative Research. Routledge, London (2017)
22. Gorla, N., Somers, T.M., Wong, B.: Organizational impact of system quality, information quality, and service quality. SIS **19**(3), 207–228 (2010)
23. International, D.: The DAMA Guide to the Data Management Body of Knowledge - DAMA-DMBOK. Technics Publications, LLC, Denville (2009)
24. Kerremans, M.: Market guide for process mining. white paper (2019). https://www.gartner.com/en/documents/3939836/market-guide-for-process-mining
25. Tushman, M.L., Murmann, J.P.: Dominant designs, technology cycles, and organization outcomes. Acad. Manag. Proc. **1998**(1), A1–A33 (1998). https://doi.org/10.5465/apbpp.1998.27643428
26. Lee, S.M., Kim, Y.R., Lee, J.: An empirical study of the relationships among end-user information systems acceptance, training, and effectiveness. MIS **12**(2), 189–202 (1995)
27. Macris, A., Papakonstantinou, D., Malamateniou, F., Vassilacopoulos, G.: Using ontology-based knowledge networks for user training in managing healthcare processes. JTM **47**(1–3), 5–21 (2009)
28. Mans, R.S., van der Aalst, W.M.P., Vanwersch, R.J.B., Moleman, A.J.: Process mining in healthcare: data challenges when answering frequently posed questions. In: Lenz, R., Miksch, S., Peleg, M., Reichert, M., Riaño, D., ten Teije, A. (eds.) KR4HC/ProHealth -2012. LNCS (LNAI), vol. 7738, pp. 140–153. Springer, Heidelberg (2013). https://doi.org/10.1007/978-3-642-36438-9_10
29. Nicolaou, A.I., McKnight, D.H.: Perceived information quality in data exchanges: effects on risk, trust, and intention to use. ISR **17**(4), 332–351 (2006)
30. Orlikowski, W.J.: Using technology and constituting structures: a practice lens for studying technology in organizations. Organ. Sci. **11**(4), 404–428 (2000)
31. Reinkemeyer, L.: Process Mining in Action: Principles Use Cases and Outlook. Springer, Cham (2020). https://doi.org/10.1007/978-3-030-40172-6
32. Rojas, E., Munoz-Gama, J., Sepúlveda, M., Capurro, D.: Process mining in healthcare: a literature review. J. Biomed. Inform. **61**, 224–236 (2016)
33. Saldaña, J.: The Coding Manual for Qualitative Researchers. Sage, Thousand Oaks (2015)
34. Tushman, M.L., Anderson, P.: Technological discontinuities and organizational environments. Adm. Sci. Q. **31**(3), 439–465 (1986)
35. Venkatesh, V., Davis, F.D.: A theoretical extension of the technology acceptance model: four longitudinal field studies. MS **46**(2), 186–204 (2000)

36. Venkatesh, V., Thong, J.Y., Xu, X.: Unified theory of acceptance and use of technology: a synthesis and the road ahead. AIS **17**(5), 328–376 (2016)
37. Wiesche, M., Jurisch, M.C., Yetton, P.W., Krcmar, H.: Grounded theory methodology in information systems research. MIS Q. **41**(3), 685–701 (2017)
38. Wynn, M.T., et al.: Grounding process data analytics in domain knowledge: a mixed-method approach to identifying best practice. In: BPM, pp. 163–179 (2019)

Digital Transformation and Business Process Innovation: A Car Rental Company Case Study

Sílvia Bogéa Gomes[1,2(✉)] ⓘ, Paulo Pinto[2] ⓘ, Flavia Maria Santoro[3] ⓘ,
and Miguel Mira da Silva[1,2] ⓘ

[1] INOV INESC Inovação, Lisbon, Portugal
silvia.bogea@inov.pt
[2] University of Lisbon, Lisbon, Portugal
{paulo.pinto,mms}@tecnico.ulisboa.pt
[3] University of the State Rio de Janeiro, Rio de Janeiro, Brazil
flavia@ime.uerj.br

Abstract. Digital innovation has forced companies to change some well-established business processes. Thus, incorporate information technology into business processes is not enough. We argue that both BPM role and related capabilities might need to be re-interpreted for the digital future. In this sense, the BPM discipline must identify relevant instruments for building on new ways to analyze, understand and support such transformations. This paper examines the results from a car rental company case in the light of DT&I-BPM-Onto, an ontology that encompasses relevant theory on the digital transformation domain. The company carried out a successful initiative by digitizing its primary end-to-end process that contributed to improvement of the company's Net Promoter Score (NPS). The focus of the case study was the process' redesign; thus we investigated: (i) the factors that led the decision to digitalize the process; (ii) how the digital transformation was conducted; (iii) the characteristics of the industry, the business itself, the transformed process, and the type of innovation implemented. The main contribution of this paper is to demonstrate that DT&I-BPM-Onto supports taking a broader picture of a process transformation case. Moreover, we provide insights and practical lessons for future projects and further research.

Keywords: Digital transformation · Case study · Car rental · Ontology

1 Introduction

Digital innovation is being incorporated into our day-to-day life, transforming business models, industries, and changing the business dynamics. Consumers' digital expectations and demands are constantly changing. Digital transformation materializes how an innovation replaces or complements existing businesses, bringing new actors, structures, practices, values, and beliefs [1]. It requires mixing people, machines, and business processes, with all of the disorderliness that these could entail [2]. Traditional companies, like airlines and banks are some well-known examples where mostly digital innovation occurs through the redesigned process [3]. Hence digital transformation is about

© Springer Nature Switzerland AG 2020
D. Fahland et al. (Eds.): BPM Forum 2020, LNBIP 392, pp. 246–262, 2020.
https://doi.org/10.1007/978-3-030-58638-6_15

processes, the Business Process Management (BPM) discipline needs to establish a background to understand and support such transformations.

The lack of a framework to support the analysis of digital transformation within the BPM discipline was first noticed by [4–6] and [7]. In a previous paper, we proposed DT&I-BPM-Onto [8], which is an ontology that encompasses most concepts related to digital transformation and innovation contextualized to the impact on the business processes. In this paper, we argue that DT&I-BPM-Onto can serve as instrument to the BPM discipline providing the theoretical background to the understanding of the steps taken to achieve a digital transformation initiative.

This paper reports the results from a successful digital transformation business case, the InterRent's® initiative, to show how DT&I-BPM-Onto [8] could be taken as a basis for the analysis in such scenarios. The global car rental industry is growing due to increasing tourism activities, globalization of operations, and the overall rise in income levels [9]. InterRent®, a company from the Europcar's group, reimagined its business by proposing a digital innovation business model named Key'n Go®, launched less than one year ago.

The focus of the case study was the process redesign project with the aim to answer the following research questions: (RQ1) What were the factors that led the decision to digitalize the process (for Key'n Go®)?; (RQ2) How was the digital transformation conducted?; (RQ3) How did the characteristics of the industry, the business itself, the transformed process, and the type of innovation implemented, considering in the ontology, contribute to the case success? In general, a qualitative case study approach uses two sources of evidence - observation and interviewing - and explores a situation where the action happens [10]. We followed the case study steps proposed by Runeson & Höst [11]: design, preparation for data collection, collecting evidence, analysis of collected data and reporting.

The remainder of the paper is organized as follows. Section 2 presents the theoretical background. Section 3 examines the InterRent® case study design. Section 4 addresses the data gathering and findings. Section 5 reports this case analysis supported by the ontology. Section 6 discusses this case through DT&I-BPM-Onto. Finally, Sect. 7 concludes the paper, highlighting the main contributions, limitations, and future work.

2 Theoretical Background

In this section, we present an overview of DT&I-BPM-Onto and related concepts such as process innovation.

The digital economy demands delivering results from the current core business of a company while simultaneously making the investments required to explore opportunities, to create and capture value. The use of digital technologies and their impact across society have, profoundly and rapidly, transformed the face of business and will continue to do so [12]. Some authors [1, 13–15] argue that digital transformation is the combined effects of several digital innovations bringing about novel actors, structures, practices, values, and beliefs that change, threaten, replace or complement existing rules of the game within organizations, ecosystems, and industries.

In previous research [8], we presented the Digital Transformation and Innovation Business Process Management Ontology (DT&I-BPM-Onto), a conceptual model that encompasses concepts and relationships of this domain. DT&I-BPM-Onto aims to share a common understanding and to establish the connection between Digital Transformation, Innovation, and BPM, by examining the interaction between these perspectives and to what extent the characteristics of the BPM initiatives are affected by, and affect, Digital Transformation and Innovation (DT&I). DT&I-BPM-Onto is composed of three sub-domains independent but complementary: (a) digital transformation and innovation, (b) BPM contextual factors, and (c) digital technology.

The digital transformation initiative is an explicit part of a business goal, turns ideas into value, through an event that occurs as a DT process, that supports the business process transformed, in one or more knowledge areas, in one or more perspectives, performed by one or more DT Agents. It results in one or more DT Outcome. The DT Drivers motivate the DT Objects. A DT Object considers a product, process, or business model that is perceived as new, requires some significant changes on the part of adopters, and is embodied in or enabled by IT [16]. The Business Process transformed involves an ecosystem that is composed by the characteristics of this Business Process, Organization and Business Environment. Business Goal is associated to Business. DT Drivers motivates DT Object, Digital Resource operationalizes DT Object, Business Organization is operationalized in Business Process.

DT&I-BPM-Onto refers to processes transformed through innovation. Process innovation is a method to align resources such as IT with the business strategies of organizations. Some studies showed that organizations consider IT a key enabler to achieve process innovation (operational and management) and process innovation initiates process reengineering, to get higher value to the business [17]. We adapted the definition of process innovation provided by [17]: an improved, renewed, or replaced end-to-end process that deliver differentiated products, services or goods, and solutions in a new way to meet the business objectives and value generation.

To achieve the process innovation, it is necessary to identify the existing problems and choose instruments to analyze the findings [18]. An approach to perception-based business process innovation is proposed by [18] segmented into four core stages: (1) engage process stakeholders - inquiries into the issues and interests related to the process under investigation, (2) collect process data - semi-structured interviews for collecting answers from key stakeholders as what, why, and who questions (3) explicate process knowledge - as a shared road map for systematically identify important observations based on the available data to aggregate them into viewpoints, and, (4) design process innovations - focused on how to innovate the process. These stages collectively form an iterative cycle of capturing, synthesizing and reconciling stakeholder perceptions to support business process innovation. More details of the concepts provided by the ontology are presented in Sect. 7 together with the discussion of results. Additionally, DT&I-BPM-Onto is fully available at https://is.gd/DTnI_BPM_Onto_Model.

3   The InterRent® Case Study Design

A case study design contains the following elements [10]: the objective to be achieved (the InterRent's® DT&I initiative case analysis taking as basis the DT&I-BPM-Onto); the case to be studied (the Key'n Go® Case); the theory of reference (the DT&I-BPM-Onto and Business Process Innovation); the research question (RQ1-3); the selection strategy (a single-case study where capture data) and method (how capture data) to collect it (as interview, media documents and observations).

This case covers the situation before and after the Key'n Go®, aiming to understand the transformation and its results. These effects were analyzed through the interpretation of this context. It is a holistic case study with a single unit of analysis which is the car rent process. The data collected was qualitative, providing better understanding of the studied phenomenon in the data analysis. The procedures and protocols for data collection are defined in Table 1. Triangulation was made using interviews and document analysis.

Table 1. The case study protocol.

Goals and scope	Data collection	Analysis	Interpretation
Exploratory	• Process models • Interviews	Analysis of the case using the DT&I-BPM-Onto	Interpretation of the results to answer the RQs

3.1 The Scenario

Europcar® is one of the main current players in the car rental industry, in 2018 it had revenue of 2.929 billion Euros and is established in over 140 countries [20]. InterRent® belongs to the Europcar®'s group since 1988. From the '80s until now, its goal has been to provide an exceptional experience to its customers. According to Hans Åke Sand, InterRent's CEO for 20 years in Sweden, *"Over ten years ago, when this industry was starting to grow here, we realized that we should not just provide our customers with a car if they ask for one. (…) We developed a service concept, according to which we provide immediately accessible transportation solutions to temporarily occurring transportation problems"* [21].

The car rental business is a relevant sector, considering the growth of mobility systems in the past years and the projection for the future [22]. There are challenges (issues of cabotage regulations, security, insurance, etc.) and opportunities (building a profitable rental operation) in this industry. InterRent®'s current strategy gives an advantage over the competitors, allowing them higher profitability and a greater number of satisfied customers [23]. However, on the other side, the forefront requires rethinking the business to improve the perceived value, both by customers and shareholders.

3.2 The As-Is Process

Most consumers choose a service based on what is grasped as the better "value for money" option, which is, in fact, the most logical approach. Brokers are responsible for the suggestions, making them the most important "bridges" between a company and the consumer. Brokers make their offers by comparing the companies' NPS (Net Promoter Score). If a company has a higher NPS, it goes up in the suggestion list; if it has a lower NPS, it goes down, or it is not even offered as an option.

The strategy of InterRent® company was to improve service quality based on processes to impact positively the NPS score. To achieve this goal, InterRent® has interacted with customers, using a Customer Focus Group initiative and reviewed the entire customer journey process. As we can see, they engage process stakeholders (including customers), collecting data, identified the existing problems, and choose components for analyses the findings (the pain points presented below). Thus InterRent® works on how to innovate the process. These stages are aligned with the collectively form an iterative cycle of capturing, synthesizing, and reconciling stakeholder perceptions proposed by [18] to support business process innovation.

Starting from the need to rent a vehicle, the Customer usually searches the Car Rental Brokers websites, on Car Rental companies' websites or simply contacts the Car Rental companies, informing, among other things, the location, date, time, type of vehicle. Then, the vehicle rental starts. The journey currently involves the following critical processes from the Customer's perspective: "Receive Offer", "Book Vehicle", "Pick Up Vehicle", "Check Vehicle Conditions", "Receive Vehicle", "Use Vehicle", "Drop-off Vehicle". Although customers are increasingly performing the first two processes ("Receive Offer" and "Book Vehicle") through Web applications, the "Pick-Up Vehicle" process was still done face-to-face manually.

InterRent® noted that the Customer usually had to wait in a queue to be helped at the car rental desk, and often is asked to review optional accessories and insurance policies, in addition to previously selected items. This procedure implied the need to increase the number customer service employees, in the store, even though it does not guarantee reduced waiting time for customers; during peak season, queues are still a pain points for customers. The "Drop-Off Vehicle" process, is also quite time sensitive as customers are usually heading to the next step of their journey; for example: the user might have to coordinate rental car's drop-off time with flight times (at the airport) and being asked to go through a careful damage inventory, together with Rent-a-car personnel, might affect their schedule. The main sources of pain on the Customer Journey, which negatively impacts on the NPS score, were identified as:

1. Waiting time in queues, to be served at the car rental desk
2. Stressing about the insurance and terms & conditions – in addition to customer service personnel's pitches for upselling opportunities
3. On picking up: Damage inventory; the need to check for inventory damage before being able to pick the car.
4. During Dropping-off: in addition to damage inventory, the customer also needs to wait to deliver the Vehicle despite time constraints

The InterRent® Customer Journey AS-IS process model is presented in Fig. 1, highlighting those issues and the need for changes.

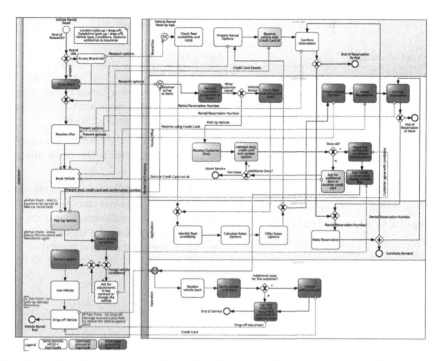

Fig. 1. InterRent® As-Is process: https://is.gd/Interrent_Customer_Journey (built using [24])

3.3   The To-Be Process: The Key'n Go® Case

As a mid-tier company, InterRent® had the ideal size to implement in, swiftly fashion, innovative solutions and still leverage the advantages of being a company of a Europcar's group company. It made InterRent® the best pilot for testing innovative solutions that, if proven to be successful, could be implemented in Europcar. Hence, InterRent® launched its "Hassle free Project", from people to people - the Keyn'Go® - and offered a complete Digital Customer Journey, for a reasonable price. The target market segment aimed is the customer who value their time and is technology friendly. Keyn'Go® project helped to address previously identified pain points (Table 2).

The benefit proposed by the InterRent® solution could be classified as an innovation in the car rental process, since it has been designed to optimize the time in the process of the renting a vehicle and the number of people involved in the process. They propose a service with fixed pre-defined conditions and insurance, aiming for the best cost benefit for their customer. Figure 2 depicts the To-Be process.

During the Key'n Go® process the customer is impacted by the Wow-factor, which in marketing means that the customer "relates to exceptional customer service in which an employee gives the customer more than they expected or something they did not expect at all", the customer feels happy and becomes "an ambassador" of the brand, sharing the service by word-of-mouth, increasing the NPS. Europcar's NPS is 50% [25]. Key'n Go®, albeit being a Mid-Tier brand reaches an average 54% with peaks over 60%

Table 2. Pain points addressed by Key'n Go®.

Pain point identified	Key'n Go®
Waiting in a queue to be served at the car rental desk	Uses automated kiosks, making it possible to deliver the car in a shifty minute while also giving the customer the power to choose a car he wishes from the fleet accessible
Need to review insurance and conditions again	Assists customer by a ground-hostess (does not sell anything) in the doubts about using the kiosk
On Picking Up damage inventory On Drop-off damage inventory, in addition to waiting time to deliver the vehicle	The Key'n Go® bundle includes the premium insurance pack, that covers this kind of situation. Now the customers going to the agreed place, to pick-up the keys at the beginning drop-off the keys at the end

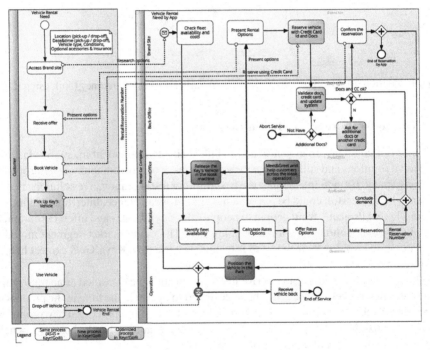

Fig. 2. Keyn'Go® To-Be process: https://is.gd/KeynGo_Customer_Journey (built using [24])

during high season, comparing who uses Key'n Go® and who waits in queue: Key'n Go® customers feel valued and return with 90–100% in NPS score.

4   Data Gathering

The project involved a team of three fully dedicated senior participants. Each member had the responsibility to interface with the other areas affected by the project. Those areas maintained their As-Is operation, participated under demand in the project, and in the end implemented the new operation. The Key'n Go was launched in May 2019 and the case study data was collected from February 2019 to February 2020. The project is already in operation for nine months. The project was carried out in a relatively short time, six months. We used multiple sources of data, including interviews, media documents and observation.

First, we reviewed company documents (e.g., internal videos presenting the project and financial reports [20]) and analyzed public documents as web magazine news [25–27], providing good starting points for the interviews and aiming at a clear insight into the company's situation. Moreover, one of the authors used the Key'n Go® App and the traditional InterRent® App to simulate a customer experience. It is worthy to mention that one of the co-authors is a member of the group responsible for strategy formulation and project development.

Second, we interviewed the three members of the team (see details in Table 3). The Director and iRPM were project members since the beginning, and the iROM joined after. Although three interviews could be considered too few for a single case study, we claim that it was enough in our case because the interviewees were the main representative of the project members and they were able to provide all the needed information about the main topics covered in the interview.

Table 3. Members of the Key'n Go® team interviewed.

Id	Role	Responsibilities
Director	InterRent® Mid-Tier segment Director	Commercial and Operations Leader at InterRent in the Mid-tier segment, holding responsibility in the Repositioning of the Brand, both in B2C and B2B2C segments, accountable for the Budget definition and execution. Besides to bethe Key'n Go® Project Director, manages all the resources needed to deliver in time the repositioning of the brand
iRPM	InterRent® Key'n Go® Product Manager	Responsible for the product features definition, managing IT team's development squads and end-to-end testing of implemented solutions
iROM	InterRent® Key'n Go® Operations Manager	In charge of the Kn'Go front-end and NPS follow-up. Feeds the iRPM with Customer's feedback on functionality drivers

Table 4. Research and Interview questions and respective summarized answers.

RQ	Question	Director	iRPM	iROM
(RQ1)	(1) Does the current InterRent® customer journey process differ from that of competitors? If so, where? Why do customers choose InterRent® today?			
		Price and convenient location of the network	Low rate	Price
	(2) What are the concerns regarding the current customer journey process, which led to the decision to evaluate the current business model with Key'n Go®?			
		Low NPS is driven by long queues and aggressive sales pitches	Quality	NPS and Aggressive ancillary sales
	(3) What do work well with the current customer journey process?			
		Fleet	Fleet	Fleet
[RQ2]	(4) Why was this project carried out?			
		Identified the need to boost sales having NPS as a barrier	Because of Broker pressure to increase NPS	To address a differentiation scenario compared to competence
	(5) Can you describe it chronologically?			
		During 2018 Europcar changed its management organization into Business Units. InterRent® started to be managed by the Low-Cost Business Unit. At this moment was identified the need and launched the DT Project	2018 Project Launched 2019 Q1 Pilot 2019 Q2 to Q4 Go Live	2019 Q1 Alicante Pilot 2019 Q2 Go Live Waves
	(6) What are the stages of the project and the main results and lessons learned in each stage?			
		Pain Points id Focus Group Solution development Solution deployment Next steps Evaluation (new opportunities)	Pain Points id Focus Group Solution development Solution deployment Next steps Evaluation (new opportunities)	Pain Points id Focus Group Solution development Solution deployment Next steps Evaluation (new opportunities)
	(7) What has changed with the launch of the Key'n Go® product?			

(*continued*)

Table 4. (*continued*)

RQ	Question	Director	iRPM	iROM
		Growing the NPS metric and RPD (Revenue Per Day)	Quality and Sales	Quality and Prices
[RQ3]	(8) What do customers praise most about the Key'n Go® product?			
		Transparency and simplicity	Simple and fast	No Queue
	(9) Are there additional inputs to improve the customer experience in the Key'n Go® process?			
		Upsell in the Kiosk	Upsell in the Kiosk	Upsell in the Kiosk
	(10) What can be automated in the Key'n Go® process that might bring significant results?			
		Wi-Fi availability	Additional driver	Wi-Fi connection
	(11) What characteristics of Key'n Go® differentiate it from the industry, InterRent®'s current business model and InterRent®'s current process, contributed most to the success of Key'n Go®?			
		No Queue, self-digitalization of the documents	One minute to Pick-up	No queue
	(12) Could you classify this case, considering how the business ecosystem transformed (before/during project/after key'n Go® launch)? See Table 4			

The interview was planned to raise general concerns based on our research questions, i.e., the topics of interest in the case study. The questions were open, using as a base the as-is and the Key'n Go® business process models, generating a broad range of answers and comments from the interviewees. The interview session begun with open questions and moves towards more specific ones. As scheduled, first the authors presented the case study goals, and explained how the data would be used. Then we started the main questions, which absorbed the largest part of the interview. Just Before the end of the interview, the key findings were summarized by the researcher, in order to get feedback and avoid misunderstandings.

During interview, the As-Is and Key'n Go® process models served as a basis to answer the questions. The interviewer took notes and consolidated them in Table 4, from the results obtained. Each respondent answered about the subject of the first 11 questions individually but question 12 was answered and discussed in group.

5 Case Study Findings Supported by DT&I-BPM-Onto

We took the DT&I-BPM-Onto as a basis to structure and analyze our findings, to further understand the success of this case and what lessons could be learned from it. In this sense, based on the interviewees' responses, media documents and observations, as cited in the previous section, we identify the main entities of this case. The terms highlighted in bold in the following paragraphs correspond to the classes (concepts) and relationships postulated by the ontology.

InterRent®'s **business goal** can be broken down into the quality increase and further process' simplification and acceleration, from a customer's perspective, in addition to the internal productivity gain. Therefore, the InterRent®'s **business goal strategy** was offering a new digital solution, the Key'n Go®, without discontinuing the current operation, which meant the coexistence in the same brand, of two very different products. The InterRent® plans to achieve its business goal offering the Key'n Go® product.

The Key'n Go® reconfigures the customer journey process, considering the pain points identified, that requires some significant changes on the part of InterRent, and it is enabled by IT. It combines effects of **digital innovation** (the Customer Journey process, the App, the Kiosk), practices (the revisited back office functions), **values** (business activities occur faster and increases its source of value creation), and beliefs (the customer at the center of the process) that change, threaten, replace or complement existing business rules within InterRent® and Car Rental Industry. The **agents** in this case they are diverse: a new **type of** actors mixing with existing actors, with a new combination's responsibilities of the existing and new **roles**: for instance, a hostess to Meet&Greet (new actor and role) and the back-office (existing actor and role) activities.

The elements of InterRent®'s **value chain** is being impacted: the key business operations and processes are transformed, and, as a consequence, the human element is impacted by the new culture. Moreover, the elements of InterRent®'s **value system** also are being impacted: The transformation of the customer experience reshaped the customer value propositions through the Key'n Go® initiative. Besides, InterRent® impact its network, or supply chains (i.e., the Brokers). The Key'n Go® transforms the **value network's** roles, relationships, value creation, value perception, etc.

Thus, Key'n Go® impacts the **industry/ecosystem** since it contributes to changing the car rental business network dynamics. Besides, the Key'n Go® impacts the **business and individual perspectives** relates to car rental experience. The **knowledge areas** involved are mainly car rental business domain, logistics, and technology.

5.1 InterRent® DT&I Process

We observed that the previous InterRent® customer journey was not that different from the competitors, and therefore not generating a high degree of competitiveness. The competition is focused on price. The concerns with the current customer journey process were that, in the customer's perception, InterRent® could offer increased quality, simplification, and speed of the process. From the company's perspective, productivity could be better. Analyzing this, they observed that the back office was not the focus. The main issue was the long queues to get the car, according to the pain points identified internally. The stages followed by InterRent® in order to redesign the business process were: Pain Points identification, Focus Group validation, Solution development, Solution deployment, Next steps Evaluation (new opportunities).

DT&I-BPM-Onto describes the main stages of the own digital transformation and innovation process [16]: discovery, development, diffusion and impact.

At the discovery stage, InterRent® used the focus group to validate the pain points previously identified from the consolidation of the NPS reports and social media complaints of their customers. Afterwards, they could come up with ideas to address those problems or needs, as for example, suggestion of extras' acquisition directly on the kiosk as a last chance. Next, they started the development stage by designing a project for a usable innovative solution. In the diffusion stage, they deployed the solution, persuaded and enabled the customers to adopt the innovation. The last stage is impact. Many questions could be raised, such as: "What transformations resulting from the digital innovation are being observed at the level of individuals, firms, markets, and society? How can adopters' appropriate maximum value from the innovation?" The last stages are currently being conducted in different physical places.

5.2   InterRent® Ecosystem Characteristics

According to the DT&I-BPM-Onto, the characteristics of the industry, the business itself, the transformed process, and the type of innovation implemented impact the DT&I Initiative and need to be considered. During the interviews, the participants discussed those topics, summarized in Table 5. According to the interviewees, the **Customer Journey process** has core value contribution, since it is essential to the business' results; it is a repetitiveness process kind; with low interdependence between others business process, and low variability. During the transformation event, the process knowledge-intensity and creativity change temporarily from low to high, in the discovery and development phases of the DT&I.

About the **Business Organization**, for the interviewees, InterRent®'s organization belongs to a large business size company group in the service industry. Currently, the business has a medium availability of resources, that change temporarily to low in the transformation event (team project), and a culture that is highly supportive of BPM (related to this kind of operation). For InterRent®, the Car Rental industry **environment**, prior to Key'n Go®'s launch, has a high degree of competitiveness. During the transformation event, it was changed to low (newness factor, no other company has something similar). The degree of uncertainty in the industry environment for InterRent®, prior to the release of Key'n Go®, before the Key'n Go® launch, was medium, but temporarily changed to high during the transformation event (associated with the project's risk) and come back to medium after the successful end of the project.

5.3   The InterRent® DT&I Object and Outcome

The **type of object** of InterRent®'s initiative was the business model, because it changed its market and organization, provided new revenue and value, produced opportunities and brought new management practices. Moreover, the **degree of newness is** incremental, with **focus** on customer journey process **exploration**. It has one or more **components**. In an DT&I initiative, not all components are digital, as the Key'n Go® App. Related to **the level of originality**, the Key'n Go® App component, is a **novel** component, and other components, as the Credit Card App, is an **existing** component. The InterRent®'s initiative carries out **new combinations of digital** (App) **and physical** (Kiosk) **components**, with new approaches of the **process** and **technologies**, to produce the Key'n

Table 5. Business ecosystem transformed.

DT&I-BPM-Onto Class	Contextual factors [28]	Data collected in the interviews from question (11)
Business process	Process Degree of Value contribution?	Core process
	Repetitiveness Process?	Repetitive
	Knowledge-intensive Process?	Low/High/Low
	Creativity-intensive Process?	Low/High/Low
	Interdependence of process participants?	High
	Process Variability?	Low
	Business Organizational Process Scope?	Inter-organizational process
Business organization	Business Industry?	Service industry
	Business Size?	Large organization
	Culture supportive of BPM?	Highly supportive
	Availability Degree of the Organizational Resources?	Medium/Low/Low
Business environment	Degree of Environment Competitiveness?	Highly/Low/Low
	Degree of Environment Uncertainty?	Medium/High/Medium

Go®. In the case of an existing component, it can be a new combination of use and/or design. Namely, this case is a **combination of use**.

The **nature** of the Key'n Go® object could be classified as a digitalization example, since the process of moving to a digital business, using digital technologies, and creating a business model that provides new revenue and value-producing opportunities. The **outcome** of the InterRent®'s initiative is the Key'n Go® itself. The **drivers** that motivate the Key'n Go® are mainly the differentiation (low-cost to mid-tier), customers (pain points) and competition (NPS). The launch of the Key'n Go® process brought Inter-Rent® a change in level (from low-cost to mid-tier) and praise from customers, reflected in the registered NPS. Customers are delighted, according to one of the testimonials, "Good price and fast check-in and checkout." There is always room for improvement, and the team is working on projects to improve the customer experience in the Key'n Go® process, including other potential automations. Currently, the characteristics of Key'n Go® set it apart in the industry, but competition is strong, and competitors have also been working towards a similar business model.

6 Discussion

The results of the case study outline what are the necessary steps taken to achieve a successful digital transformation initiative and, furthermore, how the DT&I-BPM-Onto could be taken as a basis to support the analysis of other DT&I cases in the industry. The applicability and value of an ontology in this case was to support the categorization of the entities and understand the relationships among them. Thus, the case could be framed to conceptualizations established in the literature. The research questions are more easily answered and further reflection could be done.

Concerning the (RQ1), the factors that led to the decision, to transform an existing Car Rental business through a digital process, were the business goal and business goal strategy stablished. The strategy of listening to the customer's voice and placing them at the center of the process, aiming to improve the InterRent®'s NPS, enhancing service and improving the top line. The ontology supports to organize this information through the DT&I Process stages and it facilitated the identification of what led to the decision of the Key'n Go initiative. As lessons learned, a company might consider research for opportunities, involving external people. Listening to the customer's voice was the strategy defined to achieve the business goal of increase the quality and further process' simplification and acceleration, from a customer's perspective, in addition to the internal productivity gain. Other organizations, with other Goal can define another Strategy.

Regarding (RQ2), the process' transformation was conducted in a Project Management lean structure which allowed a very close and interactive process within the team and business executives. The speed, in identifying opportunities during the project and implementing new features (lessons learned), was essential for its success. The ontology describes the concepts of value including value chain (operation, process and human resources) and value system (customer and supply chains). They were useful to learn the process end-to-end considering the whole network, moreover the DT&I Process stages. One issue to better explored here is the configuration of the project team and interfaces. It will not necessarily be the more adequate to all organizations (since there are other contexts and cultures).

Related to (RQ3), how the characteristics of the industry, the business itself, the transformed process, and the type of innovation contribute to the success of the case. The ontology helped to organize the knowledge about the goals and current results that originated the DT&I demand, what is important to customers and what the competitors are doing. It was fundamental to help explaining the success of the project. It is important consider the Industry and Organization contexts. Other industries and organizations can result in a different outcome.

All along, to support the business strategy, the DT&I InterRent® initiative focused on the business goal, developing business resources and technology resources that enabled it to be more competitive and change its level in industry, which is also supported by the concepts of the DT&I-BPM-Onto model. The business resources included a business process innovative redesign and they developed a kiosk technology to automatize part of its operation. InterRent® initiative took into account the Customer learning curve as well: a hostess to Meet & Greet is there to support them.

The DT&I-BPM-Onto describes the process characteristics considering the BPM contextual factors. Mapping this business and its industry characteristics facilitated an analysis of scenarios for the DT&I initiative. For instance, it is interesting to observe that some process' characteristics changed throughout the DT&I initiative, like the Knowledge-intensity, Creativity-intensity and Degree of Environment Uncertainty. These characteristics were also transformed with the DT&I outcome, as Availability Degree of the Organizational Resources and Degree of Environment Competitiveness. Making the process explicit and analyze it in advance is important to the strategy.

In the operation launching, the company had to convince about the advantages of the project, not only the Executives and Partners (including Brokers and B2B2C parties), but also to other product team members. Understanding all the DT&I Agents (types and roles) involvement, as the Key'n Go® initiative did, was vital to avoid problems. The DT&I Process stages of diffusion and impact are still in progress, so there is opportunity to bring even higher NPS levels, learning and refining the initiative. The ontology could also support these analyses, decision making and evolution in each of these stages. Furthermore, exploring the business value system brought more insights about success and failures in such initiatives.

The ontology enlightens that a DT&I initiative turns ideas into value, through an event that happens in one or more knowledge areas, in one or more perspectives, resulting in one or more digital innovation objects, performed by one or more agents. Agents (synonym for actors, in this case) perform activities within the DT&I initiative. Agents can be individuals, organizations, the society or software agents (as robots and algorithms). Besides, each agent plays different roles in the DT&I initiative, such as customer, user, creator, owner, partners and supplier.

Interviewee 1 explained that the brokers directly influenced the Key'n Go® initiative. However, at the launching of Key'n Go®, interviewee 1 realized that the brokers' website wasn't selling the Key'n Go® initiative because their algorithm was not prepared to equalize this new type of service with the traditional form of car rental. Then, there was a negotiation between the Key'n Go® initiative and the brokers to adjust the algorithms. This issue was very important for the success of Key'n Go® and the ontology supported us to realize that in fact it was the software agent (the algorithm) who played the role of a car rental partner. The Key'n Go® initiative had to align with the other offers on the broker's website. Thus, it is important to assess the whole network chain to ensure the success of the project.

In this case study, the focus was not on the value of the digital technology platform that supports digital transformation. So, DT&I-BPM-Onto presents concepts related to this that were not explored in this paper. Moreover, the object components were not explored in detail, as our case is still in progress. But this can be useful for other cases that are in the beginning of transformation process. The ontology supported to organize knowledge about the demand for an DT&I initiative, aiming to facilitate the interpretation of case studies of digital transformation in the industry. This can help discovering new opportunities for improvement and future work on this topic.

7   Conclusion

The InterRent's® DT&I initiative provided an opportunity to closely investigate its shift to allow a new digital customer journey. i.e. to manage the transformation of their main business process. This case covers the situations before and after the Key'n Go®, and its effects were analyzed through the participants' interpretation of context.

The main goal was to investigate deeply the steps taken to achieve the successful digital transformation initiative and, furthermore, to show how the DT&I-BPM-Onto could be taken as a basis to the analysis of this case. Therefore, the case study aimed to understand why the company decided to perform a transformation of its primary process into digital, which factors influenced this decision making, how this process transformation was conducted, and which capabilities influenced its success. To answer the research questions, the case was analyzed based on the interview results and compiled documents, grounded in the DT&I-BPM-Onto. The DT&I-BPM-Onto was used as a source of knowledge in discussions and learning process about the domain.

The main contribution of this paper is taking the perspectives of DT&I-BPM-onto, towards the InterRent® case study and thus providing a broader picture in analyzing this successful case. Although the proliferation of the use of digital technologies has brought a fundamental change in the world due to their pervasive nature, the digital business is more than a technical implementation. And the digital business value is not a synonym of benefit. It depends on aspects such as desires, goals, needs of the value-ascribing stakeholders. The theoretical implication of this research is the indication that the ontology can serve as a theoretical basis to support the evaluation of cases involving the digital transformation of business process. The main practical implications are the lessons learned from the case study per se that could be used bey companies conducting their own with digital transformation process.

The main limitation of this research refers to the use of only one case from a single car rental company. Other cases of different companies would confirm the theoretical approach examined in this paper. Future works include the refinement of the ontology as the foundation for digital transformation and innovation that will be integrated into a digital transformation and innovation business process management to guide the development of digital transformation initiatives while reducing their risk of failure.

Acknowledgements. This work has been partially supported by Portuguese National funds through FITEC - Programa Interface, with reference CIT "INOV – INESC Ino-vação - Financiamento Base".

References

1. Hinings, B., Gegenhuber, T., Greenwood, R.: Digital innovation and transformation: an institutional perspective. Inf. Organ. **28**, 52–61 (2018)
2. Davenport, T.H., Westerman, G.: Why so many high-profile digital transformations fail. Harvard Bus. Rev. **9**, 15 (2018)
3. Kirchmer, M., Franz, P.: Innovation through Business Process Management – Competing Successfully in a Digital World (2017)
4. Hess, T., Benlian, A., Matt, C., Wiesböck, F.: Options for formulating a digital transformation strategy. MIS Q. Exec. **15**, 123–139 (2016)
5. Morakanyane, R., Grace, A., O'Reilly, P.: Conceptualizing digital transformation in business organizations: a systematic review of literature. In: 30th Bled eConference Digital Transformation - From Connecting Things to Transforming Our Lives, BLED 2017, pp. 427–444 (2017)

6. Bockshecker, A., Hackstein, S., Baumöl, U.: Systematization of the term digital transformation and its phenomena from a socio-technical perspective - a literature review. In: European Conference on Information Systems: Beyond Digitization - Facets of Socio-Technical Change, ECIS 2018 (2018)
7. Bogea Gomes, S., Santoro, F.M., Mira da Silva, M.: An ontology for BPM in digital transformation and innovation. Int. J. Inf. Syst. Model. Des. **11**, 52–77 (2020)
8. Oliveira, B.B., Carravilla, M.A., Oliveira, J.F.: Fleet and revenue management in car rental companies: a literature review and an integrated conceptual framework. Omega **71**, 11–26 (2017)
9. Yin, R.K.: Case Study Research. Sage, Thousand Oaks (1994)
10. Runeson, P., Höst, M.: Guidelines for conducting and reporting case study research in software engineering. Empir. Softw. Eng. **14**, 131–164 (2009)
11. Demirkan, H., Spohrer, J.C., Welser, J.J.: Digital innovation and strategic transformation. IT Prof. **18**, 14–18 (2016)
12. Krimpmann, D.: IT/IS organisation design in the digital age – a literature review. Int. J. Soc. Behav. Educ. Econ. Bus. Ind. Eng. **9**, 1208–1218 (2015)
13. Loebbecke, C., Picot, A.: Reflections on societal and business model transformation arising from digitization and big data analytics: a research agenda. J. Strateg. Inf. Syst. **24**, 149–157 (2015)
14. Mangematin, V., Sapsed, J., Schüßler, E.: Disassembly and reassembly: an introduction to the Special Issue on digital technology and creative industries. Technol. Forecast. Soc. Change. **83**, 1–9 (2014)
15. Fichman, R.G., Dos Santos, B.L., Zheng, Z.: Digital innovation as a fundamental and powerful concept in the information systems curriculum. MIS Q. Manag. Inf. Syst. **38**, 329–353 (2014)
16. Anand, A., Fosso Wamba, S., Gnanzou, D.: A literature review on business process management, business process reengineering, and business process innovation. In: Barjis, J., Gupta, A., Meshkat, A. (eds.) EOMAS 2013. LNBIP, vol. 153, pp. 1–23. Springer, Heidelberg (2013). https://doi.org/10.1007/978-3-642-41638-5_1
17. Lewis, M., Young, B., Mathiassen, L., Rai, A., Welke, R.: Business process innovation based on stakeholder perceptions. Inf. Knowl. Syst. Manag. **6**, 7–27 (2007)
18. Robson, C.: Real World Research. Blackwell Publishing, Malden (2002)
19. Europcar: Europcar Mobility Group S.A. Consolidated financial statements for the year ended 31 December 2018
20. Christian, G., Hans, A.S.: A winning service offer in car rental. Manag. Decis. **31**(1) (1993)
21. Oliveira, B.B., Carravilla, M.A., Oliveira, J.F.: Integrating pricing and capacity decisions in car rental: a matheuristic approach. Oper. Res. Perspect. **5**, 334–356 (2018)
22. Lazov, I.: Profit management of car rental companies. Eur. J. Oper. Res. **258**, 307–314 (2017)
23. Signavio: BPM Academic Initiative [BPM Platform for Process Modeling]. Version: 14.4.2. https://www.signavio.com/news/academic-initiative-new-jersey/ (2020)
24. Europcar Mobility Group Press releases: Europcar Mobility Group Presents SHIFT 2023, Its 2023 Roadmap to Further Capture Profitable Growth, Paris, France, 25 June 2019. https://europcar-mobility-group.com/press-releases?last=1&page=5
25. Merino, N.: La Unidad Low Cost de Europcar Mobility Group presenta la nueva estrategia de marca de InterRent. Autofacil, 30 June 2019. https://www.autofacil.es/renting/2019/06/30/unidad-low-cost-europcar-mobility/51005.html
26. Silva, A.R.: Serviços de mobilidade Com uma experiência do cliente totalmente digitalizada. Revista Pontos de Vista, 28 July 2019. https://pontosdevista.pt/2019/07/28/servicos-mobilidade-experiencia-do-cliente-totalmente-digitalizada/
27. Vom Brocke, J., Zelt, S., Schmiedel, T.: On the role of context in business process management. Int. J. Inf. Manag. **36**, 486–495 (2016)

Holistic Guidelines for Selecting and Adapting BPM Maturity Models (BPM MMs)

Wasana Bandara[1]([✉]) [iD], Amy Van Looy[2] [iD], John Merideth[1], and Lara Meyers[3]

[1] School of Information Systems, Queensland University of Technology, 2 George Street, Brisbane, Australia
{w.bandara,j.merideth}@qut.edu.au
[2] Faculty of Economics and Business Administration, Department of Business Informatics and Operations Management, Ghent University, Ghent, Belgium
Amy.VanLooy@UGent.be
[3] BPM and Business Architecture Practitioner, Brisbane, Australia
lara@larameyers.com

Abstract. BPM maturity models (MMs) help organizations accomplish the BPM capabilities paramount for organizational success. Although much literature deals with how to design MMs, little knowledge exists of how organizations use BPM MMs. Moreover, the academic literature about MMs is scattered, making it hard for practitioners to learn from academia. Our purpose is to offer a holistic journey to guide organizations through three phases of BPM MM use, namely (1) choosing one out of many MMs that fits the organization's context, (2) tailoring the MM to particular needs, and (3) advising during and after a maturity assessment. Starting from a synthesis of known guidelines, a framework for BPM MM adaption is presented with evidence of its applicability when organizations are conducting maturity assessments. The analysis calls for research to derive specific guidelines for different contexts, e.g., for different levels of maturity and/or when maturity assessments are driven by consultants.

Keywords: Business process management · Maturity models · Guidelines · Case study

1 Introduction

With the rising importance of process orientation, an array of BPM maturity models (MMs) with varied focus and depth has been suggested [1–3]. A MM, also referred to as a 'stages-of-growth model' or 'stage model' [4], is a *"conceptual model that consists of a sequence of discrete maturity levels for a class of processes in one or more business domains, and represents an anticipated, desired, or typical evolutionary path for these processes"* [2]. BPM MMs help organizations to identify and evaluate their BPM capabilities, enabling organizations to create value and deliver enhanced business outcomes through the way their processes are managed. The required process-management capabilities stem from multiple dimensions (i.e., individual, organizational

© Springer Nature Switzerland AG 2020
D. Fahland et al. (Eds.): BPM Forum 2020, LNBIP 392, pp. 263–278, 2020.
https://doi.org/10.1007/978-3-030-58638-6_16

and IT), which collectively form the complex composition of BPM capabilities. Ideally, BPM MMs provide a holistic assessment, planning and execution basis for all areas relevant to BPM [5].

BPM MMs are receiving increasing attention and claim to provide meaningful answers on how organizations can and should evolve their BPM capability [1, 6–8]. As a result, many BPM MMs have been designed by practitioners and scholars alike, and more BPM MMs continue to evolve [1, 2]. Related reviews (e.g., [2] and [9]) have pointed to over 70 MMs pertaining to BPM, covering varied process aspects. Similarly, guidelines for the design of MMs are well developed, both in general [e.g. 10–12] and specifically in the BPM domain [e.g. 4, 13]. On the contrary, principles that can guide the users of BPM MMs (i.e., on MM selection, tailoring/customization and deployment) are relatively scarce, leaving the potential end users of BPM MMs unsupported [3]. This study poses the research question: *How can BPM maturity models (MMs) be selected, tailored and used?* We recognize that there can be different use-cases for BPM maturity assessments, and examine two example groups: (a) where the maturity assessment is driven and conducted solely by an organization, and (b) where the maturity assessment is driven and conducted by external consultants.

The current body of knowledge is presented in Sect. 2. Section 3 explains the applied study approach and Sect. 4 presents the final framework for BPM MM adoption. Section 5 presents a summary discussion followed by a conclusion (in Sect. 6).

2 Literature Review

2.1 Concepts Underlying BPM Maturity Models (BPM MMs)

A MM within the BPM discipline has been defined as *"a model to assess and/or to guide best practice improvements in organizational maturity and process capability, expressed in lifecycle levels, by taking into account an evolutionary road map regarding (1) process modelling, (2) process deployment, (3) process optimization, (4) process management, (5) the organizational culture and/or (6) the organizational structure'. In order to increase its usefulness, the MM design includes both a detailed assessment method and improvement method"* [14]. Pöppelbuß and Röglinger [1] distinguished two main types of MMs in the BPM context. First, 'Process Maturity Models' are concerned with the extent to which individual processes are managed. Secondly, 'BPM Maturity Models' address BPM capabilities [7], which cover all aspects of the organization pertinent to BPM, including BPM methods, tools, culture, people, etc. [15, 16].

Maturity assessments are carried out for a pre-defined set of elements, known as 'maturity factors' or 'core capabilities.' In most MMs, the maturity factors are further described and deconstructed as sub-factors or sub-capabilities. Maturity factors are often classified into separate elements such as people/culture, processes/structures, and objects/technology, which in reality are interrelated [11]. The 'maturity factors' act as the 'reference framework', against which the current state of the organization is assessed. A MM also has an 'assessment framework' which defines the elements on how the actual assessment takes place (e.g., providing ratings of the level of process quality) [2]. Both parts of the framework are needed for an organization to identify an organization's current status, gaps and possible ways to improve the BPM capabilities.

2.2   Current Guidelines Supporting BPM MM Users

BPM MMs can be applied in different contexts (e.g., by senior management and/or a BPM center of excellence, or by BPM consultants as they look for service engagement opportunities). [11] explained how MMs assist decision makers to balance divergent objectives in a comprehensive manner by incorporating formality into the improvement activities, which enables decision makers to determine whether the potential benefits have been realized. [17] have stated that: *"conflicts of interest can be avoided"* by the use of MMs, especially when the MM is developed externally to the organization, and when the assessment is done by an independent third party. MMs thus offer a framework to conduct organizational management and to decide on investments by reducing decision-making 'bias'.

A significant number of meta-analytic studies of MMs exist, but they mainly focus on comparisons or a classification of models based on MM characteristics, and thus typically focus on high-level and extrinsic characteristics of the models (e.g., MM design, number of levels, scope, etc.). However, they rarely attempt any deeper analysis or synthesis [2], in particular they are less catered towards supporting MM users. It appeared that the MM literature predominantly focuses on MM design by addressing MMs as artefacts, and defining requirements [12], design principles [4], evaluation criteria [10], steps with decision parameters [11], or phases for MM development [7]. While many authors have applied a particular MM [2], user guidelines for adapting a particular MM have only been covered to a minor extent. The latter specifically applies to the three main phases of MM adaption: (i) selecting a particular MM, (ii) tailoring it, and (iii) applying the tailored MM. For instance, regarding the first phase, the literature already refers to steps [11] or criteria [9] for selecting a MM based on the organization's needs. Before using a selected MM, it should be tailored, customized or translated into the organization's context. For instance, [5] described contingencies for BPM capability development. [11] presented steps for preparing MM deployment, while other authors referred to elements with implications for BPM MM tailoring [18] or design propositions to be adapted in a BPM MM tailoring template [18]. During BPM MM use, steps are provided for applying a MM, taking corrective actions [11], and building a strategy to support managerial actions for MM use [19].

3   Overview of Study Approach

The study design had two core phases and an additional phase (Phase 3). In Phase 1, a literature review was conducted and available guidelines for BPM MM end users synthesized and confirmed by two coders. Papers that provided direct guidance (e.g., in the form of clearly specified principles or criteria, like [9]) or indirect guidance (e.g., in the form of considerations to be mindful, like [18]) for BPM MM users were included.

Phase 2 was intended to explore and expand Phase 1. Two case studies (Cases A and B) were selected because they had: (i) adapted an existing BPM MM, (ii) at least fully prepared to deploy it, and (iii) we had good access to case data through the support of key informants. Empirical evidence was based on document analysis and interviews. Documents such as; project documents specific to the MM efforts, BPM team meeting notes, related communications to diverse stakeholders, and information available about the selected MM were used. Interviewees consisted of the BPM Centre of Excellence (CoE) leaders and members identified by the leads. All interviews were designed around reflecting on the 'what', 'how' and 'why' of adapting BPM MMs. At Case A, 2 joint-interviews with the CoE lead and deputy-lead, followed by 3 interviews with the BPM capability building program manager, senior process architect and BPM technology leader took place. Data input at Case B consisted of three interviews with the CoE lead and four group-based interviews with 13 CoE staff (who represented process architects, senior process analysts, solution architects and productivity specialists).

In Phase 3, the resulting framework was tested against two other cases (Case C and D) where the maturity assessment was actioned by external consultants who had developed their own MM and applied it to their clients to identify service opportunities for capability development. The data for Case C and D were also based on in-depth guided reflections (as in Phase 1), but they were light-weighted and were only conducted with the lead consultant who lead the design and delivery of the maturity assessment.

Case A is a large federal agency that provides social welfare support to the citizens of Australia. They work closely with a range of other federal and state government agencies and have a complex network of business processes. Case A has been adopting BPM practices to manage the myriad of complex and continuously evolving processes, with diverse BPM initiatives occurring in isolation across the agency in different sub-areas. The conduct of an organization-wide BPM maturity assessment was aimed to build awareness of the current status of BPM within the whole agency and to derive an evidence-based roadmap that proposes an 18-month and a 3-5-year plan. The agency partnered with a local university that held extensive BPM expertise and went through the selecting, tailoring and preparing stages to deploy a BPM MM. They adapted the De bruin and Rosemann's MM [20, 21], with a number of contextual customizations.

Case B is an Australian multi-national bank, and one of Australia's leading providers of integrated financial services. The bank applies many methodologies to improve operations by enhancing current business processes. Case B's enterprise-wide paradigm for BPM aimed to extend the focus to full end-to-end experiences of both customers and staff, with a specific focus on risks, controls, customer and staff experiences, quality and workforce efficiencies. Whilst these aspects became factors in Case B's creation of an in-house MM, which focused on individual process maturity, Case B also utilized the Gartner MM at the overall organizational level. This prompted Case B to look across functional hierarchies to assess 'end-to-end' processes and find opportunities for standardization and scalability. The initial assessment based on the Gartner MM was the focus for this paper.

In **Case C**, a boutique consulting company offering BPM consulting services and tools lead the maturity assessment at the client. An in-house MM derived for consulting purposes was applied to determine a roadmap of work required for consulting engagement. Sixteen maturity capabilities under four categories of BPM strategy, BPM design, BPM implementation and BPM control were assessed, applying an in-house consulting assessment tool. Input obtained from quick interviews with corporate contacts as part of pre-sales formed the basis for the assessment. Five levels of maturity with guidance on what is meant for each of 16 sections was applied, focusing both on the actual and target statuses.

Case D pertained to a large 'Big 4' Consulting Company engagement with a Utility company in North America. The Maturity assessment was done as part of the initial 'planning' component of a multistage consulting engagement to determine a work-stream focus. An in-house MM derived for consulting purposes was applied. It had 20 capabilities under four topics of Process Strategy, Process Organization, Process Management Processes and Methods and Tools, with five maturity levels. The assessment data was from a survey of selected management staff, which was designed to assess the current and aspirational states and determine the biggest gaps and priority in the context of failed previous attempts at Process Improvement (Six Sigma). This resulted in five categories of key recommendations which defined work within four work-streams.

4   Study Findings

We noticed that the guidelines found in literature were already positioned within three main stages of the MM adaption journey, namely; (i) MM *selection*, (ii) MM *tailoring*, and (iii) MM *application*. Further sub-themes (A–G; as explained below) were identified to group the content within each stage.

(A)   Guidelines pertaining to the **rationale** behind BPM maturity assessment (i.e., the *'why'* aspects that motivates the maturity assessment with a purpose and value propositions).

(B)   Guidelines pertaining to the **scope and focus** covered.

(C)   Guidelines pertaining to the **BPM capability areas**, that represent 'what' is measured in the BPM maturity assessment.

(D)   Guidelines pertaining to **'how' the measurement occurs** (i.e., the measurement items that concretize/operationalize the capability areas).

(E)   Guidelines about the **involved stakeholders** (i.e., the *'who'* aspect: respondents/users, sponsor/driver, etc.).

(F)   Guidelines about other **MM characteristics**, such as costs, reliability/validity, customizability etc.

(G)   Guidelines specifically around the **MM execution** (i.e., in Stage 3).

The extracted guidelines from Phase 1 were grouped across the MM adaption stages and re-grouped within the sub-themes (A–G). The three-numeric identification indexing each guideline captures this. For example, in '1.A.1', the first digit relates to the stage (1; MM selection), the second part relates to the sub-category (A–G, as introduced above) and the last digit is an indexed number within the sub-categorisation. Sections 4.1, 4.2 and 4.3 detail the summary findings. Given the study design, the outcomes of Phase 1 (literature) and Phase 2 (cases A and B) form the primary discussion, where insights from cases C and D are added at the end of each subsection. Tables 1, 2 and 3 present the framework, listing each guideline (column 1) and depicting supportive evidence from the literature from Phase 1 (column 2), Case A–B from Phase 2 (columns 3–4), and also Cases C–D from Phase 3 (columns 5–6). The tables represent how the literature with a priori guidelines mapped with the case study observations, depicting which guidelines were applied *as is* (denoted by an '(A)'), which guidelines were applied with extensions (denoted by an '(Ae)'), which guidelines were not considered at all (denoted by an 'X') and which guidelines were *new*.

4.1 BPM MM Selection Guidelines

All, except one (1.D.6) of the a priori guidelines were supported by either one or both of Case A and Case B (see Table 1). Some were adopted as is (see '(As)) while others were adopted with extensions where additional practices were observed (see (Aes)).

'**1.A.1**' relates to the purpose for which the MM is intended to be used, namely only raising awareness or also benchmarking and certification [9]. The driving purpose of Case A was to raise awareness and get executive buy-in for future BPM plans, hence one criterion was that the MM be simple and easy to understand by a broader community. Case B's core purpose was internal benchmarking to be able to track and show progress with the BPM Center of Excellence's (CoE) ongoing activities.

'**1.B.1**' is whether the MM addresses a specific process type or can be applied to any process (i.e., generic vs domain specific) [9]. At Case A, the intention was not to measure the maturity of any single process but was aimed at measuring overall BPM capabilities across the entire agency. In Case B the goal was to have a MM that was able to measure the maturity of generic processes. The Bank intended to formally assess 28 pre-identified 'high impact' core processes (referred to as 'HIPs'[1]) using the selected process-level MM. '**1.B.2**' captures whether the sought for MM is sourced from academia or practice [11]. This was not relevant to Case A; what did matter was that there was evidence that the MM had been rigorously designed and validated, and preferably also used in other similar government contexts. Case B had a strong preference for a MM that originated from practice. '**1.B.3**' pertains to whether the expected recommendations from the maturity assessment are problem-specific or more general in nature. The more general ones often need further contextual detailing [11]. This aspect was not considered at all by Case A or B. '**1.B.4**' refers to the number of business processes to be assessed and improved (i.e., one, more or all processes in an organization) [9]. Similar to '1.B.1', '1.B.4' this was not a consideration for Case A, as the focus was not at process-level capabilities (rather, overall organizational level). At Case B, the focus was on assessing the predefined HIPs (28 processes), from a true 'end-to-end' perspective. This was especially emphasized in areas where the bank's practices were siloed, (which had the tendency to only focus on a part of the process).

'**1.C.1**' is about identifying which capability areas should be assessed and improved (i.e., process lifecycle, process-oriented culture, process-oriented structure) [9]. Determining this upfront will help identify which MMs have the sought-after capabilities. Case A did much exploratory work to identify the required capabilities. This included looking at: (i) some of the leading MMs and seeing what capabilities were suggested there and why, (ii) related other MMs (e.g., for IT, HR, Policy, etc.) applied in government settings, and (iii) Australian Federal Government strategies and related frameworks to ascertain process capabilities that may be specific to the Australian Federal Government sector.

---

[1] 'HIPs' was an acronym used to refer to high impact processes. These are processes that are so critical, that if they failed, they would pose a significant risk (reputational, financial, regulatory) to the bank.

Table 1. BPM MM selection guidelines with summary evidence.

	Lit.	A	B	C	D
1.A.1 Determine the requested purpose of a maturity assessment (raising awareness/benchmarking/certification)	[9]	(A)	(A)	A	A
1.B.1 Determine the requested type of business processes (generic/certain domains)	[9]	X	(A)	X	X
1.B.2 Determine the requested origin (academia or practice)	[11]	(Ae)	(A)	X	X
1.B.3 Determine the requested practicality of recommendations (generic/context)	[11]	X	X	X	A
1.B.4 Determine the targeted number of business processes (all/one/more)	[9]	X	(Ae)	X	X
1.C.1 Determine the requested capabilities (BPM lifecycle/culture/structure)	[9]	(Ae)	(Ae)	A	A
1.C.2new Determine the high-level definitions of the requested Capabilities	–	new	new	X	X
1.C.3new Determine the definitions of the lower-level factors pertaining to each Capability	–	new	new	X	X
1.D.1 Determine the requested architecture type (maturity levels/capability levels)	[9]	(A)	(A)	X	X
1.D.2 Determine the requested assessment availability (public/private)	[9]	(A)	(A)	X	X
1.D.3 Determine the requested data collection technique (subjective/objective)	[9]	(Ae)	(A)	A	X
1.D.4 Determine the requested rating scales (qualitative/quantitative)	[9]	(A)	(A)	A	A
1.D.5 Determine the requested range of assessment items (number of questions)	[9]	X	(A)	A	X
1.D.6 Determine the requested assessment duration (day/week/longer)	[9]	X	X	A	A
1.E.1 Determine the requested functional role of respondents (internal/external)	[9]	(A)	(A)	X	A
1.F.1 Determine the requested reliability and validity (empirical evidence)	[9, 11]	(A)	(A)	X	X
1.F.2 Determine the requested accessibility based on costs (free/paid)	[9, 11]	(A)	(A)	X	X
1.F.3 Determine how far the MM is configurable	[11, 22]	X	(A)	X	X
1.F.4 Determine the requested architecture details (descriptive/prescriptive)	[9]	(A)	(A)	X	A
1.1new Systematically documented selection criteria		new	–	X	X

Table 2. BPM MM tailoring guidelines with supporting evidence

	Lit.	A	B	C	D
2.A.1 Determine the organization-specific purpose	[18]	(A)	(A)	N/A	N/A
2.A.2 Determine the organization-specific value preposition (i.e., effects on the organization and on business processes)	[18]	(A)	(A)	N/A	N/A
2.A.3 Determine the needed formality of realization (informal/formal)	[11]	(Ae)	(A)	N/A	N/A
2.B.1 Determine the organization-specific application area (one/multiple entities)	[11]	(X)	(A)	N/A	N/A
2.B.2 Determine the organization-specific scope (national/regional, management/operational, intra- or inter-organizational)	[18]	(A)	(A)	N/A	N/A
2.C.1 Determine the BPM capabilities that fit with the environment and organizational needs (optimal/desired levels of BPM capabilities)	[5, 18]	X	(A)	N/A	N/A
2.D.1 Determine the organization-specific measurement approach (e.g., open questions, statements on a Likert scale, matrix)	[18]	(A)	(A)	N/A	N/A
2.D.2 Determine the organization-specific measurement items (i.e., translation)	[18]	(Ae)	(A)	N/A	N/A
2.D.3 Determine the organization-specific measurement time frame	[18]	(A)	(A)	N/A	N/A
2.E.1 Define the organization-specific sponsor	[11]	(A)	(A)	N/A	N/A
2.E.2 Determine the organization-specific respondents	[11, 18]	X	(A)	N/A	N/A
2.E.3 Determine the respondents' degree of BPM knowledge (i.e., years of BPM experience, into ongoing BPM activities within the organization)	[11, 18]	(X)	(A)	N/A	N/A

At Case B, they believed that most of the leading BPM MMs covering enterprise-level BPM capabilities overlapped, hence other aspects (such as 1.B.2, 1.B.3) were prioritized over 1.C.1.

'**1.C.2**' and '**1.C.3**' were both new and observed in both cases. 1.C.2 related to paying attention to defining the capabilities that were to be assessed and 1.C.3 related to defining the lower-level factors within each capability used for the assessment. In both Case A and Case B, these definitions were derived and were also closely aligned with the organizations' current BPM terminology. Case A for example used a BPM Lexicon to standardize the definition and develop a common meaning of these across all stakeholders.

'**1.D.1**' pertained to whether the MM should support a staged (the possibility to assess and develop a single capability) and/or a continuous approach (the possibility to assess and develop overall maturity across multiple capabilities) [18]. Both Case A and B had intentions to apply a continuous approach, which aligned with their overall motivations

Table 3. BPM MM application guidelines with supporting evidence

	Lit	A	B	C	D
3.A.1 Determine the intended performance effects (e.g., financial performance, customer/employee satisfaction) and potential drivers (e.g., new business model)	[19]	(A)	(A)	X	A
3.A.2 Determine preconditions (e.g., top management support, strategy alignment)	[19]	(A)	(A)	A	A
3.D.1 Determine the frequency of application (one off/recurring)	[11]	X	(A)	A	A
3.E.1 Determine the stakeholders for corrective actions (internal/external)	[11]	X	(A)	X	A
3.G.1 Determine whether to execute an assessment (go/no-go decisions)	[11]	(A)	X	N/A	N/A
3.G.2 Determine other synergistic initiatives with the organization (e.g., integration with existing development programs)	[11, 19]	X	X	X	A
3.G.3 Determine the execution plans for corrective action (on the fly/projects)	[11]	X	X	A	A
3.G.new1 Determine the inter-relationships between the different capabilities and measurements	–	New	New	X	X
3.G.new2 Determine if/when to transfer to another MM	–	–	New	X	X

for doing maturity assessments (see 1.A.1). Again with '**1.D.2**' (whether the assessment items and level calculations are publicly available [18]), both cases were only interested in MMs that had publicly available assessment instruments. '**1.D.3**' relates to the way the information is collected during the assessment (i.e., subjectively and/or objectively) [9] and '**1.D.4**' relates to the type of data (i.e., qualitative and/or quantitative) that is collected during an assessment [9]. Case A had purely subjective intentions (where data was to be collected qualitatively) to start with, with the aim that more objective (and quantitative) measurement can be deployed after the agency gained experience and momentum on BPM MM use. Thus, the MM selection was influenced by the need to be able to make this shift in the longer term. Case B intended to use a mix of subjective and objective (and also a mix of qualitative and quantitative) measures and had the intentions of being flexible to fit the 'current-need' and "jump ship" (i.e., use alternative MMs if desired). '**1.D.5**' represented the maximum number of questions to be answered during the assessment [9]. The precise number of questions was not a direct consideration at Case A but given their goals to maintain a 'simple' and "straight forward" assessment, MMs with a more concise set of questions were more attractive. Similarly, for Case B, the number of questions was not a direct determinant, as this was a function of other aspects such as 1.C.1 and 1.D.3. '**1.D.6**' related to the maximum duration of an assessment [9]. Again, this was a not a direct consideration by Case A or B as it depended on other aspects (such as 1.D.3, 1.D.4 and 1.D.5) and how the overall maturity assessment efforts were to be project managed.

'**1.E.1**' captures whether to include people from within or outside the assessed areas, as respondents [9]. This was considered by both cases, but more at the MM adoption stage than the MM selection stage, where potential respondents were carefully hand-picked in each case; considering stakeholders internal (within) and external (outside) to the assessed area. At Case B the principles of having an end-to-end process influenced the respondent selection.

'**1.F.1**' relates to how well the MM has been evaluated (i.e., evidence that the MM is able to assess maturity and helps to derive more enhanced and effective business processes), with prior evidence [9, 11]. As mentioned with 1.B.2, Case A was highly inclined to selecting a 'rigorous' MM. One reason for this was the aim to communicate the assessment results to the higher government authorities (i.e., Minister and steering committee) and the need to have a robust model on which the assessment was based on. Case B also intended to select a reliable MM 'in theory' but did not go through any formal process to ascertain the Model's validity and reliability. '**1.F.2**' relates to if the MM is free for use or not. And if not free, what the direct costs for accessing and using the MM are [9, 11]. Both Cases opted for a free MM; to maintain an overall low cost. Note that for Case B, the Gartner assessment was included as part of another engagement the two entities had. '**1.F.3**' is about how far the MM elements can be customized and their ease of integration into existing organizational contexts [11, 22]. This aspect was not considered in Case A, but was as an important consideration in Case B. Not only was this relevant for the first implementation of the maturity assessment, but they also considered the ease of transitioning from one MM to another over time. '**1.F.4**' refers to the degree of guidance the MM gives to achieve higher maturity levels (i.e., descriptive, implicit prescriptive, explicit prescriptive) [9]. Both cases looked at this at MM selection and found that hardly any gave prescriptive guidelines.

A novel aspect (see '**1.1new**') that related to and overlaid a number of the detailed guidelines was identified within Case A, namely the systematic derivation and application of MM selection criteria (SC). Case A derived an initial set of high-level SC consolidating agency-specific needs and anticipated MM features, which were then subjected to a detailed SC with sub-criteria, ranking and weights. Both stages were documented for transparency, as a means to justify selecting the 'best-fitting' MM.

The use of these guidelines within Case C and Case D was different, due to the selector being a consulting organization who was selecting and designing a maturity model, and not the company that was applying it. Therefore, the goals of a consulting organization will be at the forefront – selecting and selling defined services, favoring competitive strengths, utilizing existing assets and intellectual property. Also, the consulting company will be more focused on outputs (plans and roadmaps) depending on when in their methodology they recommend using the maturity assessment. It is understandable that there will be less focus on factors within Category F (related to MM characteristics such as costs and reliability) as the company designs once and applies to multiple clients, and has an advantage in having a unique MM product on the market that fits with their methodology and recommended deliverables. These cases were included as even though they demonstrate a key variation in the selection and applicability of a model, they are common examples of the use of maturity assessments in the environment. It is common for organizations not to have existing skills in this area. And thus, they seek assistance from consultants in applying BPM principles and making recommendations for BPM capability uplift.

4.2 BPM MM Tailoring Guidelines

All a priori MM tailoring guidelines were supported by either one or both of Case A and Case B (see Table 2), with some extensions observed. Similar to the MM selection stage, both Case A and B revisited the rationale and motivation for maturity assessment at the MM tailoring stage as well. '**2.A.1**' relates to determining the purpose of assessment and how it will fit to the target group (e.g., to identify a need for BPM methods or knowledge, benchmarking, action plans) [18]. '**2.A.2**' is about the expected outcomes of the maturity assessments and how to cater the maturity assessment to meet these needs (i.e., effects on the organization and on business processes) [18]. These two facets remained the same (as the prior MM selection stage) for both cases. '**2.A.3**' determines whether the assessment will be an informal appraisal or a formal assessment. Case A chose a quasi-formal option, where the actual assessment would be done by an internal sub-committee with advisory members of BPM MM researchers from a local university. At Case B, the assessment was informal and performed by the newly appointed General Manager of BPM, who focused on the capabilities within the 28 HIPs.

'**2.B.1**' gets one to think about which areas will be subjected to the assessment (i.e., specific entity or multiple entities) [11]. This aspect was not specifically considered at Case A but given the enterprise-wide BPM capability assessment goals the areas where BPM had progressed and was led by clear leadership was aimed for (yet ill defined). Case B targeted Australian sited processes. '**2.B.2**' relates to the specific scope of assessment; at a geographical level (i.e., national, regional or local), organizational level (i.e., management or operational) or process level (i.e., intra- or inter organizational) [18]. Case A was set at an inter-organizational level. With Case B, the process level assessments were aimed at the HIPs which were often intra-organizational. The Enterprise-wide BPM capability assessment was aimed at a national scope.

'**2.C.1**' is about deciding the optimal levels of BPM capabilities, as dependent on environmental and organizational characteristics. The dynamic market conditions and the users' purposes will help determine relevant BPM [5, 18]. Case A did not attempt to predefine optimal levels of capabilities. Case B did this, by utilizing a pre-existing Enterprise Capability Model recently produced within its Enterprise Services unit, and then assessing all human resource capabilities directly involved in bank-wide BPM.

'**2.D.1**' is about how the assessment will look like (e.g., open questions also or only statements on a 5-point Likert scale, of a matrix to be filled out) [18]. At Case A, this was considered in detail earlier (as explained with 1.D.3-1. D.5) and designed (ready for execution) in the tailoring phase. In Case B this was completed via responses to questions on a 5-point scale. '**2.D.2**' relates to translating the MM measurement using concepts appropriate to the target respondents [18]. Case A was very conscious about making the assessment process and results meaningful to the non-BPM specific audience (i.e., ministerial leaders/senior agency executives) and derived a detailed lexicon to support this. Like Case A, Case B translated the results into funding proposals submitted to the Executive Committee to support the BPM improvement work to be performed by the Chief Process Office of the Bank. **'2.D.3'** Defines the time frame planned for the assessment [18]. At Case A, this was planned to proceed immediately. At Case B, the intention was to proceed at the originally-set point in time, and thereafter repeat the assessment in an annual frequency.

'2.E.1' relates to identifying the main person responsible for driving the maturity assessment [11]. At Case A, this was one of the BPM-leads who was trying to use the maturity assessment as a means to 'communicate-upwards'. Case B was sponsored by the newly created and appointed Chief Process Officer (CPO). **'2.E.2'** defines the pool of respondents who will contribute to the assessment as input providers (i.e., employees, management, business partners or a combination of such) [11, 18]. At Case A this was rather ad-hoc and ill-defined. At Case B, this was performed by Chief Process Office, with input from relevant heads of business units **'2.E.3'** Determine the respondents' degree of BPM knowledge (i.e., years of BPM experience, insights into ongoing BPM activities within the organization) [11, 18]. This was not considered at Case A. Case B was performed within the Chief Process Office by individuals with decades of experience in BPM and/or bank specific operations.

Most of the tailoring guidelines were not applicable for Case C and D as both cases related to consulting companies creating maturity models which were applied at client sites. These MMs were originally designed to be of use with multiple clients so either, did not require many adjustments or would be designed to be easily adapted. In these cases, the former was applicable, and no tailoring was required and they were used directly as designed, with little consideration of any need to adapt to specific client needs. Matters outside of the formalized assessment were still considered, informally. Example factors considered informally included; the history of BPM within the organization, the sponsor's focus, and the context of the capabilities. The context included such factors as understanding the dependencies between related capabilities and timelines required for foundational baseline capabilities (such as tools and awareness) being in place before considering capabilities of a more advanced nature.

4.3   BPM MM Application Guidelines

One of the a priori MM application guidelines was supported by all cases and two were supported by both Case A and B. Two new guidelines emerged (Table 3). The motivations ('**3.A.1**') were revisited again in the model application phase, with the aim to answer why the organization needs a process orientation (i.e., financial, customer/employee satisfaction and/or operational performance, new business models, innovations, support sustainable society) [19]. The motivations were consisted for Case A across all three phases. For Case B, during the application phase, customer and staff experience were top priorities, followed by risk and cost/income drivers. '**3.A.2**' confirms that the planned assessment is aligned with the organization's strategic plans and also supported by the top management [19]. This was observed within Case A, where the assessment goals were refined by input from the senior executives as the maturity assessment roll out was been planned. In Case B, this took place as a natural outcome of the creation of the Chief Process Office.

'**3.D.1**' was to decide on the frequency of MM application; ideally if it is one off (non-recurring) or repeated, and if repeated in what intervals [11]. In both Case A and B, this highly related to the decisions made in 2.D.3 and remained as-is.

'**3.E.1**' identifies who (i.e., specific staff, line organization, or externals) will be involved when executing BPM capability developments resulting from the maturity assessment, including what roles they will play [11]. As in 2.E.2, this was overlooked in Case A. In Case B, this was determined by each Process Owner in conjunction with that business's Master Black Belt, Black Belt and Green Belt community.

'**3.G.1**' relates to the 'go' or 'no go' options of executing the actual assessment. The maturity assessment may proceed or halt after all the detailed preparation is done [11]. Case A halted the initiative after an initial informal assessment (and did not proceed with a capability development phase; also, commonly known as a road map). This was not considered at Case B as progressing till the end was a commitment made from the outset. '**3.G.2**' relates to decisions on whether or not to couple new/enhanced BPM capability building with existing developmental programs. And if yes, to identify existing development initiatives to integrate with [11, 19]. This was seen in both Case A and B (e.g., piggy backing on agency-wide IT implementations) (e.g., integrating with existing initiatives determined). '**3.G.3**' relates to determining if the identified 'gaps' when planned to be addressed should have dedicated well defined projects or occur on the fly [11]. In Case A, given the assessment discontinued the capability enhancements were not thought through in detail. At Case B, the different capability enhancements did form a mix of new projects (e.g., the creation of 'Customer Led Simplification (CLS)' projects), and 'on-the-fly' continuous improvement initiatives.

Two new guidelines were identified. '**3.G.new1**' relates to looking at the different capabilities as a 'whole system' rather than isolated silos. Essentially it is to identify how one capability may relate to others. In Case A, a comprehensive mapping of these interrelationships was done. For example, a process architecture that facilitates 'strategic alignment', will require training (a 'people' factor) and may also be a 'tool' and appropriate procedure that guides and governs its use (thus touching on 'methods' and 'governance'). This was implicitly in Case B as well, but not directly designed for such.

3.G.new2 captures the consideration of when to swap MMs, especially in reoccurring MM applications. This was strongly emphasized in Case B where they argued that as an organisation's BPM maturity increase over time and need for maturity assessment evolves, they may wish to 'move-on' to other MMs. For example, the more sophisticated the adoption and utilisation of BPM within an organisation, the use of the more detailed and sophisticated MMs should follow – as the benefits of less detailed MMs will hit 'the law of diminishing returns' for the organisation.

With Phase 3, all but one (3.G.1) a priori MM application guidelines were observed in Case C and Case D. Given that both cases intended to encourage sales (Case C) and work (Case D), the recommendation was always going to be to proceed and a go/no go decision would not be applicable. With '3.A.1', Case D looked closely at performance effects as a tailored response was developed as an outcome from the maturity assessment that took into account history and context. Case C was a very simple assessment for one purpose only, to get an initial evaluation, and so was not concerned with the outcomes, just providing a reading of where the organization was at. Case C and D although quite different did take into account the existing state of the organization and the factors that would ensure organizational readiness such as top management support which are incorporated into any consulting engagement relating to capability uplift ('3.A.2'). With '3.D.1' Case C was designed as a one-off assessment and Case D was designed to be used annually but would be determined on success of initial work to see if momentum could be maintained. Case D looked at stakeholders as part of implementing a roadmap '3.E.1'. Case C was a standalone assessment and Case D tried to incorporate existing past and future projects and priorities to a limited extent '3.G.2'. The focus on Case C and D was to foresight future work so considered the approach for building BPM capabilities, and selecting which ones to focus on, and was designed to be a more defined planned approach ('3.G.3').

5 Summary Discussion

Figure 1 presents a visual summary of the resulting framework for BPM MM adaption, which combines three stages, including 20 guidelines for *BPM MM selection*, 12 guidelines for *BPM MM tailoring* and nine guidelines for *BPM MM application*. The guidelines are divided into seven sub-themes (A-G) pertaining to the: (A) rationale behind maturity assessment, (B) scope and focus covered, (C) capabilities measured, (D) operationalization of the measures, (E) involved stakeholders, (F) MM meta-characteristics, and (G) MM execution.

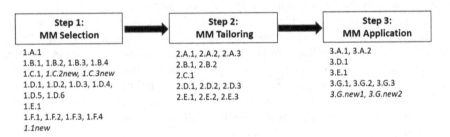

Fig. 1. A BPM selection and adaption guidelines

The guidelines were interrelated in a nested manner across the stages, i.e., decisions made in Stage 1 would influence the decisions and options in Stage 2 and Stage 3. Themes appearing across the stages have specific guidelines for decisions in that point of time of the journey (as enumerated in Fig. 1 and summarized in Table 4). These sub-themes occur in varying degrees across the three phases (Table 4). The framework synthesizes otherwise scattered literature-based guidelines and demonstrates how this synthesized knowledge is validated and at times extended by our case insights. Compared to the literature, five new guidelines have been added, namely three selection guidelines and two application guidelines. The analysis indicated dependencies between the sub themes and phases. For example, the rationale behind maturity assessment (A) can influence the scope and focus (B) for the assessment, which capability areas are selected (C), and also how the MM is eventually executed (G).

Table 4. Summary overview of the extracted themes

	Sub-categories of themes						
Main phases	A	B	C	D	E	F	G
(1) For selecting/choosing a BPM MM	X	X	X	X	X	X	
(2) For tailoring or customizing a BPM MM	X	X	X	X	X		
(3) For the actual application of the BPM MM	X			X	X		X

6 Conclusion

This paper has provided detailed guidelines for the selection, tailoring and application of BPM MMs, from a user perspective. The framework was built with a literature synthesis and further supported by case study insights. The framework is a useful artifact for practitioners with 41 detailed guidelines, which are categorized into seven categories and grouped in three phases. Meanwhile, the framework can support multiple journeys (e.g., possibly by combining MMs) to inspire future MM adaptations rather than stimulating extra BPM MMs.

The current framework has limitations and leaves room for further development and validation. First, while the literature typically differentiates between the levels of BPM maturity assessments (e.g., at the process level and organizational level [1]), the framework guidelines do not consider this differentiation more deeply. Also, the different MM user-types with diverse motivations and means of applying BPM MMs in practice need further inquiry. An attempt to generalize the findings by different MM use-case scenarios (i.e., where the maturity assessment is driven by consultants, in Case C and Case D) proved that the guidelines are not fully applicable to this context. Future research can derive a typology of BPM MM use-cases to refine the framework with evidence obtained from representatives of such different scenarios.

References

1. Pöppelbuß, J., Röglinger, M.: What makes a useful maturity model? A framework of general design principles for maturity models and its demonstration in business process management. In: Proceedings of the European Conference on Information Systems, pp. Paper28 (2011)
2. Tarhan, A., Turetken, O., Reijers, H.A.: Business process maturity models: a systematic literature review. Inf. Softw. Technol. **75**, 122–134 (2016)
3. Van Looy, A., Poels, G., Snoeck, M.: Evaluating business process maturity models. J. Assoc. Inf. Syst. **18**, 1 (2017)
4. Röglinger, M., Pöppelbuß, J., Becker, J.: Maturity models in business process management. Bus. Process Manag. J. **18**, 328–346 (2012)
5. Niehaves, B., Poeppelbuss, J., Plattfaut, R., Becker, J.: BPM capability development – a matter of contingencies. Bus. Process Manag. J. **20**, 90–106 (2014)
6. Bucher, T., Winter, R.: Taxonomy of business process management approaches. In: vom Brocke, J., Rosemann, M. (eds.) Handbook on Business Process Management 2. International Handbooks on Information Systems, pp. 93–114. Springer, Heidelberg (2010). https://doi.org/10.1007/978-3-642-01982-1_5

7. De Bruin, T., Freeze, R., Kaulkarni, U., Rosemann, M.: Understanding the main phases of developing a maturity assessment model. In: Proceedings of the Australasian Conference on Information Systems (ACIS), pp. 8–19 (2005)
8. BPM&O: Status Quo Prozessmanagement 2010/2011 (2011)
9. Van Looy, A., De Backer, M., Poels, G., Snoeck, M.: Choosing the right business process maturity model. Inf. Manag. **50**, 466–488 (2013)
10. Salah, D., Paige, R., Cairns, P.: An evaluation template for expert review of maturity models. In: Jedlitschka, A., Kuvaja, P., Kuhrmann, M., Männistö, T., Münch, J., Raatikainen, M. (eds.) PROFES 2014. LNCS, vol. 8892, pp. 318–321. Springer, Cham (2014). https://doi.org/10.1007/978-3-319-13835-0_31
11. Mettler, T.: Maturity assessment models: a design science research approach. Int. J. Soc. Syst. Sci. **3**, 81–98 (2011)
12. Becker, J., Knackstedt, K., Pöppelbuß, J.: Developing maturity models for IT management: a procedure model and its application. Bus. Inf. Syst. Eng. **1**, 213–222 (2009)
13. De Bruin, T., Rosemann, M.: Towards a business process management maturity model. In: ECIS 2005 Proceedings of the Thirteenth European Conference on Information Systems (2005)
14. Van Looy, A., De Backer, M., Poels, G.: Defining business process maturity. A journey towards excellence. Total Qual. Manag. Bus. Excellence **22**, 1119–1137 (2011)
15. Hammer, M.: The process audit. Harv. Bus. Rev. **85**, 111 (2007)
16. Rohloff, M.: Case study and maturity model for business process management implementation. In: Proceedings of the 7th International Conference, BPM 2009 (2009)
17. Fraser, M.D., Vaishnavi, V.K.: A formal specifications maturity model. Commun. ACM **40**, 95–103 (1997)
18. Christiansson, M.-T., Van Looy, A.: Elements for tailoring a BPM maturity model to simplify its use. In: Carmona, J., Engels, G., Kumar, A. (eds.) BPM 2017. LNBIP, vol. 297, pp. 3–18. Springer, Cham (2017). https://doi.org/10.1007/978-3-319-65015-9_1
19. Christiansson, M.-T., Rentzhog, O.: Lessons from the "BPO journey" in a public housing company: toward a strategy for BPO. Business Process Management Journal (2019)
20. de Bruin, T.: Business process management: theory on progression and maturity. Queensland University of Technology (2009)
21. de Bruin, T., Doebeli, G.: An organizational approach to BPM: the experience of an Australian transport provider. In: vom Brocke, J., Rosemann, M. (eds.) Handbook on Business Process Management 2. IHIS, pp. 741–759. Springer, Heidelberg (2015). https://doi.org/10.1007/978-3-642-45103-4_31
22. Mettler, T., Rohner, P.: Situational maturity models as instrumental artifacts for organizational design. In: Proceedings of the Design Science Research in Information Systems and Technology (DESRIST) Conference. (2009)

Standardization, Change and Handoffs

A Theoretical Model for Business Process Standardization

Bastian Wurm[✉] and Jan Mendling

Vienna University of Economics and Business, Vienna, Austria
{bastian.wurm,jan.mendling}@wu.ac.at

Abstract. Process standardization is for many companies a matter of strategic importance. Process standardization enables companies to provide consistent quality to customers and to realize returns of scale. Research in this area has investigated how process standardization impacts process outcomes, such as cycle time, quality, and costs. However, there are only limited insights into antecedents that lead to process standardization. Furthermore, it is not clear which contextual elements play a role when standardizing business processes. In this paper, we address this research gap by developing a theoretical model for business process standardization. The model is relevant for academics and practitioners alike, as it helps to explain and predict business process standardization by various antecedents and contextual factors.

Keywords: Process standardization · Theoretical model · Theory development

1 Introduction

Business process management (BPM) is a management paradigm that allows companies to improve their performance by managing their business processes in an end-to-end fashion [17]. Among others, associated advantages of BPM are increased customer satisfaction [25], product quality [34] and profitability [19]. One of the main challenges companies face when adapting a process-oriented management view is determining the trade-off of process standardization versus process variation [18]. When too many variants of the process are allowed to coexist, the variability in the process leads to inefficiencies [10]. A mechanism to cope with this is business process standardization (BPS) [42].

Empirical research about BPS has been mainly centered around the relation of process standardization to process performance [35,41]. [53] also investigated the effect process complexity and standardization effort have on BPS. There is, however, a lack of research on antecedents of business process standardization that explains how process standardization can be achieved. In particular, it remains unclear how different contextual factors [7], e.g. the knowledge-intensity of the business process or environmental uncertainty, mediate the effect of process standardization on business process performance. This is a problem as it

© Springer Nature Switzerland AG 2020
D. Fahland et al. (Eds.): BPM Forum 2020, LNBIP 392, pp. 281–296, 2020.
https://doi.org/10.1007/978-3-030-58638-6_17

limits our understanding of how process standardization can be achieved and when it might be disadvantageous.

In this paper, we address this research gap by developing a theoretical model for business process standardization. Our research question thus reads as follows:

What are antecedents of business process standardization?

We address this question by conducting a structured literature review [32] and multiple expert interviews. The result of our work is a theoretical model that identifies and interrelates antecedents of business process standardization and connects them to different context elements. The model is relevant for research and practice, as it summarizes and integrates existing knowledge on building blocks that are required when standardizing business processes. This way, managers can understand what it takes to standardize processes, and what are important factors that they may not be able to influence, but should take into account as well.

The remainder of this paper is structured as follows. First, we elaborate on the current state of the literature on process standardization and socio-technical systems theory that we use as a theoretical lens. Second, we present the method that we used to derive our theoretical model. Third, we outline our findings and how this leads us to propose a theoretical model for business process standardization. We then discuss implications and limitations of our work. Last, we conclude the paper by summarizing its key contribution.

2   Research Background

In this section we discuss the research background against which we position our work. We first present the literature on BPS. Afterwards, we elaborate on socio-technical systems theory that serves us as an important theoretical lens in this paper.

2.1   Business Process Standardization

From a procedural perspective, process standardization can be understood as the alignment of a set of process variants towards a defined meta-process [42]. By contrast, different process variants are generated from a meta-process in order to comply with legislative requirements of different countries [40], to serve specific customer needs [16,18], or to tailor services and products to different markets [64,65]. Figure 1 visualizes process standardization and process variant generation schematically. Accordingly, we define BPS as *the unification of business process variants and their underlying actions* [9,51]. Similarly to BPS, process harmonization [45] aims at the alignment of business processes, but leaves more degrees of freedom with respect to the exact implementation. Consequently, we understand process harmonization as a specific instance of process

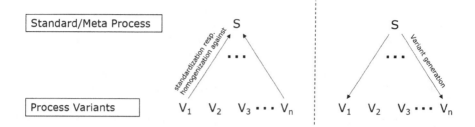

Fig. 1. Process standardization/variant generation (extended from [42])

standardization. Importantly, not all standardization initiatives lead to implementation of *standard processes* that are carried out exactly the same in each and every part of the organization.

To facilitate process standardization, researchers have proposed different procedure models [31,38,42]. The reference model by [31], for example, comprises a total of three phases. In the first phase, a process owner and his or her team is appointed. In the second phase, the actual standard process is developed. This is done by comparing, merging existing processes and by positioning the newly created standard process. The comparison of processes allows to benchmark processes against one another and identify best practices. Then, and the standard processes is defined and modeled. The second phase is completed with the identification of implementation barriers and the estimation of the return on investment expected from implementing the process. Third, the standard process is implemented and substitutes existing process variants in place. [31] suggest to introduce the standard process gradually at more and more locations. Before the process is released, the process owner is trained to manage and improve the process in a centralized manner.

According to [36,37], processes form a continuum ranging from standard, over routine to non-routine processes. Standard processes, on one side of the continuum, are very effective and efficient, as they make use of asset specificity. However, they can only deal with predefined scenarios and schemata. Non-routine processes, on the other end of the continuum, are non-repetitive and cannot be described prior to execution. For this type of process, task accomplishment serves as the most important criterion to evaluate execution success (compare [21,54]).

Especially for multinational enterprises, process standardization serves as an essential mechanism to improve operational efficiency [41]. It helps organizations to realize returns of scale and reduce costs [35,65]. BPS enables the reduction of organizational complexity while increasing transparency [29], ultimately leading to enhanced control over large corporations. Yet, there are significant drawbacks and barriers to BPS. First, process standardization comes at the cost of limiting *flexibility* [22,58]. Second, [49] found, that process standardization should not target pockets of creativity [55], i.e. *creative* parts of processes. Additionally, slack is emphasized as an important element, which allows routine and non-routine processes to cope with *uncertainty* and *ambiguity* [36]. This is one of the reasons, why very *complex* processes are said to be roadblocks for process standardization [36,53].

Empirical research on business process standardization has primarily focused on the relation between process standardization and process performance [35,41]. In particular, these studies explored the resulting effect of BPS on efficiency gains in cycle time, process quality, and process cost. However, there are at least four important limitations. First, there seems to be only little clarity on the delimitation of the concept and construct of business process standardization and other concepts that are related to BPS. E.g. [52,53] operationalize BPS mainly from an execution perspective, but also include one item concerned with process documentation. [44] additionally consider Information Technology as one of the facets of process standardization. Second, antecedents of process standardization have not been examined, yet. Even though multiple authors mention the connection of BPS with governance [58], documentation [60], and strategic focus [7], the relations are neither described in detail nor hypothesized. Third, while there is substantial research on the effects of process standardization, there is limited work on how to actually achieve it. It seems needless to say that only once something has been achieved, its true effects can be realized. Fourth, research on process standardization is largely acontextual. However, to allow for precise predictions and managerial implications, we need to consider mediating factors, and context [7]. In summary, research on BPS would benefit from explicating the nomological network in which it is embedded.

2.2   Business Processes as Socio-Technical Systems

Socio-technical systems theory (STST) [4,5] has influenced information systems research to a great extend. The theory as captured in Fig. 2 essentially states that the technical sub-system, comprised by technology and tasks, and the social sub-system, formed by structure and people, are interdependent. Still today, many IS projects fail due to the exclusive focus on either one of the sub-systems, while neglecting the other.

Instead of designing and managing each sub-system individually, the technical and the social sub-system should be jointly designed and managed to guarantee its functioning. The socio-technical systems perspective can be applied to business process as well. In business processes, participants (social system) carry out tasks by use information technology (technical system) [10]. Business processes comprise certain authority relations (structure) that define who makes decisions and what processes are ought to be executed. Thus, business processes are socio-technical systems and are in turn part of more complex and larger socio-technical systems within organizations and beyond. This view is also represented by work system theory [1].

In terms of business process standardization, many authors have focused the characteristics of process execution or its effects on process performance (e.g. [45] and [41]). However, the broader socio-technical system that business processes are part of and that plays a role for process standardization has received limited

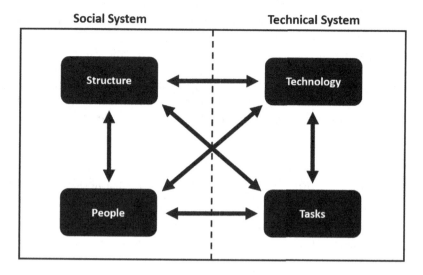

Fig. 2. The socio-technical systems view [5]

attention. In a similar way, [7] argue for a contingent approach to BPM. Consequently, a theoretical model that explains and predicts [14] the antecedents of business process standardization has to consider socio-technical elements, like culture [33] and the knowledge-intensity [22] of business processes.

3 Method

To develop our theoretical model, we carried out a structured literature review [32] and conducted several expert interviews within a multinational organization. We describe the details of our methodological approach in the following.

3.1 Literature Review

We conducted a systematic literature review to identify, evaluate, and interpret the literature [32,33] on business process standardization and harmonization. As the primary source to retrieve literature, we used four databases that cover a wide range of publication outlets: ProQuest, ScienceDirect, EBSCOhost, and IEEE Xplore. Each database is widely recognized and covers publications from Information Systems, Business Process Management, Information Technology, and other research fields. This inclusive scope in literature search helped us to avoid a bias. We applied the search terms "Process Harmoni*" OR "Process Standard*" to search titles, abstracts, and keywords of publications without any restrictions to specific periods of time.

The literature search yielded more than 250 publications (journal articles and conference papers) matching the search criteria. Table 1 shows how the search hits are distributed over the different databases that we used. Additionally, we carried out backward and forward search [63] to identify literature that was relevant, but did not show up in our direct search. All papers that we found were assessed using the same rigorous selection process [13]. First, we considered and retrieved only articles in English or German. Second, all duplicate papers were removed. For all remaining papers, we read title, abstract, and conclusion in order to evaluate whether they should be subject to full text assessment. Those papers that appeared to be relevant were then studied in detail.

Table 1. Search results

Database	URL	Number of search hits
Science Direct	http://www.sciencedirect.com	93
Proquest	http://search.proquest.com	51
EBSCOHost	http://search.ebscohost.com	58
IEEE Xplore	http://ieeexplore.ieee.org	53
Sum	**255**	

3.2 Interviews

Furthermore, we carried out expert interviews with process experts of a multinational company to complement our literature analysis. In total, we conducted five semi-structured in-depths interviews with experienced subject matter experts in the area of BPM. The goal of the interviews was to triangulate our findings from the literature review, to gather concrete illustrations of how BPS is achieved in practice, as well as to detect and address any blind spots that may have been present.

All interview partners were highly experienced employees of a manufacturing company in the construction industry. The case company has its head quarters in a German-speaking country and operates subsidiaries in more than 120 countries worldwide. Overall, the company has a workforce of more than 22,000 people. The case company is one of the market leaders and its products mainly address the premium segment. Our case company is very process-oriented; it operates a Center of Excellence for process management and a dedicated community of process managers on a global, regional, and on country level. In more general terms, our case company is well known for its highly mature IT capabilities and IT landscape. All this makes it a very interesting and promising case to study.

In order to collect different perspectives on process standardization, we purposefully sampled interview participants that were responsible for the management of processes on a country and on a global level. Table 2 lists all interview partners, including the 'head of process excellence and IT governance', the 'head of Human Resources (HR) management processes and systems team', a global process manager for the marketing process as well as a 'local process coordinator and process expert'. When quoting parts of interviews as means of illustrative evidence, we will point to the respective interview. The interviews lasted between 1 and 3 h. All interviews were recorded and transcribed, followed by a detailed analysis.

Table 2. Interview participants

Interview	Position
Interview 1	Head of process excellence and IT governance
Interview 2	Release manager for core systems and coordinator for ISO standardization activities
Interview 3	Local coordinator for processes and process experts, France
Interview 4	Head of HR management processes and systems team
Interview 5	Global process manager for a marketing process

4 A Theoretical Model for Business Process Standardization

Based on the literature and interviews we derived the theoretical model depicted in Fig. 3. For ease of understanding, we will first discuss the different constructs that we identified as relating to business process standardization, followed by the propositions that result form their inter-relation.

Strategic Focus. Strategic alignment between the management of business processes and corporate strategy is one of the key principles of BPM [48]. Pursuing process standardization is first and foremost a matter of strategic focus. The weight that an organization puts on profit versus growth influences extend to which companies will aim at process standardization versus local innovation and variation of business processes [64]. This also relates to the dual definition of quality [37]. Quality can be either perceived as the conformance to ex-ante specified requirements or the tailoring of products and services to specific customer needs.

Further, the implementation of standard processes is a strategic question, since it requires top management support and management involvement [30, 31, 62]. Only with necessary top-level project sponsors, crucial organizational change can be carried out and governance mechanisms can be implemented that promote continuous process standardization efforts.

Governance. The design and implementation of organizational structures, roles, responsibilities, and performance measurements are essential parts of process governance [56, p. 223]. Governance of business processes affects all parts of the process life-cycle, from design and implementation, over day-to-day management activities up to redesign and optimization of business processes [39].

Process standardization requires the centralization of governance mechanisms and instruments. For all stages of the process life-cycle, there is a clear shift of competencies and authority from subsidiaries and business units towards headquarters compared to a localized business process [44]. Process design is not elaborated in single business units, but by the headquarters from where it is distributed to all business units for execution [58]. Where processes are not designed from the scratch, internal or external best practices can serve as a template for the process standard [42]. Process decisions are mandated centrally, which market organizations need to adhere to. Clear roles and responsibilities for all process stakeholders are defined [59]. Often this is combined with periodical process audits, checking whether defined practices are followed on a local level [3,40]. On top of that, process improvements are coordinated centrally. This requires a sophisticated schema including the assessment of process optimization and redesign proposals from business units for the overall organization as well as the adaption of the global standard and the distribution of eventual process changes back to the process in each business unit [31,58].

Documentation and Training. Without rigorous documentation and training, employees are left to act on their own discretion. This can lead to each employee executing the business process differently. In a standardized process, documentation will guide employees. Even though the target of good documentation should not be to overstrain employees, documenting the business process without leaving room for interpretation helps a uniform process performance [8, p. 6]. To achieve a high level of process standardization, there should be only one set of documentation materials available [18], which is regularly updated [59] and employees can refer to.

Information Technology. IT is an integral component of comprehensive BPM. It has not anymore only supportive functionality, but has become a business driver itself [23]. The Strategic Alignment Model of [24] stresses the importance of strategic and functional integration of business and IT. IT can never be an end in itself, but contributes to process change and redesign to improve business performance [6]. Thus, IT and business (processes) are highly interrelated and should not be treated independently; on the contrary a tight integration should be pursued [43]. This is especially true with regards to BPS.

Variations in processes will most often go hand in hand with variations on IT side and vice versa. By use of an integrated and common information system irrespective of location, consistent processes can be assured [61, p. 1008]. Wherever differences in processes are needed, information systems have to be customized accordingly (Interview 4; Interviewee 5) or different software has to be employed [50]. In a similar manner, legacy systems in different business units can call for the adoption of business processes [41]. With process standardization, information systems can be much more standardized and leading than recording. In fragmented information systems, manual workarounds via Excel or other tools are needed connecting various applications. Not only is this frequently a source of error, but it also for process variation (Interviewee 5). On the contrary, IT systems in highly standardized processes are not only well integrated [57], but can help leading employees through the process. Finally, BPS builds the basis for and makes significant contribution towards process automation [41].

Knowledge-intensity. The more a business process relies on experience and judgment of employees rather than rules that can be pre-specified, the more knowledge-intensive the process is [28,68]. Knowledge-intensive processes are hard to impossible to model, as it the exact steps of the process cannot be specified before its actual execution [20]. This also affects the design of information Technology. IT that is designed around knowledge-intensive processes can only provide rough specifications for process execution. E.g. information technology in hospitals often provides only very rough guidance, as patients' symptoms and according treatments vary considerably. Instead, doctors need to rely on their medical education and the experience they gained [37].

Legal. Regulatory Compliance Management (RCM) is a field in BPM, which has been gaining more and more interest recently [11]. Legal requirements and external regulations, e.g. ISO norms (ISO, n.d.), and hence need to be considered for standard building and process execution. External requirements in Compliance Management can be divided in 'must comply' and 'can comply' scenarios. Laws, tax requirements, and import/export regulations fall under the 'must comply' scenario, as non-compliance will cause immediate penalties from national or international actors. Accordingly, process standardization is influenced as described by one of the interviewees:

> "When we build a standardized process, legal directives have to be checked [...]. If each country has different legal [directives] and rules, we will have difficulties to have standardized processes." (Interviewee 3)

Culture. Based on our literature analysis and interviews, we found that particularly two of the cultural dimensions derived by Hofstede [26] are relevant for BPS. First, uncertainty avoidance relates to the attitude of members of a society towards uncertain situations and their tolerance for ambiguity. In countries with high uncertainty avoidance, novel and surprising situations are perceived as a threat, while ease and the acceptance of uncertainty in life is a sign of low uncertainty avoidance. In turn, societies with high levels of uncertainty avoidance will try to minimize the exposure to uncertainty by establishing rules and routines [26]. This indicates that in cultures shaped by high uncertainty avoidance, employees will be more likely to follow a prescribed and standard process, then in countries facing low uncertainty avoidance. [27] reported high levels of uncertainty avoidance for German speaking countries and Japan, and lower values for English speaking areas and Chinese cultures. Second, individualism vs. collectivism is another important contextual element with regards to BPS. These societal opposites reflect whether people of a society are keen to emphasize independence and self-reliance over belonging to groups. In collectivistic societies, there are strong ties not only among the immediate family, but also beyond.

People seek for consensus not to endanger harmony, even if that means hiding his or her opinion [26]. It is thus valid to assume, that in collectivistic cultures employees do not question orders from superiors and execute processes as requested. In most developed countries individualism prevails, while the opposite is the case for less developed nations [26].

Interviewed experts primarily stressed corporate culture and different business habits across countries as an influential factor when standardizing business processes. Shared values among employees and subsidiaries can be seen as an indirect means of control, which helps to enforce headquarters objectives [65]. [12] point out that the diversity of cultures can be a hindrance to knowledge sharing. To overcome this barrier a culture of trust needs to be developed, which serves as a common foundation. Having said this, multiple senior executives from our case company also emphasized that often culture was used as an empty argument to push standardization incentives away from business units.

Process Standardization. As outlined in Sect. 2, process standardization and process variation are two sides of the same coin. Generally speaking, the enactment of a business process is standardized, if the process is executed the same way in each and every part of the organization (Interviewee 1; Interviewee 7). Process flow is composed of the activities executed in course of each process instance, the sequence of those activities and the resources used or consumed for each activity. The more variation and exceptions we find in activities, process sequence, and process resources, the less standardized a process is [2]. In turn, such a process can face high levels of uncertainty as every process instance might differ and require different resources [52].

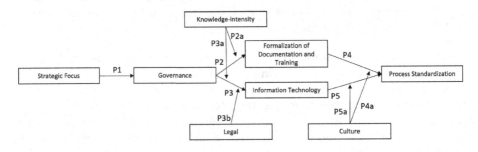

Fig. 3. Model

Proposition 1: A strategic focus that is oriented towards profitability (as opposed to a growth strategy) will result in tighter and more central process governance.

Proposition 2: The more tight governance mechanisms are the more formalized will be process documentation and process training.

Proposition 2a: The effect of governance mechanisms on documentation and training is mediated by the knowledge-intensity of the respective business process. The higher the level of knowledge-intensity of the business process, the less strong will be the effect of governance mechanisms on documentation and training.

Proposition 3: The more tight governance mechanisms are the more constraining will be employed information technology.

Proposition 3a: The effect of governance mechanisms on information technology is mediated by the knowledge-intensity of the respective business process. The higher the level of knowledge-intensity of the business process, the less strong will be the effect of governance mechanisms on employed information technology.

Proposition 3b: Legal aspects mediate the effect of governance on information technology. The more consistent (diverse) legal requirements for business processes are, the more constraining (enabling) will be employed information technology.

Proposition 4: The more formalized process documentation and process training the more standardized will the business process be.

Proposition 4a: The effect of process documentation and training on process standardization is mediated by culture. The more the culture of process participants is characterized by uncertainty avoidance and collectivism, the stronger is the positive effect of formalization on process standardization.

Proposition 5: The more constraining information technology is the more standardized will the business process be.

Proposition 5a: The effect of information systems on process standardization is mediated by culture. The more the culture of process participants is characterized by uncertainty avoidance and collectivism, the stronger is the positive effect of constraining information systems on process standardization.

5   Discussion

5.1   Implications

Our theoretical model has implications for research on BPS, which we discuss in turn. First, we integrate scattered insights into an overarching nomological network that explicates the antecedents of business process standardization. We have reviewed more than 250 academic journal articles and conference papers that relate to process standardization and our theoretical model integrates this existing literature. Our interviews with process experts complement the literature analysis and provide further illustrative evidence. In particular, the model integrates the different perspectives and shows how process standardization relates to the larger picture in organizations by using the socio-technical systems theory. Thus, our model extends prior work by making the relationships between different concepts explicit, e.g. the relationships between information technology [44] and process variation or between documentation [53] and process variation.

The model helps managers to understand antecedents of business process standardization and how they are interrelated. In this way, the model can be used to identify necessary conditions to achieve process standardization, on the one hand. On the other hand, it helps to understand how certain contextual factors, such as legal requirements, culture, and the knowledge-intensity of business process, moderate the impact the effects that management can take to standardize business processes. For researchers, the model provides an overview of the literature that future research can build on. For example, the model can be used as a point of reference to modify and extend existing procedure models for business process standardization.

5.2 Limitations

There are several limitations of our research. First and most important, the focus of this paper was the development of a theoretical model for business process standardization. We derived the model from existing literature and carried out expert interviews to gain additional insights from practice. However, we did not empirically test our model, yet. For many of the constructs in our model there are already measurement scales available (e.g. [67,68]). Future research can make use of these scales to provide evidence for or falsify the propositions we derived. With the increasing availability of digital trace data, not only traditional survey methods can be used to test our propositions, but also process mining techniques become increasingly useful for theory development and theory testing [15]. Second, all of our interview partners work for the same organization. While we have purposefully interviewed process experts in different positions, with varying levels of seniority, and located in different countries, our choice of interview partners may have affected our findings. Third, business process standardization is not a means by itself, but is used to achieve a certain purpose. Business process standardization is generally associated with an improvement of cycle time, costs and quality [41]. However, we know very little about the mutual strength of contextual factors [7] that mediate these effects. Last, we can observe an emerging body of research that aims to broaden the design space of business processes by means of explorative BPM [46]. The focus of these studies is the design of processes for trust [47] or individual customer needs [66]. To sustain in the long run, companies have to balance exploitation of existing processes with the exploration of new processes.

6 Conclusion

In this paper, we developed a theoretical model for business process standardization. The model summarizes and integrates the state of the art on process standardization research; it shows how process standardization is influenced by antecedents and which conditions mediate their impact. With this we provide an overview of key drivers of process standardization that helps process managers to consider the different factors that come into play when standardizing business processes. Fellow researchers are invited to further build on and extend or test our model.

Acknowledgements. The work of Bastian Wurm has received funding from the EU H2020 program under the MSCA-RISE agreement 645751 (RISE_BPM).

References

1. Alter, S.: Work system theory: overview of core concepts, extensions, and challenges for the future. J. Assoc. Inf. Syst. **14**(2), 72 (2013)
2. Balint, B.: Standardization frameworks in services offshoring: the relationship between process implementation thoroughness, task complexity, and performance improvement. In: Proceedings of the 45th Hawaii International Conference on System Sciences (HICSS 2012), pp. 4366–4375 (2012)
3. Bass, J.M., Allison, I.K., Banerjee, U.: Agile method tailoring in a CMMI level 5 organization. J. Int. Technol. Inf. Manage. **22**(4), 77–98 (2013)
4. Bostrom, R.P., Heinen, J.S.: MIS problems and failures: a socio-technical perspective. Part I: the causes. MIS Q. **1**(3), 17–32 (1977)
5. Bostrom, R.P., Heinen, J.S.: MIS problems and failures: a socio-technical perspective. Part II: the application of socio-technical theory. MIS Q. **1**(4), 11–28 (1977)
6. vom Brocke, J.: In-memory value creation, or now that we found love, what are we gonna do with it? BPTrends **10**, 1–8 (2013)
7. vom Brocke, J., Zelt, S., Schmiedel, T.: On the role of context in business process management. Int. J. Inf. Manage. **36**(3), 486–495 (2016)
8. Curran, P., Undheim, T.A.: The java community process standardization, interoperability, transparency. In: Proceedings of the 7th International Conference on Standardization and Innovation in Information Technology (SIIT 2011) (2011)
9. Davenport, T.H.: The coming commoditization of processes. Harv. Bus. Rev. **83**(6), 101–108 (2005)
10. Dumas, M., La Rosa, M., Mendling, J., Reijers, H.A.: Fundamentals of Business Process Management, 2nd edn. Springer, Heidelberg (2018). https://doi.org/10.1007/978-3-662-56509-4
11. El Kharbili, M.: Business process regulatory compliance management solution frameworks: a comparative evaluation. In: Proceedings of the 8th Asia-Pacific Conference on Conceptual Modelling (APCCM 2012), pp. 23–32 (2012)
12. Finestone, N., Snyman, R.: Corporate South Africa: making multicultural knowledge sharing work. J. Knowl. Manage. **9**(3), 128–141 (2005)
13. Grant, M.J., Booth, A.: A typology of reviews: an analysis of 14 review types and associated methodologies. Health Inf. Libr. J. **26**(2), 91–108 (2009)
14. Gregor, S.: The nature of theory in information systems. MIS Q. **30**(3), 611–642 (2006)
15. Grisold, T., Wurm, B., Mendling, J., Vom Brocke, J.: Using process mining to support theorizing about change in organizations. In: Proceedings of the 53rd Hawaii International Conference on System Sciences (2020)
16. Hall, J.M., Johnson, E.M.: When should a process be art, not science? Harv. Bus. Rev. **87**(3), 58–65 (2009)
17. Hammer, M.: What is business process management? In: vom Brocke, J., Rosemann, M. (eds.) Handbook on Business Process Management 1, pp. 3–16. Springer, Heidelberg (2010). https://doi.org/10.1007/978-3-642-00416-2_1
18. Hammer, M., Stanton, S.: How process enterprises really work. Harv. Bus. Rev. **77**(6), 108–118 (1999)
19. Hammer, M.: The 7 deadly sins of performance measurement (and how to avoid them). MIT Sloan Manage. Rev. **48**(3), 19–28 (2007)

20. Harmon, P.: Alternative approaches to process analysis and modeling. BP Trends **4**(13) (2006)

21. Harmon, P.: Business Process Change: A Guide for Business Managers and BPM and Six Sigma Professionals. Morgan Kaufmann, Burlington (2007)

22. Harmon, P.: Artistic processes. BPTrends **7**(9) (2009)

23. Harmon, P.: The scope and evolution of business process management. In: vom Brocke, J., Rosemann, M. (eds.) Handbook on Business Process Management 1, pp. 37–81. Springer, Heidelberg (2010). https://doi.org/10.1007/978-3-642-00416-2_3

24. Henderson, J., Venkatraman, N.: Strategic alignment: leveraging information technology for transforming organizations. IBM Syst. J. **32**(1), 472–484 (1993)

25. Hinterhuber, H.H.: Business process management: the European approach. Bus. Change Re-engineering **2**(4), 63–73 (1995)

26. Hofstede, G.: Dimensionalizing cultures : the Hofstede model in context. Online Read. Psychol. Cult. **2**(1), 1–26 (2011)

27. Hofstede, G., Hofstede, G.J., Minkov, M.: Cultures and Organizations: Software of the Mind, 3rd edn. McGraw-Hill, New York (2010)

28. Işik, Ö., Mertens, W., Van den Bergh, J.: Practices of knowledge intensive process management: quantitative insights. Bus. Process Manage. J. **19**(3), 515–534 (2013)

29. Kampker, A., Maue, A., Deutskens, C., Foerstmann, R.: Standardization and innovation: dissolving the contradiction with modular production architectures. In: Proceedings of the 4th International Electric Drives Production Conference (EDPC 2014) (2014)

30. Kettenbohrer, J., Beimborn, D.: What you can do to inhibit business process standardization. In: 20th Americas Conference on Information Systems (AMCIS 2014), pp. 1–11 (2014)

31. Kettenbohrer, J., Beimborn, D., Kloppenburg, M.: Developing a procedure model for business process standardization. In: Proceedings of the 34th International Conference on Information Systems (ICIS 2013), pp. 1–11 (2013)

32. Kitchenham, B.: Procedures for performing systematic reviews. Ph.D. thesis, Joint Technical report Keele University (TR/SE-0401) and National ICT Australia Ltd. (0400011T.1) (2004). https://doi.org/10.1.1.122.3308

33. Kummer, T.F., Schmiedel, T.: Reviewing the role of culture in strategic information systems research: a call for prescriptive theorizing on culture management. Commun. Assoc. Inf. Syst. **38**(1), 122–144 (2016)

34. Küng, P., Hagen, C.: The fruits of business process management: an experience report from a Swiss bank. Bus. Process Manage. J. **13**(4), 477–487 (2007)

35. Laumer, S., Maier, C., Eckhardt, A.: The impact of business process management and applicant tracking systems on recruiting process performance: an empirical study. J. Bus. Econ. **85**(4), 421–453 (2015). https://doi.org/10.1007/s11573-014-0758-9

36. Lillrank, P.: The quality of standard, routine and nonroutine processes. Organ. Stud. **24**(2), 215–233 (2003)

37. Lillrank, P., Liukko, M.: Standard, routine and non-routine processes in health care. Int. J. Health Care Qual. Assur. **17**(1), 39–46 (2004)

38. Manrodt, K.B., Vitasek, K.: Global process standardization: a case study. J. Bus. Logist. **25**(1), 1–23 (2004)

39. Markus, M.L., Jacobsen, D.D.: Business process governance. In: vom Brocke, J., Rosemann, M. (eds.) Handbook on Business Process Management 2, pp. 201–222. Springer, Heidelberg (2010). https://doi.org/10.1007/978-3-642-01982-1_10

40. Mocker, M., Weill, P., Woerner, S.L.: Revisiting complexity in the digital age. MIT Sloan Manage. Rev. **55**(4), 73–81 (2014)

41. Muenstermann, B., Eckhardt, A., Weitzel, T.: The performance impact of business process standardization: an empirical evaluation of the recruitment process. Bus. Process Manage. J. **16**(1), 29–56 (2010)

42. Muenstermann, B., Weitzel, T.: What is process standardization? In: Proceedings of the International Conference on Information Resources Management (CONF-IRM 2008), pp. 1–17 (2008)

43. Rahimi, F., Møller, C., Hvam, L.: Business process management and IT management: the missing integration. Int. J. Inf. Manage. **36**(1), 142–154 (2016)

44. Romero, H.L., Dijkman, R.M., Grefen, P.W.P.J., van Weele, A.J.: Factors that determine the extent of business process standardization and the subsequent effect on business performance. Bus. Inf. Syst. Eng. **57**(4), 261–270 (2015). https://doi.org/10.1007/s12599-015-0386-0

45. Romero, H.L., Dijkman, R.M., Grefen, P.W.P.J., van Weele, A.J., de Jong, A.: Measures of process harmonization. Inf. Softw. Technol. **63**, 31–43 (2015)

46. Rosemann, M.: Proposals for future BPM research directions. In: Ouyang, C., Jung, J.-Y. (eds.) AP-BPM 2014. LNBIP, vol. 181, pp. 1–15. Springer, Cham (2014). https://doi.org/10.1007/978-3-319-08222-6_1

47. Rosemann, M.: Trust-aware process design. In: Hildebrandt, T., van Dongen, B.F., Röglinger, M., Mendling, J. (eds.) BPM 2019. LNCS, vol. 11675, pp. 305–321. Springer, Cham (2019). https://doi.org/10.1007/978-3-030-26619-6_20

48. Rosemann, M., vom Brocke, J.: The six core elements of business process management. In: vom Brocke, J., Rosemann, M. (eds.) Handbook on Business Process Management 1, pp. 107–122. Springer, Heidelberg (2010). https://doi.org/10.1007/978-3-642-45100-3_5

49. Rosenkranz, C., Seidel, S., Mendling, J., Schaefermeyer, M., Recker, J.: Towards a framework for business process standardization. In: Rinderle-Ma, S., Sadiq, S., Leymann, F. (eds.) BPM 2009. LNBIP, vol. 43, pp. 53–63. Springer, Heidelberg (2010). https://doi.org/10.1007/978-3-642-12186-9_6

50. Ross, J.W.: Creating a strategic IT architecture competency: learning in stages. MIS Q. Exec. **2**(5), 31–43 (2003)

51. Schaefermeyer, M., Grgecic, D., Rosenkranz, C.: Factors influencing business process standardization: a multiple case study. In: Proceedings of the 43rd Hawaii International Conference on Systems Sciences (HICSS 2010), pp. 1–10 (2010)

52. Schaefermeyer, M., Rosenkranz, C.: "To standardize or not to standardize?" - Understanding the effect of business process complexity on business process standardization. In: Proceedings of the European Conference on Information Systems (ECIS 2011), p. 32 (2011)

53. Schaefermeyer, M., Rosenkranz, C., Holten, R.: The impact of business process complexity on business process standardization: an empirical study. Bus. Inf. Syst. Eng. **5**, 261–270 (2012). https://doi.org/10.1007/s12599-012-0224-6

54. Seidel, S.: A theory of managing creativity-intensive processes. Dissertation, The University of Muenster, Muenster, Germany (2009)

55. Seidel, S., Mueller-wienbergen, F., Becker, J.: The concept of creativity in the information systems discipline: past, present, and prospects. Commun. Assoc. Inf. Syst. **27**(1), 217–242 (2010)

56. Spanyi, A.: Business process management governance. In: vom Brocke, J., Rosemann, M. (eds.) Handbook on Business Process Management 2, pp. 223–238. Springer, Heidelberg (2010). https://doi.org/10.1007/978-3-642-01982-1_11

57. Steinfield, C., Markus, M.L., Wigand, R.T.: Through a glass clearly: standards, architecture, and process transparency in global supply chains. J. Manage. Inf. Syst. **28**(2), 75–108 (2011)

58. Tregear, R.: Business process standardization. In: vom Brocke, J., Rosemann, M. (eds.) Handbook on Business Process Management 2, pp. 307–327. Springer, Heidelberg (2010). https://doi.org/10.1007/978-3-642-01982-1_15

59. Trkman, P.: The critical success factors of business process management. Int. J. Inf. Manage. **30**(2), 125–134 (2010)

60. Ungan, M.C.: Standardization through process documentation. Bus. Process Manage. J. **12**(2), 135–148 (2006)

61. de Vries, M., van der Merwe, A., Kotzé, P., Gerber, A.: A method for identifying process reuse opportunities to enhance the operating model. In: Proceedings of the IEEE International Conference on Industrial Engineering and Engineering Management (IEEM 2011), pp. 1005–1009 (2011)

62. Wagner, H.T., Weitzel, T.: How to achieve operational business-IT alignment: insights from a global aerospace firm. MIS Q. Exec. **11**(1), 25–36 (2012)

63. Webster, J., Watson, R.T.: Analyzing the past to prepare for the future: writing a literature review. MIS Q. **26**(2), xiii–xxiii (2002)

64. Weill, P., Ross, J.W.: A matrixed approach to designing IT governance. MIT Sloan Manage. Rev. **46**(2), 26–34 (2005)

65. Williams, C., van Triest, S.: The impact of corporate and national cultures on decentralization in multinational corporations. Int. Bus. Rev. **18**(2), 156–167 (2009)

66. Wurm, B., Goel, K., Bandara, W., Rosemann, M.: Design patterns for business process individualization. In: Hildebrandt, T., van Dongen, B.F., Röglinger, M., Mendling, J. (eds.) BPM 2019. LNCS, vol. 11675, pp. 370–385. Springer, Cham (2019). https://doi.org/10.1007/978-3-030-26619-6_24

67. Wurm, B., Schmiedel, T., Mendling, J., Fleig, C.: Development of a measurement scale for business process standardization. In: 26th European Conference on Information Systems (ECIS 2018) (2018)

68. Zelt, S., Recker, J., Schmiedel, T., vom Brocke, J.: Development and validation of an instrument to measure and manage organizational process variety. PloS One **13**(10), e0206198 (2018)

A Causal Mechanism Approach to Explain Business Process Adoption and Rejection Phenomena

Andreas Brönnimann$^{(\boxtimes)}$ (iD)

Edith Cowan University, Perth, Australia
a.broennimann@ecu.edu.au

Abstract. Introducing change to organizational business processes is an inherently social event. People perform process activities to realize corporate and personal goals. When confronted with changes to their daily routine environments, people, being social actors, reflect critically on the changes presented to them. Collective interactions may lead to acceptance or rejection decisions about the process change, which could be why process change projects regularly fail during their implementation phase. Managing the complexity of interacting social mechanisms during the process implementation phase may be decisive in determining the success or failure of process change projects.

The lack of social mechanism models in this field indicates a business domain less managed. Problems during the process implementation phase may have financial implications, and can cause delays that reduce customer satisfaction.

This paper presents a research approach, consisting of a conceptual ontology model together with a mechanism discovery method. The approach seeks to uncover causal social mechanisms underlying adoption and rejection phenomena during business process implementation. It aims to strengthen research seeking to explain 'why things happen' during change initiatives.

The impact of this research envisions a central mechanism repository to further advance BPM practices. The mechanism repository together with the change ontology model could assist with the analysis of social dynamics from the people perspective to improve the management of process implementation projects.

Keywords: Business process change management · Ontology · Realism · Causal mechanism

1 Introduction

Business Process Management (BPM) aims to continuously improve business processes to keep up with changing market demands [23]. BPM ensures transparency and alignment between strategy and technology across the end-to-end

© Springer Nature Switzerland AG 2020
D. Fahland et al. (Eds.): BPM Forum 2020, LNBIP 392, pp. 297–312, 2020.
https://doi.org/10.1007/978-3-030-58638-6_18

processes that organizations execute to deliver goods and services. Most BPM life cycle phases have matured to a degree where advanced method and tool support is present [13]. Process analysis follows detailed qualitative and quantitative methods, process models are developed following the BPMN notation and stored in large process repositories, where they can be monitored using process mining applications [36]. But despite these advancements, business process change projects are still facing problems during their implementation phase.

Many companies manage to analyze and design improved process models, but then experience difficulties transitioning successfully to their new processes. Empirical studies reveal that 46% of change projects experience problems and 16% fail entirely [7,26,35]. It has been argued that the neglect of people, collectives and culture contributes significantly to process implementation failures [8,11]. For example, implementation projects that apply the wrong change management approach [22], or present a requirements mismatch between current and future processes [25], may disempower employees and consequently lose their support. Employees rejecting business processes prevents organizations from realization the benefits of redesigned processes [22].

Social dynamics may trigger resistance to change at the individual and the collective level. At the individual level, reasons for change presented by management may appear illogical to employees causing dissatisfaction about the decision. The introduction of restructured processes creates job insecurities, as BPM may still be perceived as a management style to reduce staff. Fear of future interpersonal conflicts may lead to communication problems and low job satisfaction breeds resistance [29]. Collective resistance is oftentimes linked to mismatching cultural values and beliefs [37]. Merging different organizational cultures may also create political or ideological resistance. The requirement to adopt new values and practices by collectives is frequently met with rejection or general change incomprehension.

Past research has identified BPM change success [35] and failure factors [1]. These studies provide a degree of insight into "*what*" affects business process change, and yet quantitative covariance-based approaches struggle to explain "*why things happen*" as process change unfolds. This paper contributes to the call for research to progress BPM as a behavioral science, more specifically the social aspects of the process implementation phase [30].

Knowledge about adoption and rejection phenomena during business process change implementation at a deeper level is necessary for its effective intervention management. Therefore, this paper approaches the process change domain from a realist research perspective to uncover underlying social mechanisms that are at play during business process change projects.

Various ontologies have been developed within the BPM domain to foster knowledge and information sharing [13,14,21,24,28]. While these BPM related ontologies focus on technological implementations and automation, Ariouat et al. [6] discusses the ontological importance of the people perspective during the life cycle's design and analysis phase. However, research into the causalities of social business process change mechanisms during process implementations have received only little research.

This paper first outlines a realist ontology to identify and relate important entities involved during process change. Then, a research method, supported by an illustrative example, applies the ontology model as a research tool to discover social mechanisms. The final section discusses the value impact of having social mechanism knowledge within BPM change to support the development of more effective change interventions.

2   Developing the Process Change Ontology

The advancement of discipline knowledge depends on a shared understanding about existing entities within that discipline. Such shared understanding can be established and communicated in form of an ontological model, which ensures that research is directional and comparable. An ontology allows researchers to see and understand the entities it explains with the same sense and meaning.

This section outlines the development of an ontology model that combines relevant entities, concepts, and relations. The ontology represents a research instrument assisting in the discovery of hidden social mechanisms governing business process change phenomena. The ontology creation follows the specification and conceptualization steps defined in the Methontology framework [15]. Information about the business process change domain was collected from the literature. The resulting ontology model is shown using UML notation in Fig. 1.

2.1   Ontology Scope and Goals

The scope coves socially relevant business process change aspects such as individuals, collectives, and their interactions with each other. Additionally, organizational as well as individual culture affect the inner attitude of people and collectives towards proposed process change. Organizational structures such as rules and regulations may encourage or suppress people's actions for change support or objection.

The ontology's goal is to guide research to develop causal explanations in form of mechanism descriptions of social interactions between people, organizational structures and culture that lead to the overall event of change acceptance or rejection. It addresses questions like

- How do goals affect process change behavior in process stakeholders?
- How do powers and liabilities of collectives lead to adoption or rejection?
- What causes different social mechanisms to emerge during a change project?
- Under which conditions are powers actualized?
- How and under what conditions does culture affect agent actions?
- Under what conditions do organizational artifacts allow or prevent power relations?
- How are the change events caused by mechanisms during process change?
- Why are some mechanisms successful in causing a change event, but others are not?

2.2   Specifying the Ontology Perspective

Different theories of explanations exist on causal relations between consecutive events. According to the positivist view, given a situation, where a certain event X has happened, it can be deduced that it will always be followed by the same event Y. Positivists derive from the repeated series of events a sense of statistics based predictability, which may provide answers to '*What caused Y to happen?*'. But positivist views fall short to answer the more valuable, deeper lying questions of '*Why and how did X cause Y to happen?*' [10]. This requires an approach that looks beneath the empirically surfaced events.

Critical realists reject event correlation as a form of causal explanation and emphasize the importance of causal knowledge as a means to expand scientific understanding. The concept of causation critically hinges on the existence of underlying, usually hidden, causal mechanisms [10].

Critical realists perceive reality as three nested, layered domains. These are the *empirical*, the *actual*, and the *real*. The empirical domain resides within the actual domain, which in turn sits within the real domain [10, 17, 32]. At the top sits the empirical domain, where events occur that have been observed by humans. In the middle is the actual domain, where events occur that may or may not have been observed. A plethora of events occur without ever being perceived by an observer. The real domain as the lowest domain holds all existing things. They interact in causal mechanisms, which are the source of emerging events into higher domains.

Reality is made up of things, which have particular dispositions to act in certain ways. A disposition is a particular *power* or a *liability* [18]. Powers allow a thing to actively initiate externally directed forces of action. A liability is a passive power representing a susceptibility to be acted upon by powers [10]. Powers may mature as they mutually interact with other dispositions along a causal chain from their initial state of being *possessed* to *exercised*, and eventually *actualized*, when they cause an event to occur [16].

The interaction patterns of powers and liabilities between things constitute a particular mechanism. Mechanisms may or may not cause a particular event to happen, because of other intervening mechanisms [2].

Mechanisms have the potential to cause certain events. An event is considered as a form of change that has been inflicted upon by another thing's acting power. Events may either be observed or they remain hidden from the human mind [10].

To summarize, critical realism provides guidance by defining the concept of causation using mechanisms. Understanding the causalities of process change acceptance and rejection is key to be able to mange process change projects.

2.3   Conceptualizing the Ontology

While critical realism defines properties of fundamental, yet domain unspecific things, the domain of business process change is a social environment, categorized by the intense interactions between people, their culture and the organization that are required to make change happen. Social realism extends the critical realism concept by providing socio-cultural and socio-structural analysis factors to explain actions of social entities [2]. As such the ontology shown in Fig. 1 combines both philosophical foundations.

Social realists define *Structure*, *Culture*, and *Agency* as extensions of critical realist things. This means that they also have powers and liabilities through which they may interact. Agency describes the domain of human collectives with their behavior and social interactions. Social structure emerges from human interactions over time, such as organizations or business processes. Culture represents the domain of ideas and beliefs; these are social things with their own properties, powers and liabilities.

Social Agency. Humans are considered as *Agents* with an action capacity. Actions involve motivation, rationalization and reflexive monitoring [19].

Agents evolve through different phases during their lifetime [4]. They begin as *Primary Agents*, who become entangled in the present network of society. During this ongoing entanglement, some agents may develop the desire to change the existing structures and cultures to get ahead in life. But primary agents cannot drive change, because their actions are uncoordinated.

Personal and corporate goals cause primary agents to form coordinated, goal-oriented collectives. They become *corporate agents*, who achieve their personal and corporate goals through the power of collective movements. Their power actualization is conditional and depends on the powers and liabilities of surrounding culture, structures and people, which have powers to resist, block, abstain, and avoid imposed force [4].

Humans reflect critically through internal conversations with themselves on changes in their surroundings [5]. This inner voice deliberates decisions about future courses of actions to satisfy personal and organizational goals. Autonomous, communicative, meta, and fractured are distinct modes of reflexivity inherent in human beings, which affect their personal and collective behavior upon change confrontation.

For example, given a process change situation, an autonomous reflexive BPM project manager would drive the change for personal reasons to achieve a promotion or family happiness, and regardless of how it affects other agents. A member of the process participant collective may act communicatively, therefore seeking advice from others before forming a personal opinion. Another member may evaluate the change from a meta perspective, meaning the thinking process considers how the group is affected first. Fractured reflexives have a troubled life situation, where they cannot make decisions by themselves and would usually agree with any proposed change.

Social Structures. *Social Structures* emerge from human interactions, who occupy roles within the structure [2]. Structures can be nested to form a hierarchy. Examples of structures are social clubs, governments, corporations with their different departments, as well as business processes and sub-processes. While these structures are not directly observable, they possess causal powers to

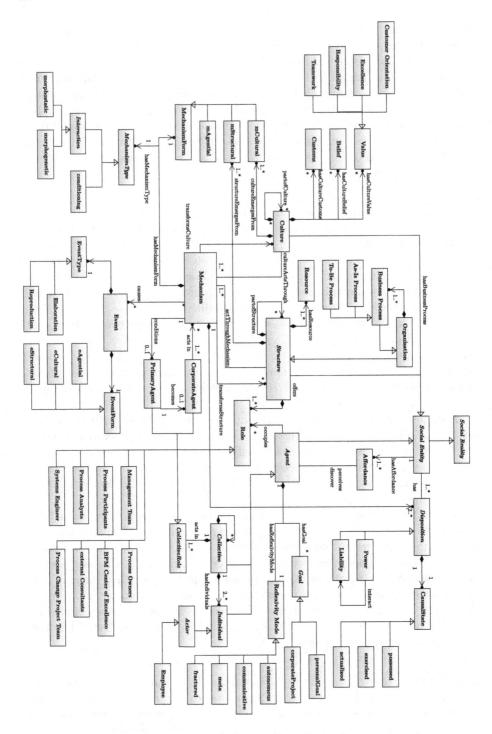

Fig. 1. Social business process change ontology for implementation phases

influence people. Structures are not considered entirely rigid, because they have liabilities through which agents can change them. Structures within the domain scope are the business process itself and the organization.

The powers give structures the ability to act upon agents. For instance, an organization operates according to certain policies and rules, which dictate decisions for particular situations. The structure of a business process is often formally represented in a business process model that explains how work is to be executed and in what order. The business process has the power to form a new employee to become part of the existing process team in ways that the process as it is will continue to exist. This could be through training sessions and social interactions of explanation. This conditioning affects an agent's possibilities for change consideration. High familiarity with a process may prevent an agent from recognizing change potential.

Culture. While there are many different definitions of culture, the common theme of having shared basic assumptions emerges [11]. Realists understand *Culture* as a set of social values, beliefs, and customs. Similar to structure, culture also follows a similar interaction cycle with agency of influence, reshaping or reconstruction. Culture, formed by past human generations, conditions current agency, who may or may not desire to change the present culture. Agency's change powers may generate a new culture or reuse the existing culture [3].

Culture with respect to BPM is considered an under-researched area, yet it is agreed that culture plays an important role as culture is one factor in the BPM maturity model and if neglected will most likely result in project failures [31].

Culture, similar to social structure, is considered to be layered. Cultural levels exist at national, organizational, which impacts BPM projects the most, and work group level. Domain independent properties of culture are *artifacts* like company symbols and logos, *espoused values* like norms, rules and principles, and *basic underlying assumptions* representing complex value relations around work, family and the personal self. The result of a delphi study identified cultural values distinct for BPM culture. These values are referred to as CERT: *Customer Orientation, Excellence, Responsibility,* and *Teamwork* [33].

The BPM-Culture-Model captures these cultural roles within BPM, where values will manifest through actions in structure [11], while the BPM Context Framework classifies organizational level culture as highly-, medium-, and non-supportive of BPM [12]. A very recent study using the CERT values confirms the causal influence of preexisting culture on agents as well as that BPM culture is non-rigid and can be changed [34]. This causal link represents the cultural conditioning of agents, who will act through mechanisms attempting to improve the prevailing BPM Culture. The ways of how culture manifests in events through mechanisms affecting process change are yet to be determined.

Affordances. An *affordance* is the relation between the dispositions of goal driven agents and structures with mutually compatible dispositions, which allow the agents to achieve their goals, if a mechanism is invoked to realize the affordance relation. Affordances exist outside of human knowledge, but they can be discovered [20]. Thus, the perception of an affordance by an agent prior to the invocation of a mechanism is a necessity.

Agents in different roles and social states will perceive different affordances between the current and proposed process states. For instance, the current process may afford a process participant, who values family time, a short commute to work, while the proposed process may afford the process owner, who seeks career development, a job promotion. Similarly, while an experienced change manager may also perceive a job promotion affordance in a complex change project, an inexperienced change manager may not perceive an affordance at all.

Social Mechanisms and Events. Structure, culture, and agency are mediated by underlying causal mechanisms that generate social events. These are classified as *elaboration*, representing change, while the alternative called *reproduction* represents no change. A proposed to-be process being successfully adopted by a collective of process participants, for instance, denotes a structural elaboration event, while the rejection of the process is a structural reproduction event.

Morphostatic Mechanisms preserve the continuation of structures and culture, thus resulting in structural and cultural reproduction events. *Morphogenetic Mechanisms* are mechanisms seeking change and resulting in elaboration events in structure and culture. An active mechanism is not sufficient to cause its related event, for there may be other counteracting mechanisms that are blocking or preventing the event from being generated. It is this dynamic flux of interacting mechanisms, enacted through the powers and liabilities of agents, that causes elaboration or reproduction events in the social reality. It is about these mechanisms that critical realists seek knowledge to explain why things happen.

3   Method for Social Mechanism Discovery

While the ontology defines relevant entities and places them in relation to each other, it's instantiation relies on actionable steps to become applicable to conduct social research analysis. Wynn and Williams [38] present a step-wise approach to elicit social mechanisms. The data collection steps are *Event Explication* and *Structure, Culture and Context Explication*, while the analysis happens during *Retroduction*, and *Empirical Corroboration*. The application of triangulation and multimethods during data collection increases validity.

Because the research mode is explanatory, data collection is preferably performed after the change project, but it may also start with the beginning of the change phase. The proposed methods for retrospective data collection after the change project are individual and collective interviewing, questionnaires [5] as well as project documentation reviews, if accessible. The aim is to identify particular behavioral tendencies for actions when considering personal goals, cultural values, or structural entities.

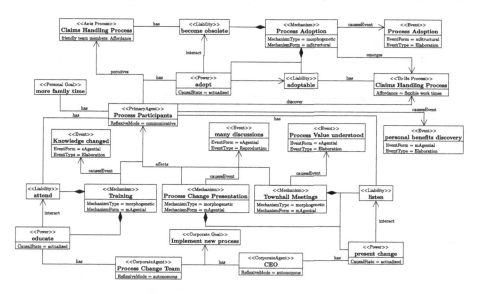

Fig. 2. Example of a business process ontology instance

When data can be captured during the change project, observation is the preferred choice as it is less stressing for the project. Research data can be collected from primary sources like process change stakeholders or from secondary sources like company documents. While it is preferential to collect data from all change stakeholders involved, it may not always be possible due to time and project limitations.

3.1 Method Steps

The application of the method is illustrated for clarification with an example explaining a change of an insurance claims handling process, shown in Fig. 2.

A top-down directed change starts when the organization's strategies change or new government policies and regulations need to be implemented. While certain changes are welcomed by employees, oftentimes this type of change is perceived as being forced onto process participants due to the necessity of its nature. Neglecting the voice of affected employees may lead to dissatisfaction and resulting in refraining from the change. Similarly, the prevailing cultural beliefs in a work group can define the level of collective resistance toward change.

Event Explication. The researcher seeks to extract empirical events from experiences. Descriptions of events can be collected from participants or personal experiences, if the researcher was present at the time. Events can be of agential, cultural, and structural form according to the ontology defined in the preceding section. The event descriptions capture minute actions or interactions of individuals and collectives with structures and culture. For instance, the cognitive realization of a particular process affordance by a process participant constitutes an event. The richer the event description are, the more rigorous understanding can be explicated from them. The aim is to identify the sequential or concurrent temporal order of events. The possibility of ordering such events suggests the existence of a causal chain.

Structure, Culture and Context Explication. The inquiry moves from the observed events and asks what powers and liabilities are required to be actualized to cause that particular event. The inquiry seeks to identify potential structural, cultural or agential entities that may have exposed those powers and liabilities within a given temporal context. Certain event explanation may require considering the involvement of multiple entities. The explications of contextual conditions, under which actions have occurred, allows to define when actions can and cannot trigger.

In the example, process participants may describe the following events and experiences. They may state that the new process was accepted overall, but that the beginning of the change phase caused a lot of discussion among co-workers as they had been skeptical. Yet, the skepticism changed with the occurrence of certain events. For instance, the moment when the process participants knowledge changed and the affordance of the new process had been revealed. Or when they had eventually understood the process value.

The researcher now explicates the following entities in relation to the events. The CEO presented the process change project and later on a townhall meeting, while the change team conducted a training session. Process participants are the recipients of the change. It could be assumed that the training session caused the knowledge change of the participants or allowed them to understand the value of the new process. Next, the analysis phase begins.

Retroduction. The data collected may contain multiple possible explanations for the phenomenon under investigation. Retroduction attempts to discover the most likely mechanism explanation by linking identified power and liability capacities of explicated entities and events caused by them. The researcher engages in complex thought processes, contemplating different mechanism scenarios that may explain sections of or the entire causal chain leading up to the final event.

Empirical Corroboration. While multiple mechanism explanations result from the retroductive thinking step, empirical corroboration seeks conformance of these explanations with the data collected and is a form of validation against the source data. In cases, where the most likely explanation is not conform with the source data, the researcher may have to re-interview participants to verify or reject the assumption made. It is important to confirm any assumptions made as any assumption left unconfirmed jeopardizes the validity of the mechanism description.

In the example, the process participant collective values family time and they tend to discuss work related changes among each other during times of uncertainty. This behavior of seeking input and confirmation from close people is associated with communicative reflexives. The current claims handling process affords them to work with friendly team members. Communicative reflexive do not object change from the beginning, but also critically evaluate change through reflection as a way of mediation.

The process change team and the CEO, who are actively driving the change project, have the corporate goal to implement the new process, thus they may act as autonomous reflexives. Autonomous reflexives act as an adoption driver and use techniques to influence others through mechanisms. Research data may reveal that the initial change presentation did not contain detailed enough information or that the CEO's popularity is low among process participants. Hence, the participants initially lean towards change rejection.

The data records may show that training sessions happened directly after the CEO's change presentation. The process change team has the power to educate others, while the process participants have a liability to attend these meetings as part of their employee responsibilities. Assuming, that the process participant collective does not act rebelliously by blocking the education power or non-attendance, process participants may have raised their concerns about change conflicting with their personal goals. This may be a form of collective action. The change team now reveals the previously unperceived affordance of the new process to allow more flexible work times. Hence, the training mechanism has affected the collective causing the knowledge change event.

The CEO addresses the process participants next in an informal townhall meeting to achieve collective adoption. The process participants have an obligation to listen to the presentation. Assuming that the town-hall mechanism can operate uninterruptedly and nothing else interferes with the presentation power of the CEO, the mechanism would also change the participants understanding by stressing the affordance of work flexibility. Further forms of dialogue also exist [27]. The process participants are now ready to adopt the new process, which causes an elaboration event indicating the occurrence of structural change.

Adhering to this narrative form allows researchers to present a causal description that links events with their underlying mechanisms to the best of their knowledge given the current information available.

4   Limitations

Applying the ontology as part of the described research steps does not come without limitations. Causal logic defines the strength of mechanism descriptions, which rely on accurate, relevant and timely data collected from an ethnographic data collection. Some data collection obstacles may appear when social analysis is applied during a change project and when applied retrospectively after the project has finished.

When used during the change project, passive data collection through observation and document analysis is limited to the number of researchers. There is a higher likeliness of missing important event developments in a larger project, when only a small number of data collectors are active. While increasing the number of researchers is an option, it also increases data collection costs. However, the researcher's own observations and experiences as the change project develops also contribute valuable insight.

Active data collection, using interviews and questionnaires, may be perceived as an additional time consumer and perceived by change directors and participants as not directly related to project goals. Furthermore, while structures and cultural aspects are less likely to undergo rapid change, people's opinions and experiences with respect to the process change are more volatile. People interact and influence each other over the course of the project causing minute chains of events to occur, which are hard to collect.

Retrospective data analysis represents the main method of data collection for causal analysis. Data collection related problems after the project has finished are related to the availability of people involved in the project as well as the amount of time past between data collection and project end. Individuals and entire collectives may not be available past the designated project end date due to contractual obligations, for instance hired consulting groups. Other individual people may have left the company after the project ended.

5   Impact on BPM Research and Practices

While BPM research has grown within the information systems discipline, it can also be addressed from a behavioral science point. BPM as a behavioral science investigates human, organizational, and cultural aspects to achieve better management of business processes [30]. The approach presented in this paper contributes to the advancement of behavioral BPM research by describing an ontology and research method to discover social processes during business process change. While some of the concepts in the social domain have been studied before [8,9,34], they have not been combined to present an overarching ontology model, yet the strength of the ontology emerges only from the integration of constituent, individually researched areas.

The ontology shall not be perceived as a static, engineered artifact, but its true value for research and practitioners emerges eventually through continued application and revision in process change projects over time. The ontology model in its current maturity state represents a research instrument striving to understand adoption and rejection phenomena during the social phase process change implementation through the uncovering of mechanisms. To be of value to the BPM research community as well as BPM analysts, the ontology needs to mature as a research instrument first.

When used to conduct research after the process change project has ended, the ontology provides a direction for research to investigate causal links during process change. The more past projects are analyzed, the more knowledge about the interactions of mechanisms is generated. Ongoing case research will mature the ontology and identified mechanisms. Eventually, mechanisms leading to change adoption and those leading to change rejection will emerge. Identified mechanisms are to be stored in a mechanism repository. Once a mechanism

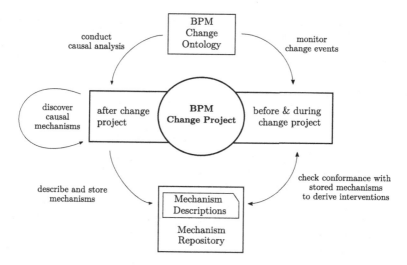

Fig. 3. Ontology applications in BPM research and practice

knowledge repository has been established, the ontology together with discovered mechanisms becomes valuable and applicable as a management tool for BPM practitioners to improve the manageability of BPM change projects (Fig. 3).

The ontology in combination with the knowledge from stored mechanisms can assist during future and current running process change projects to analyze social dynamics from the people perspective. While planning the change, the project change team would start an analysis of the cultural, social, and structural entities and their tendencies of acting in favor or against the change. Standard data collection methods can be applied such as individual and collective interviewing techniques, reviewing past tendencies as well as personal development goals, although HR data access is oftentimes restricted. The analysis results provide insight into the preferential conditions for the change implementation phase.

During the implementation phase, the change team will continue to observe emerging events and deduce currently active mechanisms through comparison with the stored ontology and mechanisms. If observations reveal strong acting of change rejecting mechanisms, the change team can plan and intervene in situations otherwise headed towards rejection. Seeking confirmation about any event or mechanism assumptions made becomes important, because any neglected confirmation may lead to the triggering of wrong or unnecessary intervention actions.

Ideally, the intervention mechanism prohibits the change rejecting mechanisms and supports the change adoption mechanisms. Mechanism knowledge becomes valuable for BPM analysts during the business process change implementations when suitable response interventions can be developed to counteract process implementation rejection mechanisms. Applying the knowledge about the causal chain during change allows the BPM change team to activate specific intervention mechanisms at the right time and under the appropriate conditions, rather than attempting to satisfy lists of critical success factor criteria. The development of appropriate intervention responses needs to be addressed in future research.

6   Conclusion

Business process change projects are critical for successful organizations and therefore demand detailed knowledge about the underlying mechanisms. Only when it becomes possible to explain how causal powers as part of hidden mechanism have caused the adoption or rejection of a business process, will the management of process change become more effective.

This paper presents an approach to discover these social mechanisms during business process change projects. First, it develops an ontology model founded on realist understanding of reality and society. Then an instantiation method is shown to develop case specific ontologies that assist researchers with the retroduction of causal mechanism during process change. Lastly, the envisioned impact of the approach for researchers and practitioners of BPM is discussed.

While the approach is complex and subject to limitations, it presents a viable alternative to existing research to investigate the problem of failing business process change implementations because of social resistance.

The scope of this paper is limited to the extent that it presents an inception of an ontological model for the business process change domain. In order to increase credibility in the causal research approach using an overarching ontology, future research may benefit from case study based data collection and analysis.

References

1. Antony, J., Gupta, S.: Top ten reasons for process improvement project failures. Int. J. Lean Six Sigma **10**(1), 367–374 (2019). https://doi.org/10.1108/ijlss-11-2017-0130
2. Archer, M.S.: Realist Social Theory: The Morphogenetic Approach. Cambridge University Press, Cambridge (1995)
3. Archer, M.S.: Culture and Agency: The Place of Culture in Social Theory. Cambridge University Press, Cambridge (1996)
4. Archer, M.S.: Being Human: The Problem of Agency. Cambridge University Press, Cambridge (2000)
5. Archer, M.S.: The internal conversation: mediating between structure and agency: full research report ESRC end of award report. Technical report, RES-000-23. ESRC, Swindon (2008). http://www.researchcatalogue.esrc.ac.uk
6. Ariouat, H., Hanachi, C., Andonoff, E., Benaben, F.: A conceptual framework for social business process management. Procedia Comput. Sci. **112**, 703–712 (2017). https://doi.org/10.1016/j.procs.2017.08.151
7. Bandara, W., Alibabaei, A., Aghdasi, M.: Means of achieving business process management success factors. In: Proceedings of the 4th Mediterranean Conference on Information Systems. Department of Management Science & Technology, Athens University of Economics and Business, Athens (2009). https://eprints.qut.edu.au/30074/

8. Baumöl, U.: Cultural change in process management. In: vom Brocke, J., Rose-mann, M. (eds.) Handbook on Business Process Management 2: Strategic Alignment, Governance, People and Culture. INFOSYS, pp. 487–514. Springer, Berlin (2010). https://doi.org/10.1007/978-3-642-01982-1_23

9. Bernhard, E., Recker, J., Burton-Jones, A.: Understanding the actualization of affordances: a study in the process modeling context. In: International Conference on Information Systems (ICIS 2013): Reshaping Society Through Information Systems Design, vol. 5, pp. 1–11. AIS Electronic Library (AISeL), Milan, December 2013. https://eprints.qut.edu.au/63052/

10. Bhaskar, R.: A Realist Theory of Science. Harvester Press, Sussex (1978)

11. vom Brocke, J., Sinnl, T.: Culture in business process management: a literature review. Bus. Process Manag. J. 17(2), 357–378 (2011). https://doi.org/10.1108/14637151111122383

12. vom Brocke, J., Zelt, S., Schmiedel, T.: Considering context in business process management: the BPM context framework. bptrends.com (2015)

13. Dumas, M., La Rosa, M., Mendling, J., Reijers, H.A.: Fundamentals of Business Process Management, 2nd edn. Springer, Heidelberg (2018). https://doi.org/10.1007/978-3-662-56509-4

14. Fahad, M.: BPCMont: business process change management ontology (2016)

15. Fernández-López, M., Gómez-Pérez, A., Juristo, N.: Methontology: from ontological art towards ontological engineering. Technical report SS-97-06. The AAAI Press, Menlo Park (1997)

16. Fleetwood, S.: Causal laws, functional relations and tendencies. Rev. Polit. Econ. 13(2), 201–220 (2001). https://doi.org/10.1080/09538250120036646

17. Fleetwood, S.: An ontology for organisation and management studies, pp. 25–50. Routledge, London (2004)

18. Fleetwood, S.: Powers and tendencies revisited. J. Crit. Realism 10(1), 80–99 (2011). https://doi.org/10.1558/jcr.v10i1.80

19. Gibson, E.J., Walker, A.S.: Development of knowledge of visual-tactual affordances of substance. Soc. Res. Child Dev. 55(2), 453–460 (1984). http://www.jstor.org/stable/1129956

20. Gibson, J.J.: The theory of affordances, pp. 56–60. Routledge, New York (1979)

21. Green, P., Rosemann, M.: Applying ontologies to business and systems modelling techniques and perspectives. J. Database Manag. 15(2), 105–117 (2004). https://doi.org/10.4018/jdm.2004040105

22. Grisdale, W., Seymour, L.F.: Business process management adoption: a case study of a South African supermarket retailer. In: Proceedings of the South African Institute of Computer Scientists and Information Technologists Conference on Knowledge, Innovation and Leadership in a Diverse, Multidisciplinary Environment - SAICSIT 2011 (2011). https://doi.org/10.1145/2072221.2072234

23. Harmon, P.: The scope and evolution of business process management. In: vom Brocke, J., Rosemann, M. (eds.) Handbook on Business Process Management 1. IHIS, pp. 37–80. Springer, Heidelberg (2015). https://doi.org/10.1007/978-3-642-45100-3_3

24. Hepp, M., Leymann, F., Domingue, J., Wahler, A., Fensel, D.: Semantic business process management: a vision towards using semantic web services for business process management. In: IEEE International Conference on E-Business Engineering (ICEBE 2005), pp. 535–540 (2005). https://doi.org/10.1109/ICEBE.2005.110

25. Jeston, J.: BPM and change management: two perspectives (2011). https://www.bptrends.com/

26. Jørgensen, H.H., Albrecht, J., Neus, A.: Making change work: Erfolgsfaktoren für die einführung von innovationen. Technical report, IBM Global Business Services (2007). http://www-935.ibm.com
27. Klewes, J., Langen, R.: Change 2.0: Beyond Organisational Transformation. Springer, Heidelberg (2008). https://doi.org/10.1007/978-3-540-77495-2
28. Pedrinaci, C., Domingue, J., Alves de Medeiros, A.K.: A core ontology for business process analysis. In: Bechhofer, S., Hauswirth, M., Hoffmann, J., Koubarakis, M. (eds.) ESWC 2008. LNCS, vol. 5021, pp. 49–64. Springer, Heidelberg (2008). https://doi.org/10.1007/978-3-540-68234-9_7
29. Pejić Bach, M., Bosilj Vukšić, V., Suša Vugec, D.: Individual's resistance regarding BPM initiative: case study of the insurance company. Naše gospodarstvo/Our Econ. **63**(4), 29–39 (2017). https://doi.org/10.1515/ngoe-2017-0021
30. Recker, J., Mendling, J.: The state of the art of business process management research as published in the BPM conference. Bus. Inf. Syst. Eng. **58**(1), 55–72 (2015). https://doi.org/10.1007/s12599-015-0411-3
31. Rosemann, M., vom Brocke, J.: The six core elements of business process management. In: vom Brocke, J., Rosemann, M. (eds.) Handbook on Business Process Management 1. IHIS, pp. 105–122. Springer, Heidelberg (2015). https://doi.org/10.1007/978-3-642-45100-3_5
32. Sayer, A.: Realism and Social Science. Sage, London (2000)
33. Schmiedel, T., vom Brocke, J., Recker, J.: Development and validation of an instrument to measure organizational cultures' support of business process management. Inf. Manag. **51**(1), 43–56 (2014). https://doi.org/10.1016/j.im.2013.08.005
34. Schmiedel, T., Recker, J., vom Brocke, J.: The relation between BPM culture, BPM methods, and process performance: evidence from quantitative field studies. Inf. Manag. **57**(2), 103175 (2020). https://doi.org/10.1016/j.im.2019.103175
35. Trkman, P.: The critical success factors of business process management. Int. J. Inf. Manag. **30**(2), 125–134 (2010). https://doi.org/10.1016/j.ijinfomgt.2009.07.003
36. Van Der Aalst, W.: Process Mining: Discovery, Conformance and Enhancement of Business Processes, vol. 2. Springer, Heidelberg (2011). https://doi.org/10.1007/978-3-642-19345-3
37. Very, P.: The Management of Mergers and Acquisitions. Wiley, West Sussex (2005)
38. Wynn, D., Williams, C.K.: Principles for conducting critical realist case study research in information systems. MIS Q. **36**(3), 787–810 (2012). https://doi.org/10.5555/2481655.2481663

Analyzing a Helpdesk Process Through the Lens of Actor Handoff Patterns

Akhil Kumar[(⊠)] and Siyuan Liu

Pennsylvania State University, University Park, PA 16802, USA
{akhilkumar,sxl168}@psu.edu

Abstract. In this study, we analyze the activity logs of fully resolved incident management tickets in the Volvo IT department to understand the handoff patterns i.e., how actors pass work from one to another using sequence analytics, an approach for studying activity patterns from event log sequences. In this process the process model itself is rather simple, but a large amount of variety is present in it in terms of the handoff patterns that arise. Hence, process modeling is not so helpful to gain a deeper understanding of the performance of the process. We offer an alternative approach to analyze such processes through the lens of organizational routines. A generic actor pattern here describes the sequence in which actors participate in the resolution of an incident. We characterize actor handoff patterns in terms of canonical sub-patterns like straight, sub- and full-loop, and ping-pong. Then, we predict incident resolution duration with machine learning methods to understand how actor patterns affect duration. Finally, the evolution of patterns over time is analyzed. Our results shed light on emergence of collaboration and have implications for resource allocation in organizations. They suggest that handoff patterns should be another factor to be considered while allocating work to actors along with position, role, experience, etc.

Keywords: Workflows · Routines · Handoffs · Sequence analytics · Actor patterns · Pattern variety · Machine learning · Social network analysis

1 Introduction

Any business or healthcare process can be viewed as a series of handoffs between task actors (or workers) who perform successive tasks until the process instance is completed. After an actor completes her task, she hands off the process instance or case to the next actor. Such behavior is observed in various kinds of application areas ranging from medicine and software development to insurance claim processing. Some handoffs also occur in a *ping-pong pattern* such that an actor A hands off a task to actor B only to have it returned later, either after some work is done or just untouched. This leads to an actor handoff pattern represented by the sequence A-B-A, A-B-C-D-A, etc. Such ping-pong behavior arises from an alternating pattern in which the same actor appears more than once.

Most of the research in the Business Process Management (BPM) and workflow literature has been devoted to the discovery of process models, conformance checking

© Springer Nature Switzerland AG 2020
D. Fahland et al. (Eds.): BPM Forum 2020, LNBIP 392, pp. 313–329, 2020.
https://doi.org/10.1007/978-3-030-58638-6_19

and process enhancement. This research assumes that historical process execution logs of completed process instances are available for analysis. Thus, say, we have a log like:

T1 – T4 – T5
T1 – T3 – T5 – T4
T1 – T2 - T4 – T5
T1 – T2
T1 – T5

By applying a process mining algorithm [26] we may discover a process model that always starts with task T1; next tasks T2 and T3 appear in a choice (or alternative) structure and are followed by a parallel structure of T4 and T5 that can appear in any order. Moreover, it is also possible to skip the T2–T3 or T4–T5 substructure, but only one skip is allowed not both. The discovery of process models in this manner is useful for it helps us to understand the control flow of a process. The drawback with process mining approaches is that when considerable variety is inherently present in a process, capturing it in excruciating detail leads to a spaghetti model that is unreadable. This diminishes its real value.

Researchers in the area of *organizational routines* define routines as "repetitive, recognizable patterns of interdependent actions, carried out by multiple actors." [5] They accept that variety exists in real world processes and have developed approaches to quantify routine variation, by posing questions like: Does the process in Unit A have more variety than the one in Unit B? [17] Moreover, researchers in organizational routines ascribe greater agency and tacit knowledge to actors of various tasks in terms of their interpretations of how the task should be done rather than treating actors as fully interchangeable. Their focus is on interdependencies among actions, people and technology in contrast to the control flow, data flow and resource perspectives of BPM.

Our approach is in part inspired by previous work in the context of routines [17, 18], but our work fits into the broader area of work distribution and resource allocation in BPM (e.g. [9, 22]). Consequently, we ask: what are the patterns of interaction among generic actors (as opposed to actions) in a large real log and how can they inform us? A pattern like 1-2-1-3-1 shows the order of involvement of actors 1, 2 and 3 in the completion of a process instance through four handoffs among themselves. We can interpret this pattern as actor 1 dividing some work between actors 2 and 3 and finally integrating the two pieces of work for resolving the ticket. Alternatively, this pattern could arise when 2 was unable to complete the work sent by 1 and returned it, thus 1 had to instead turn to 3 to perform it. The incident resolution process we consider is simple because multiple workers take turn on solving the same problem, yet it creates an enormous amount of variety. The significance of our work lies in that, by examining the actual worker interaction patterns and their durations in detail, we expect to gain a deeper understanding of good collaboration practices, and to use these insights in developing better work allocation policies in organizations.

Our study was made possible by access to a large log from the incident resolution process at the Volvo automotive company. This data set is public and hence the results can be verified. By correlating the most frequent types of actor sequence patterns in this dataset with the duration for resolving the incident we were able to gain many useful insights about the significance of actor patterns. By analyzing this data, we hope to address questions like:

- RQ1. What are the (frequent) actor patterns found in resolving tickets?
- RQ2. Are some actor patterns better than others (in terms of time) and why?
- RQ3. Can we develop an algorithm to classify patterns as good or bad?
- RQ4. What features of actor patterns affect resolution time of the tickets?
- RQ5. How does the social interaction pattern among workers evolve with time?

In this way, we can shed more light on the resource perspective in a business process. This perspective has implications for assignment of resources to a process in an efficient way.

This paper is organized as follows. Section 2 gives some preliminaries about handoffs and sequence analytics. Then, Sect. 3 describes our log data and the main results of our analysis. Next, Sect. 4 develops our method for classifying patterns as good versus bad. Later, Sect. 5 examines process evolution, while Sect. 6 offers a discussion and covers related work. The paper concludes with Sect. 7.

2 Sequence Analytics

A handoff is a transfer of work from one actor to another. Research in healthcare has shown that communication breakdowns among medical professionals can lead to adverse effects on surgical patients [15]. These breakdowns result from poor handoffs involving verbal communications and ambiguities about responsibilities. There is ample evidence to suggest that the nature of social interactions and interdependencies among participants (or resources) who collaborate on a routine or a process does have an impact on the outcome and performance of the process in terms of quality, failure rate, etc. [12].

Sequence analytics refers to the concept of analyzing the sequences of actions or elements to detect similarities and differences across the sequences [17]. For example, in biology this concept is used to detect evolutionary patterns, rate of mutation and any genetic modifications that occurred in time. This concept was later adopted by sociologists and more recently in information systems to detect socio-material entanglement in work processes [6]. An *action* or *task sequence pattern* is a series of possible orderings of related tasks to complete a process or a workflow. Some examples are:

T1-T2-T3-T4
T1-T3-T4-T1

Similarly, it is possible to also consider actor sequences. An actor sequence would define sequences of specific actors such as: A1-A2-A3-A4, or A3-A2-A4-A1, etc. Each sequence denotes the order in which various actors perform tasks to complete a workflow or routine. In contrast to these two notions, in this paper we are interested in studying actor patterns. By a pattern, we mean an ordering or sequencing in which *generic actors* perform tasks to complete a workflow or a routine. Thus, an actor pattern like 1-2-3-4 means that some actor 1, handed over the work to actor 2 who in turn passed it along to

3 and so on. We call this a *straight pattern*, i.e. 2 or more actors appear in a sequence. Another pattern 1-2-3-1 is a combination of *straight* and *full-loop sub-patterns* where the work is returned *at the end* to the same actor who started it. Yet another pattern may be 1-2-1-2-1-3. This is a combination of *straight* and *ping-pong* sub-patterns since the incident returns to an actor after one step. It is important to note here that 1, 2, 3 are generic placeholders for actors, and not specific names of actors. See Fig. 1 for examples of these sub-patterns that are the elements of any full pattern. Finally, a sub-loop pattern falls in between a ping-pong and a full-loop wherein the sub-loop extends a distance that is longer than a ping-pong pattern, but it falls short of a full-loop. Note that in a loop or ping-pong pattern the incident returns to the same worker for further work.

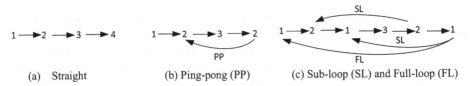

(a) Straight   (b) Ping-pong (PP)   (c) Sub-loop (SL) and Full-loop (FL)

Fig. 1. Different kinds of canonical actor sub-patterns

Our goal is to analyze such generic patterns to determine the kinds of patterns appear most frequently and to understand if some patterns are better than others in terms of incident resolution times.

These sub-patterns are complete in that they can accurately describe a work sequence. Also note that the ping-pong is a special case of the sub-loop. The variation between two work sequences can be quantified as the edit distance between two strings, since each sequence is a string. One common metric for the edit distance is the Levenshtein distance [16] or the number of edit operations such as insert, delete, move, etc. required to transform one sequence into another. Aggregating the edit distance among all pairs of work sequences and dividing by the total number of pairs can give a notion of variation or variety across all the work sequences or patterns [17].

3   An Analysis of Volvo IT Incident Management

3.1   Data Set Description and Analysis

This process log data is publicly available and was initially released as a part of the Business Process Intelligence (BPI) Challenge in 2013 [1]. The dataset contains a log of incidents or cases to be resolved. Each incident has a unique serial number. Typically, there are many log records for each incident to reflect any status or owner change. A log record captures the status of the incident and includes information like serial number, date, status, sub-status, impact, product, country, owner, support team, and organizational line (see the partial log shown in Table 1). There are 7554 cases or incidents and 65553 events or records in the log for an average of 8.7 log records per incident. The period of this data set extends from the end of March 2010 until middle of May 2012. A process diagram for this helpdesk process is shown in Fig. 2. After a customer reports a problem, an incident is created or opened for it. Then it is assigned to an agent who works on it and passes it to another agent who works on it further until it is solved; or else it is transferred to a new owner. Thus, most of the work for resolving an incident is performed in the inner loop at the center of Fig. 2. To understand such a process we apply the approach of organizational routines to uncover the amount of variety present in the different handoff patterns.

Table 1. A snapshot of the incident log

SR number	Date	Status	Impact	Product	Country	Owner name
1-364285768	2010-03-31	Accepted	Medium	PROD582	France	Frederic
1-364285768	2010-03-31	Accepted	Medium	PROD582	France	Frederic
1-364285768	2010-03-31	Queued	Medium	PROD582	France	Frederic
1-364285768	2010-04-06	Accepted	Medium	PROD582	France	Anne Claire

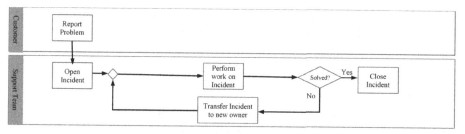

Fig. 2. Process model for the incident resolution process

The owner attribute in the log record denotes the actual actor who performs a task. When two successive (in time sequence) log records have different owners it indicates a handoff of work from the previous owner to the new one. If two successive records have the same owner, it means that there is a status change and not a handoff. We were only interested in the incidents where at least one handoff occurred.

The dataset was loaded in a MySQL database for the analysis. We first removed the log records for the owner 'Seibel' because this is the information system, and not a human owner. Our focus was on studying the effect of handoffs among human actors only. After all, a resource allocation algorithm can only select a specific human from a set of alternatives. Then we removed all records for incidents where only one human owner was involved, i.e. no handoffs occurred, and for incidents that were not resolved. This left us with 4375 incidents - 1755 tickets of low impact, 2413 of medium impact, 204 of high impact and 3 of major impact. Queries were written in MySQL using constructs like JOIN, GROUP BY, ORDER BY, COUNT, CONCAT, etc. to categorize the help tickets by status, extract help ticket country patterns, calculate metric values and determine traffic flows. Results from MySQL were further analyzed in Excel. Statistical methods, and machine learning and social network analysis techniques were also employed.

We observed that the largest number of incidents (about 93% of all) was concentrated in the last part of the dataset, within a short period from April 1 until May 15 of the year 2012. There were 4064 incidents during this period out which the largest number by impact were 2204 medium impact incidents. There were 1670, 187 and 3 of low, high and major impact, respectively. This was attributed to a learning curve effect in the initial period that stabilized later. By concentrating on this period, we were able to eliminate any learning curve effect by removing just a small fraction of the total number of incidents. By plotting resolution time for incidents against time it was observed that resolution times declined with time (see Fig. 3).

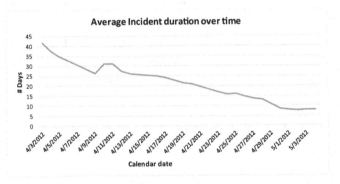

Fig. 3. Average duration over time

Next, we created generic or abstract patterns from the sequences of handoffs for each incident in the log in the following manner. For example, the ticket #1-523391859 has 8 events recorded in the log and contained a series of operations in resolving the case. The incident went through multiple owners, 'Elaine-Elaine-Elaine-Elaine-Elaine-Rafael-Rafael-Siebel', before resolution. As we were interested in the abstract handoff patterns among actors, we removed consecutive repetitions with the same owner name (and also owner Siebel) resulting in, 'Elaine-Rafael'. By coding first actor as '1' and second actor as '2' and so on, we were able to generate patterns to convert the sequence of owners' names to numbers 1, 2, 3, ... These numbers give the order in which an actor appears in the incident resolution process. For each incident we generated the actor pattern along with other information like frequency, average duration, etc. These results are shown in Tables 2 and 3 for medium and low impact incidents, respectively.

3.2 Preliminary Analysis of Dataset

Tables 2 and 3 show the Top-10 most frequent actor patterns that appear in the data set for low and medium impact incidents, respectively. The top 5 actor patterns account for about 70% and the top 10 for about 80% of the incidents in both tables. The last column of Tables 2 and 3 shows the rank by duration time of the various patterns. We have excluded from both tables patterns that had a frequency of less than roughly 1% of the total number of incidents in their category. From these patterns one can easily determine the number of unique owners that took part in resolving the corresponding incident and also the number of handoffs.

Table 2. Top 10 most frequent actor patterns for **Low** Impact Incidents

	Actor pattern	Frequency	Average duration	Rank
1.	1-2	577	10.2	1
2.	1-2-3	317	10.8	3
3.	1-2-3-4	177	14.3	7
4.	1-2-3-4-5	83	15.0	8
5.	1-2-3-4-5-6	48	15.5	9
6.	1-2-3-1	33	10.5	2
7.	1-2-1	31	10.9	4
8.	1-2-3-4-5-6-7	27	17.8	10
9.	1-2-3-2	25	13.0	5
10.	1-2-3-4-1	19	13.8	6

Table 3. Top 10 most frequent actor patterns for **Medium** Impact Incidents

	Actor pattern	Frequency	Average duration	Rank
1.	1-2	747	9.3	5
2.	1-2-3	492	10.4	6
3.	1-2-3-4	195	13.8	8
4.	1-2-1	92	9.0	3
5.	1-2-3-4-5	90	12.9	7
6.	1-2-3-1	78	7.6	1
7.	1-2-3-4-5-6	35	14.3	9
8.	1-2-3-4-1	30	8.0	2
9.	1-2-3-2	19	16.2	10
10.	1-2-3-4-5-1	18	9.2	4

Note that 9 out of 10 patterns are common to both tables. Further, the top 3 patterns are identical, and 4 out of the top 5 are common as well. This suggests that similar patterns are used to resolve incidents of both low and medium impact.

One interesting effect found in Table 3 is that a pattern like 1-2 is more frequent than a similar pattern 1-2-1, but the latter takes smaller duration. Similarly, we find that pattern 1-2-3 is more frequent than 1-2-3-1, but the latter has a shorter duration by

27%. In fact, many such loop patterns have a shorter duration than their straight pattern counterparts. This raises the question, why does an instance with one additional handoff to loop around take a shorter duration than without it? Further investigation showed that in many incidents there was a large lag time between the last two log entries of "Completed-Resolved" and "Completed-Closed". The last step was performed by the system. In many cases with the straight pattern it increased to 8 days but was lower in the loop pattern.

The helpdesk incidents relate to many products of the company, but the largest number of incidents are for PROD424 that has 520 incidents. We examined all 111 incidents of medium impact for 'PROD424' and show the results for the Top-10 patterns in Table 4. Notice from the last column for pattern type that only two of the 10 patterns are pure straight patterns though they account for 80% incidents, while the other patterns for 20% incidents. Moreover, among the top-10 least duration patterns of the 406 low impact incidents for this product not one was a straight one. These observations suggest that effective collaboration tends not to happen always in a straight chain, rather it is more helpful if an incident is exchanged among various actors who work on it.

Table 4. Top 10 smallest duration actor patterns for **Medium** Impact Incidents for **PROD424**

	Pattern	Frequency	Average duration	Sub-patterns
1.	1-2-3-4-2	1	6.0	Straight, sub-loop
2.	1-2	24	6.8	Straight
3.	1-2-1-2-3	1	8.0	Straight, sub-loop
4.	1-2-3-1-2-4-5-6	1	8.0	Straight, sub-loop
5.	1-2-1	5	8.8	Straight, ping-pong
6.	1-2-3-4-5-2-5	1	9.0	Straight, ping-pong, sub-loop
7.	1-2-3-2	1	9.0	Straight, ping-pong
8.	1-2-3-1	2	9.0	Straight, full-loop
9.	1-2-3-4-2-5-6-1	1	9.0	Straight, sub-loop, full-loop
10.	1-2-3	26	9.6	Straight

Preliminary results with OLS (Ordinary Least Squares) regression showed that there is a significant relationship between attributes start time, impact and Handoffs, and the dependent variable Duration at the 0.001 level. This OLS produces a high R-squared of 0.98. Clearly, these main factors can explain the duration well. In the next section, we will develop an algorithm that classifies duration as low or high in terms of certain attributes of each pattern.

4 Characterizing Patterns and Explaining Duration

As observed above, variety relates to the number of unique handoff patterns that are present in a log. We found that there were 545 patterns in 4375 incidents, i.e. on average *a single pattern applies only to 8 incidents*. Next we tried to model each pattern more formally in terms of its constituents, such as the number of handoffs, owners, straight, ping-pong, full-loop and sub-loop elements present in them. Some example characterizations are shown in Table 5.

Table 5. Coding the frequent workflow sequences patterns for low impact incidents

Actor handoff pattern	Hand-offs	Owners	Straights	Full- loop	Ping-pong	Sub-loop
1-2-3-4-5-6-7	6	7	1	0	0	0
1-2-3-1-4-5-6	6	6	2	0	0	1
1-2-3-4-5-1	5	5	1	1	0	0
1-2-3-4-2-5	5	5	1	0	0	1
1-2-3-4-2	4	4	1	0	0	1
1-2-3-4-2-3	5	4	2	0	0	1
...	...	...	...	...	...	...

We selected only the patterns with an occurrence frequency of 5 or more. There were 23 such low impact patterns. Next, we coded this information to classify the duration in terms of these 6 elements as attributes. Thus, *straights* attribute measures the # of times continuous sequences of length two or more appear in the pattern. The full-loop attribute can only be 0 or 1 since a pattern can, by definition, only have one full-loop such that actor 1 appears in the end of the pattern. The ping-pong and sub-loop attributes measure the respective occurrences of these two behaviors. The average duration for the 23 patterns ranged from 10.2 to 22.0 days. The 12 instances with duration <15 were coded as low duration (1) and the rest with duration ≥ 15 were coded as high (or 0).

Various rule and tree-based machine learning techniques were applied to this dataset using Weka [27, 28] to see how accurately we could predict the duration as low (1) or high (0) based on the sequence information. The results for J48, Random Forest and One-Rule methods are shown in Table 6 for low and medium impact patterns. For low impact patterns, the best classification accuracy of 82.61% on the full learning set was achieved by the Random Forest algorithm. However, the other two methods were close behind. When 10-fold cross-validation was performed on the data, the performance of Random Forest decreased considerably to 52.17%. However, the One-rule method produced the same result of 78.26% in both scenarios. The One-rule found by Weka for low impact patterns is:

owners:
$< 4.5 \rightarrow 1$
$>= 4.5 \rightarrow 0$
(18/23 instances correct)

Table 6. Classification accuracy using different methods

	Low impact		Medium impact	
	Full training set	10-fold cross-validation	Full training set	10-fold cross-validation
One rule	78.26%	78.26%	66.67%	45.83%
J48	73.91%	60.87%	91.67%	66.67%
Random forest	82.61%	52.17%	95.83%	75%

This rule is quite intuitive in that it finds that fewer number of owners leads to a low duration with the cut-off being at 4. J48 did slightly worse than One-rule and found the cutoff to be at 5; thus, if the number of owners was 5 or less, then the pattern was classified as low duration. Now, turning to medium impact incidents, we repeated the above coding approach for the 24 instances with occurrence frequency of 5 or more. The cut-off for duration was 22.8 with low duration patterns in the range [11.9, 20.1] and high-duration patterns in [25.5,119.6]. The results show that now Random Forest method dominates on both metrics, with J48 close behind and One-rule is much worse. J48 is easier to visualize as a tree (see Fig. 4). The leaf nodes give the outcome (0 – high duration; 1- low duration) and the number of correct/wrong classifications. It shows that duration is low when a full-loop is present. Otherwise, ping-pong behavior should be present and # handoffs should be small. We tried other rule-based methods in Weka such as ZeroR and Decision Table as well but they did not do well. In the next section, we will delve deeper into some of the reasons for the findings here by looking at how patterns evolve over time.

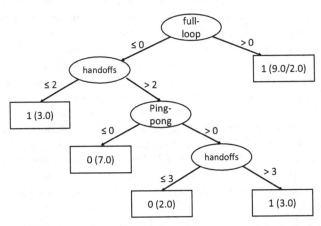

Fig. 4. J48 decision tree for predicting duration from patterns (medium impact)

5 Pattern Evolution Analysis

Pattern evolution shows how the work sequence patterns change with time. In this section, we analyzed the data in terms of its temporal pattern to see how certain key metrics like duration, #owners, #handoffs, etc. varied with time. This analysis was performed during the time range from April 1 through May 15, 2012 (a 45-day period) when the bulk of the incidents were performed as discussed in Sect. 3. Figure 5 shows the plot of handoffs and owners v. date for incidents that were initiated on a given date. This figure shows that both #owners and #handoffs decrease steadily with time. Another interpretation for this trend of decreasing numbers of owners and handoffs is that the company needed to compromise on incident resolution. Given, say, the same amount of resources available and more work to be done, you tend to do things quicker, and maybe less carefully with less testing, double-checking, etc. This would suggest that the decrease in duration may have occurred at the expense of quality of the solution.

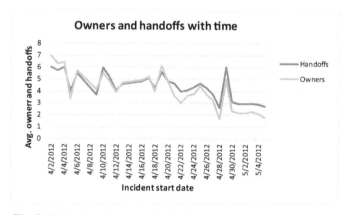

Fig. 5. Evolution of average handoffs and average owners with time

This means that the patterns are becoming shorter since the pattern length = #handoffs + 1. Moreover, the number of owners per incident is decreasing thus indicating that the incident requires less effort to resolve. In addition, Fig. 3 shows that duration is declining steadily as well which is consistent with shorter handoff patterns. This is corroborated by Table 7 where the 10 most frequent patterns and their frequencies for all incidents that start on three different dates that are successively 10 days apart from each other within the 45-day period (i.e. start date in the dataset for the incidents is April 13, April 23 and May 3, respectively). The number of incidents that start on these three dates increases rapidly from 33 to 710. The patterns produced on the incidents that started on a later date are shorter and simpler with both fewer owners and handoffs.

Variety calculations were performed for the top 10 handoff sequences in Table 8 using the formula:

$$\text{Variety} = \frac{1}{n(n-1)} \sum_{i=1}^{n} \sum_{j=1}^{n} d(i,j)$$

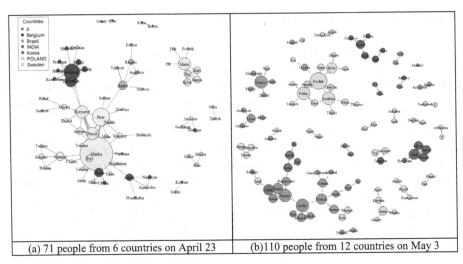

| (a) 71 people from 6 countries on April 23 | (b)110 people from 12 countries on May 3 |

Fig. 6. Evolution of social network on two different dates

Table 7. Most frequent patterns for two different dates for all incident types

April 13 (33 incidents)	Freq.	April 23 (143 incidents)	Freq.	May 3 (710 incidents)	Freq.
1-2	9	1-2	26	1-2	306
1-2-3	7	1-2-3	24	1-2-3	158
1-2-3-4-5-6	3	1-2-3-4	22	1-2-3-4	61
1-2-3-4-5	2	1-2-3-4-5	12	1-2-1	26
1-2-3-1	1	1-2-3-4-5-6-7	5	1-2-3-1	23
1-2-3-4-1	1	1-2-3-4-5-6-7-8	4	1-2-3-4-5	20
1-2-3-4-3-5-6	1	1-2-1	3	1-2-3-4-1	8
1-2-3-4-5-6-5-7	1	1-2-3-4-3	3	1-2-3-4-5-6	5
1-2-3-4-5-6-7-8-9	1	1-2-3-1	2	1-2-3-4-5-1	5
1-2-3-1-4-5-6-7-5	1	1-2-3-2	2	1-2-3-2	5

Table 8. Most frequent patterns for two different dates

	Deg. centr. (min,max)	Close centr. (min, max)	Betw. centr.	Diameter
April 23	2.82 (1,19)	0.029 (0.014,0.39)	7.29e−04	14
May 6	1.82 (0,7)	0.01 (0.009,0.011)	5.41e−05	24

In the above expression, $d(i, j)$ is the edit distance between sequence i and sequence j, and n is the number of sequences. We use Levenshtein distance [16] for the edit distance. On April 13 variety was 1.67; on April 23 it was 1.53, while on May 3 it was 1.36. Thus, we can see a declining trend in variety even though the number of incidents is rising sharply from 33 to 143 to 710 across these dates. One can also notice from Table 7 that the number of owners and handoffs are both smaller on later dates than the earlier ones. This suggests that as workers become more adept at resolving incidents of a certain type, they take less time to complete them. Moreover, notice that on the third date (May 3), 9 of the 10 patterns are either short, pure straight, or straight with full-loop.

We also examined the evolution of the social network over time. Table 7 summarizes key metrics obtained from social network analysis on April 23 and May 3. It shows that all three metrics of centrality decrease as handoff patterns evolve. Figure 6 shows the social graphs for the top 100 interaction pairs (by frequency) on the two later dates (the number of incidents on the first date was very small). This graph was produced in R with the **igraph** package. The different colors of the nodes represent different countries, and the edge width is proportional to the number of interactions between a pair of owners. A pair of owners is said to interact once if both of them work on an incident. The size of the nodes is proportional to the centrality of the target.

We can see both from Table 8 and Fig. 6 that the network on April 23 is much denser with higher values for all the three centrality metrics and a smaller diameter. The sparser network of May 3 suggests that fewer handoffs and owners are involved as the actors get more adept at resolving the incidents and each incident needs fewer hands. We can also notice that there are fewer interactions across countries as the network evolves. This is consistent with the finding in [10] based on an extensive analysis of multi-country incident resolution that when a larger number of countries is involved it hurts the resolution duration.

This analysis gives us a deeper insight into the evolution of handoff patterns. Moreover, now it is also possible to relate these results with those found in Sect. 4 by analyzing the characteristics of the respective patterns on the three successive dates. By applying the rules for low and medium impact patterns discovered in Sect. 4, we can see that, while only 3 out of 10 patterns on April 13 would be classified as low duration patterns, there are 7 and 8 patterns on April 23 and May 3, respectively, that would fall in this category. This again suggests that perhaps a learning process is taking place leading to the selection of better work sequences.

6 Discussion and Related Work

Several interesting results emerge from our study. By decomposing actor patterns into straight, loop and ping-pong elements, and applying basic machine learning methods, we found that pure straight patterns are often dominated by loop and ping-pong patterns with the same (or even more) number of owners.

The superior performance of loop and ping-pong patterns suggests that rather than one actor holding onto an incident, frequent exchanges among actors are better. A second feature of loop and ping-pong patterns is that a single actor may take ownership of the incident and monitor it. This accelerates its progress leading to faster resolution. In fact, the actor who appears more than once may be playing the role of a "coordinator" to facilitate the smooth transfer of work among others. Evidence of this is also found in the work of Liu, et al. [13] who have constructed social network analysis to develop an enhanced organizational model. In their model, various actors play social roles like team leader, coordinator, etc. arguably leading to superior team formations.

In the literature, studies have looked at how routines evolve over time in the context of university housing routines [4] (e.g. for the student moving-in process) and invoice processing routines [19]. These studies find that even over a short time period the patterns of action are significantly different. Our results agree with theirs in this respect. However, while these studies seemed to find that process outcomes did not improve systematically, our results suggest that learning is taking place that leads to a kind of emergent collaboration. This means that over time, the actors can learn to collaborate in ways that are more efficient. There is evidence in literature that complex problem solving requires a mix of independent and collaborative work [8]. A straight chain of new actors would lead to each actor incurring additional "set-up" time to understand the problem and it may be an inefficient way of solving it.

Although the loop and ping-pong patterns perform better, yet the pure straight patterns are predominant. It may well be that for relatively easy incidents pure straight patterns are perhaps the best. What our results also indicate is that resource allocation algorithms should be designed to take handoff patterns into consideration. In terms of pattern variety, out of 111 incident cases pertaining to PROD424, there were 32 actor patterns. Among these, 28 had a frequency of 5 or less (24 patterns only 1). The longest pattern had 18 handoffs and the smallest 1, showing the large diversity among patterns.

Our results can be incorporated into work allocation algorithms by, for instance, limiting the number of owners that work on an incident. After a cut-off value is reached, it would be desirable to bounce the incident back to one of the previous owners who is already familiar with the case rather than to a new one who will have to make a fresh start on it. There is a large body of work on resource allocation in BPM (see, e.g. [9, 22]) that we can review only briefly here. However, one part of this work relates to resource assignment languages [2], resource preference models [3], etc. There is also some valuable work in the area of organizational mining that helps to understand resource profiles and assignment patterns and relates to how the involvement of resources influences the control flow of a process [14, 21, 23, 24, 29]. Related work has also examined affordance networks that combine actors, actions and artifacts [20]. Event interval analysis [25] is also relevant in the context of actor patterns to understand the nature of handoff intervals. Finally, an effort to identify handoffs within and across process and organizational boundaries with a view to optimizing them is described in [11].

It is important to point out some key limitations of our work. First, the dataset does not provide any information about the nature and content of an incident other than its impact. Thus, we are not able to compare the level of difficulty of incidents within the same impact level. Second, there is no contextual information for an incident. Third, we do not have knowledge of the worker skill levels across countries or the methods used for assignment of incidents to them. We expect to address some of these issues in future work.

7   Conclusions

In this paper, we presented an approach for analysis and classification of actor work sequences that can help to improve our understanding of the resource perspective of a process and give fresh insights into work design practices. This study was conducted in the context of a rich data set from an incident resolution process. We found that the pure straight patterns were often dominated by work sequences where the ping-pong, sub-loop and full-loop sub-patterns were present because they led to interaction among the various owners of an incident in a more efficient manner. In a straight pattern, it appears there was a tendency for a worker to "hang-on" to an incident and thus delay its progress.

In future work, we would like to examine other data sets to see if this approach for pattern classification can be generalized more broadly. The link between various sub-patterns within a pattern and its effect on effective collaboration and teamwork needs to be studied more carefully. Hence, there is a need to understand sub-patterns and interactions among them in more detail. Moreover, it would be very helpful to study how handoff patterns and sub-patterns can be more tightly integrated into resource allocation methods. A resource allocation algorithm should be able to learn to distinguish good handoff patterns from bad ones and then make resource assignments accordingly. Finally, it would be worthwhile to investigate whether notions of task complexity along the lines of [7] can be incorporated into this framework as another factor that affects duration of an incident.

References

1. BPI Challenge (2013). http://www.win.tue.nl/bpi/doku.php?id=2013:challenge
2. Cabanillas, C., Resinas, M., Ruiz-Cortés, A.: Defining and analysing resource assignments in business processes with RAL. In: Kappel, G., Maamar, Z., Motahari-Nezhad, H.R. (eds.) ICSOC 2011. LNCS, vol. 7084, pp. 477–486. Springer, Heidelberg (2011). https://doi.org/10.1007/978-3-642-25535-9_32
3. Cabanillas, C., García, J.M., Resinas, M., Ruiz, D., Mendling, J., Ruiz-Cortés, A.: Priority-based human resource allocation in business processes. In: Basu, S., Pautasso, C., Zhang, L., Fu, X. (eds.) ICSOC 2013. LNCS, vol. 8274, pp. 374–388. Springer, Heidelberg (2013). https://doi.org/10.1007/978-3-642-45005-1_26
4. Feldman, M.S.: Organizational routines as a source of continuous change. Organ. Sci. 11(6), 611–629 (2000)
5. Feldman, M.S., Pentland, B.T.: Reconceptualizing organizational routines as a source of flexibility and change. Adm. Sci. Q. 48(1), 94–121 (2003)
6. Gaskin, J., Berente, N., Lyytinen, K., Yoo, Y.: Toward generalizable sociomaterial inquiry: a computational approach for 'zooming in & out' of sociomaterial routines. MIS Q. 38(3) (2014)
7. Hærem, T., Pentland, B., Miller, K.: Task complexity: extending a core concept. Acad. Manag. Rev. 40(3), 446–460 (2015)
8. Hagemann, V., Kluge, A.: Complex problem solving in teams: the impact of collective orientation on team process demands. Front. Psychol. 8, 1730 (2017). Published 2017 Sep 29. https://doi.org/10.3389/fpsyg.2017.01730

9. Kumar, A., van der Aalst, W.M.P., Verbeek, E.M.W.: Dynamic work distribution in workflow management systems: how to balance quality and performance. J. Manag. Inf. Syst. **18**(3), 157–193 (2002). Winter

10. Kumar, A., Liu, S.: Analyzing performance of a global help desk team operation – country handoffs, efficiencies and costs. In: Proceedings of the 53rd Hawaii International Conference on System Sciences, pp. 353–362 (2020)

11. Leyer, M., Iren, D., Aysolmaz, B.: Identification and analysis of handovers in organisations using process model repositories. Bus. Process Manag. J. (forthcoming) (2020). https://doi.org/10.1108/BPMJ-01-2019-0041

12. Lindberg, A., Berente, N., Gaskin, J., Lyytinen, K.: Coordinating interdependencies in online communities: a study of an open source software project. Inf. Syst. Res. **27**(4), 751–772 (2016)

13. Liu, R., Agarwal, S., Sindhgatta, R.R., Lee, J.: Accelerating collaboration in task assignment using a socially enhanced resource model. In: Daniel, F., Wang, J., Weber, B. (eds.) BPM 2013. LNCS, vol. 8094, pp. 251–258. Springer, Heidelberg (2013). https://doi.org/10.1007/978-3-642-40176-3_21

14. Ly, L.T., Rinderle, S., Dadam, P., Reichert, M.: Mining staff assignment rules from event-based data. In: Bussler, C.J., Haller, A. (eds.) BPM 2005. LNCS, vol. 3812, pp. 177–190. Springer, Heidelberg (2006). https://doi.org/10.1007/11678564_16

15. Mazzocco, K., et al.: Surgical team behaviors and patient outcomes. Am. J. Surg. **197**(5), 678–685 (2009)

16. Miller, F.P., Vandome, A.F., McBrewster, J.: Levenshtein distance: Information theory, computer science, string (computer science), string metric, damerau? Levenshtein distance, spell checker, hamming distance (2009)

17. Pentland, B.T.: Sequential variety in work processes: organization science, vol. 14, no. 5, pp. 528–540 (2003)

18. Pentland, B.T., Haerem, T., Hillison, D.W.: Using workflow data to explore the structure of an organizational routine. Organizational routines: Advancing empirical research (2009)

19. Pentland, B.T., Hærem, T., Hillison, D.W.: The (n) ever-changing world: stability and change in organizational routines. Organ. Sci. **22**(6), 1369–1383 (2011)

20. Pentland, B.T., Recker, J., Wyner, G.: A thermometer for interdependence: Exploring patterns of interdependence using networks of affordances, manuscript (2015)

21. Pika, A., Leyer, M., Thandar Wynn, M., Fidge, C.J., ter Hofstede, A., van der Aalst, W.: Mining resource profiles from event logs. ACM Trans. Manag. Inf. Syst. **8**(1), 1:1–1:30 (2017)

22. Russell, N., ter Hofstede, A.H.M., Edmond, D., van der Aalst, W.M.P.: Workflow data patterns: identification, representation and tool support. In: Delcambre, L., Kop, C., Mayr, Heinrich C., Mylopoulos, J., Pastor, O. (eds.) ER 2005. LNCS, vol. 3716, pp. 353–368. Springer, Heidelberg (2005). https://doi.org/10.1007/11568322_23

23. Schönig, S., Cabanillas, C., Jablonski, S., Mendling, J.: A framework for efficiently mining the organisational perspective of business processes. Decis. Support Syst. **89**, 87–97 (2016)

24. Schönig, S., Cabanillas, C., Di Ciccio, C., Jablonski, S., Mendling, J.: Mining team compositions for collaborative work in business processes. Softw. Syst. Modeling **17**(2), 675–693 (2016). https://doi.org/10.1007/s10270-016-0567-4

25. Suriadi, S., Ouyang, C., van der Aalst, W.M.P., ter Hofstede, A.H.M.: Event interval analysis. Decis. Support Syst. **79**(C), 77–98 (2015)

26. van der Aalst, W.M.P.: Process Mining: Data Science in Action, 2nd edn. Springer, Heidelberg (2016)

27. Weka website. https://www.cs.waikato.ac.nz/ml/weka/
28. Witten, I.H., et al.: Data Mining: Practical Machine Learning Tools and Techniques, 4th edn. Morgan Kaufmann, San Francisco (2016)
29. Yang, J., Ouyang, C., Pan, M., Yu, Y., ter Hofstede, A.H.M.: Finding the "liberos": discover organizational models with overlaps. In: Weske, M., Montali, M., Weber, I., vom Brocke, J. (eds.) BPM 2018. LNCS, vol. 11080, pp. 339–355. Springer, Cham (2018). https://doi.org/10.1007/978-3-319-98648-7_20

Author Index

Printed in the United States
By Bookmasters